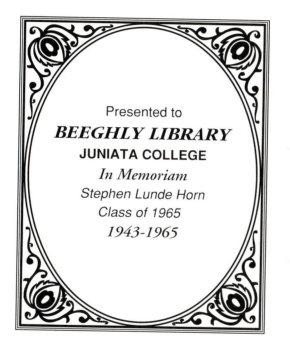

GROUNDBREAKING SCIENTIFIC EXPERIMENTS, INVENTIONS, AND DISCOVERIES OF THE MIDDLE AGES AND THE RENAISSANCE

GROUNDBREAKING SCIENTIFIC EXPERIMENTS, INVENTIONS, AND DISCOVERIES OF THE MIDDLE AGES AND THE RENAISSANCE

ROBERT E. KREBS

**Groundbreaking Scientific Experiments, Inventions
and Discoveries through the Ages**
Robert E. Krebs, Series Adviser

GREENWOOD PRESS
Westport, Connecticut • London

Library of Congress Cataloging-in-Publication Data

Krebs, Robert E., 1922–
  Groundbreaking scientific experiments, inventions and discoveries
of the Middle Ages and the Renaissance / Robert E. Krebs.
    p. cm.—(Groundbreaking scientific experiments, inventions and
  discoveries through the ages)
  Includes bibliographical references and index.
  ISBN 0–313–32433–6 (alk. paper)
    1. Science, Medieval. 2. Science, Renaissance. 3. Technology—
History. I. Title. II. Series.
Q124.97.K73 2004
509.4′0902—dc22   2003060075

British Library Cataloguing in Publication Data is available.

Library of Congress Catalog Card Number: 2003060075
ISBN: 0–313–32433–6

First published in 2004

Greenwood Press, 88 Post Road West, Westport, CT 06881
An imprint of Greenwood Publishing Group, Inc.
www.greenwood.com

Printed in the United States of America

∞™

The paper used in this book complies with the
Permanent Paper Standard issued by the National
Information Standards Organization (Z39.48–1984).

10 9 8 7 6 5 4 3 2 1

To my mother, Anna E. Krebs, who inspired me to seek
my own drummer

CONTENTS

List of Figures   ix

Series Foreword   xi

Introduction   xv

1.   Astronomy   1

2.   Geography and Exploration   33

3.   The Biological Sciences: Botany and Zoology   67

4.   Medicine, Disease, and Health   85

5.   Mathematics   123

6.   Physics and Chemistry   155

7.   Inventions and Innovations   189

8.   Weapons and War   243

Glossary   291

Selected Bibliography   297

Name Index   303

Subject Index   309

LIST OF FIGURES

1.1	Stonehenge	2
1.2	Aristotle's Spherical Crystal "Shells" of the Universe	6
1.3	Ptolemy's Universe	8
1.4	Early Weight-Driven Clock	11
1.5	Chinese Sighting Tube	11
1.6	Indian Observatory	15
1.7	Egyptian Clock Sticks	16
1.8	Kepler's Law of Areas	27
2.1	T-O Maps	56
2.2	Cylindrical Projection	61
2.3	Mercator Projection	63
3.1	Aristotle's Ladder of Life	69
3.2	Monocots vs. Dicots	73
4.1	A Trephined Skull	86
5.1	Conic Sections	126
5.2	Evolution of Arabic Numerals	130
5.3	Ancient Chinese and Arabic Numerals	138
5.4	Mayan Numerals	141
5.5	Stevin's Equilibrium Demonstration	153
6.1	Kepler's Law of Areas	164
6.2	Earth's Magnetic Poles	167
6.3	Electromagnetic Spectrum	169
6.4	Four Elements of Alchemy	179

6.5	Alchemic Symbols	182
6.6	Alchemists' Classification of Substances	187
7.1	Three Types of Middle Ages Abaci	194
7.2	Medieval Arch with Keystone and Flying Buttresses	195
7.3	Early 15th-Century Eyeglasses	208
7.4	Heavy Plow of the Renaissance	224
7.5	Ptolemy's Device for Measuring Small Angles and Triquetrum	226
7.6	Tycho Brahe's 11-Foot Quadrant	227
7.7	6th-Century c.e. Roman Siege Ladder	228
7.8	Low-Whorl Drop Spindle	231
7.9	Four Types of Waterwheels	237
7.10	14th-Century European Wheelbarrow	239
7.11	Persian Vertical Windmill	240
8.1	Mace	245
8.2	Protective Armor	247
8.3	Byrne Scale Armor	249
8.4	Armor of the Middle Ages and Renaissance	252
8.5	"Loophole" for Firing Crossbows	257
8.6	Pentagon Shape of Fortification for Castle	258
8.7	Windsor Castle—Main Tower	261
8.8	Windsor Castle—Round Towers	262
8.9	Chinese Fire Lance	263
8.10	Early Types of Islamic Grenades	263
8.11	Early Chinese Cannon	264
8.12	Sling-Type Trebuchet	265
8.13	Siege Engine	267
8.14	Spear-Shooting Cannon	269
8.15	Chinese Bronze Cannon	269
8.16	Ancient Crossbow	277

SERIES FOREWORD

The material contained in five volumes in this series of historical groundbreaking experiments, discoveries, and inventions encompasses many centuries from the pre-historic period up to the 20th century. Topics are explored from the time of prehistoric humans, the age of classical Greek and Roman science, the Christian era, the Middle Ages, the Renaissance period from the years 1350 to 1600, the beginnings of modern science of the 17th century, and great inventions, discoveries, and experiments of the 18th and 19th centuries. This historical approach to science by Greenwood Press is intended to provide students with the materials needed to examine science as a specialized discipline. The authors present the topics for each historical period alphabetically and include information about the women and men responsible for specific experiments, discoveries, and inventions.

All volumes concentrate on the physical and life sciences and follow the same historical format that describes the scientific developments of that period. In addition to the science of each historical period, the authors explore the implications of how historical groundbreaking experiments, discoveries, and inventions influenced the thoughts and theories of future scientists, and how these developments affected people's lives.

As readers progress through the volumes, it will become obvious that the nature of science is cumulative. In other words, scientists of one historical period draw upon and add to the ideas and theories of earlier periods. This is evident in contrast to the recent irrationalist philosophy of the history and sociology of science that views science, not as a unique, self-correcting human empirical inductive activity, but as just another social or cultural activity where scientific knowledge is conjectural, scientific laws are contrived, scientific theories are all false, scientific facts are fickle, and scientific truths are relative. These volumes belie postmodern deconstructionist assertions that no scientific idea has greater validity than any other idea, and that all "truths" are a matter of opinion.

For example, in 1992 the plurality opinion by three jurists of the U.S. Supreme Court in *Planned Parenthood v. Case* restated the "right" to abortion by stating: *"at the heart of liberty is the right to define one's own concept of existence, of meaning of the universe, and of the mystery of human life."* This is a remarkable deconstructionist statement, not because it supports the right to abortion, but because the Court supports the relativistic premise that anyone's concept of the universe is whatever that person wants it to be, and not what the universe actually is based on: what science has determined by experimentation, the use of statistical probabilities, and empirical inductive logic.

When scientists develop factual knowledge as to the nature of nature they understand that "rational assurance is not the same thing as perfect certainty." By applying statistical probability to new factual data this knowledge provides the basis for building scientific hypotheses, theories, and laws over time. Thus, scientific knowledge becomes self-correcting as well as cumulative.

In addition, this series refutes the claim that each historical theory is based on a false paradigm (a methodological framework) that is discarded and later is just superseded by a new more recent theory also based on a false paradigm. Scientific knowledge is of a sequential nature that revises, adds to, and builds upon old ideas and theories as new theories are developed based on new knowledge.

Astronomy is a prime example of how science progressed over the centuries. Lives of people who lived in the pre-historical period were geared to the movement of the sun, moon, and stars. Cultures in all countries developed many rituals based on observations of how nature affected the flow of life, including the female menstrual cycle, their migrations to follow food supplies, or adaptations to survive harsh winters. Later, after the discovery of agriculture at about 8000 or 9000 B.C.E., people learned to relate climate and weather, the phases of the moon, and the periodicity of the sun's apparent motion to the Earth as these astronomical phenomena seemed to determine the fate of their crops.

The invention of bronze by alloying first arsenic and later tin with copper occurred in about 3000 B.C.E. Much later, after discovering how to use the iron found in celestial meteorites and still later, in 1000 B.C.E.

when people learned how to smelt iron from its ore, civilization entered the Iron Age. The people in the Tigris-Euphrates region invented the first calendar based on the phases of the moon and seasons in about 2800 B.C.E. During the ancient and classical Greek and Roman periods (about 700 B.C.E. to A.D. 100) mythical gods were devised to explain what was viewed in the heavens or to justify their behavior. Myths based on astronomy, such as the sun and planet gods as well as Gaia the Earth mother, were part of their religions affecting their way of life. This period was the beginning of the philosophical thoughts of Aristotle and others concerning astronomy and nature in general that predated modern science. In about 235 B.C.E. the Greeks first proposed a heliocentric relationship of the sun and planets. Ancient people in Asia, Egypt, and India invented fantastic structures to assist the unaided eye in viewing the positions and motions of the moon, stars, and sun. These instruments were the forerunners of the invention of modern telescopes and other devices that made modern astronomical discoveries possible. Ancient astrology was based on the belief that the positions of bodies in the heavens controlled one's life. Astrology is still confused with the science of astronomy, and it still is not based on any reliable astronomical data.

The ancients knew that a dewdrop of water on a leaf seemed to magnify the leaf's surface. This led to invention of a glass bead that could be used as a magnifying glass. In 1590 Zacharias Janssen, a spectacle-maker, discovered that two convex lenses, one at each end of a tube, increased the magnification. In 1608 Hans Lippershey's assistant turned the instrument around and discovered that distant objects appeared closer, thus the telescope was discovered. The telescope has been used for both navigation and astronomical observations from the 17th century up to the present time. The inventions of new instruments, such as the microscope and telescope, led to new discoveries such as the cell by Robert Hooke and the four moons of Jupiter by Galileo, who made this important discovery that revolutionized astronomy with a telescope of his own design and construction. These inventions and discoveries enabled the expansion of astronomy from an ancient "eyeball" science to an ever-expanding series of experiments and discoveries leading to many new theories about the universe. Others invented and improved astronomical instruments, such as the reflecting telescope combined with photography, the spectroscope, and Earth-orbiting astronomical instruments resulting in the discovery of new planets, galaxies, and new theories related to astronomy and the universe in the 20th century. The age of "enlightenment" through the 18th and 19th centuries culminated in an explosion of new knowledge of the universe that continued through the 20th and into the 21st centuries. Scientific laws, theories, and facts we now know about astronomy and the universe are grounded in the experiments, discoveries, and inventions of the past centuries, just as they are in all areas of science.

The books in the series *Groundbreaking Scientific Experiments, Discoveries and Inventions* through the Ages are written in easy to understand language with a minimum of scientific jargon. They are appropriate references for middle and senior high school audiences, as well as for the college level nonscience major, and for the general public interested in the development and progression of science over the ages.

Robert E. Krebs

INTRODUCTION

This volume encompasses the science of the periods in history known as the Middle Ages and Renaissance. It is the second reference book in the Greenwood Press series *Groundbreaking Scientific Experiments, Inventions and Discoveries through the Ages,* and it encompasses the science-related material that is often neglected in the philosophy and history works of these periods. The terms Prehistoric, Ancient, Classical, Dark Ages, Medieval/Middle Ages (early and late), Byzantine, Renaissance, Age of Enlightenment, Pre-Modern, Modern, and Post-Modern Ages are all rather relative and dependent upon the context in which they are used, as well as the dates that identify them. The book is intended as a reference for high school and college/university students, as well as for the general reader who seeks information about science, discoveries, inventions, and technologies and the men and women involved in these activities for the centuries between 500 and 1600 C.E.

Technical and important words are highlighted in bold type when first used in the text and are included in the glossary. A bibliography of selected references and separate name and subject indexes are included.

Note: The following abbreviations and conventions are used in this book:

- **B.C.E.** denotes events "Before the Common Era," or "Before the Christian Era."
- **C.E.** represents dates during or after the "Common Era" or "Christian Era."
- **B.P.** means years "Before the Present Time or Year."
- **ca.** (for circa) is used when an exact date is not known (approximate date).

- **fl.** means "flourished" for this time in history.
- **mya** stands for "Millions of Years Ago."
- ~ is placed in front of an estimated quantity.

A Short Background and History

Historians often have problems establishing the dates for periods that encompass important historical events, trends, and developments. An example is the period of history extending from the fall of Rome to the Age of Enlightenment, better known as the Middle Ages and Renaissance (from about 500 to 1600 c.e.). This period is a slice of history extending from darkness to an awakening of human potential when men and women sought new knowledge. After an interval of intellectual stagnation, the later years during this period are known for the advancement of science and new pragmatic inventions that led to improved technologies and great discoveries that significantly advanced civilization. Obviously, a tremendous amount of history preceded this age of rebirth of literature, art, philosophy, and science, when there were no sharp delineations between historical periods. Dates that designate the beginning and ending of an historical period have been assigned as a paradigm for pedagogical purposes.

Following is an attempt to set nonspecific dates for identified historical periods important to scientific and human developments:

- **Prehistoric Ages**—ca. 5,000,000 to about 12,000 b.p.:

This period of prehistory ranges from approximately the time in the past when the ancient hominid *Australopithecus* roamed central Africa to about 40,000 b.p. when modern humans known as *Homo sapiens sapiens* evolved. At the end of this long period of prehistory, at about 8,000 to 9,000 b.c.e., humans became settled farmers and began to domesticate animals. This led to the settlement of communities and the division of labor that, over time, provided more leisure for the development of the arts, technologies, and science.

- **Early civilizations**—about 6000 to 500 b.c.e.:

During this period people from the Far East (China, India), Middle and Near East (Mesopotamia, Semite regions), and the Mediterranean region (Egypt, Greece, Rome) communicated with written language; wrote poetry; developed calendars and other astrological devices and theories; wove fabrics; made pottery; worked with metals; used oil lamps; designed simple boats and sails; and improved spears, bows and arrows, and other weaponry. These developments seem to have spurred further advancements in agriculture, innovative manufacturing, art, war, and to some extent, science. This was the time when civilization, as we think of it, began.

- **Classical Age**—ca. 500 B.C.E. to ca. 500 C.E.:

During this time in history the Eastern cultures of India and China, as well as those of Greece and Rome, flourished. This was also the period when Western cultures exhibited innovations and creative approaches to solving the mysteries of nature that differed from Eastern civilizations. In addition, information, knowledge, and technology from China, India, and Arabia were transported to the Mediterranean region. The literature, art, science, and intellectual developments of the Classical Period influenced later historical periods, including contemporary approaches to science and philosophy.

- **Dark Ages**—ca. 450 to ca. 1300 C.E.:

This is a somewhat derogatory term that has been consistently used to describe the years after the fall of Rome ca. 450 C.E. It also coincides with the period often referred to as the Early Middle Ages. Although there are evidences of culture and enlightenment during these years, this period represents a decline in general learning and innovative technologies. However, it is a mistake to assume that there was a dearth of contributions to civilization during this interval. For example, China built boats with waterproof bulkheads, invented paper (later manufactured in Europe), devised block printing (later improved in Europe), constructed complex canals with waterwheels, and manufactured explosive devices using black powder (also later improved in Europe). The main deficit of the so-called European "Dark Ages" was the reluctance, and, in many cases, the refusal to look beyond the theological dogma and scholasticism of the early European Catholic Church, which dominated most areas of life on the European continent during this period.

- **Middle Ages**—ca. 500 to about 1300 C.E.:

The period referred to as the Middle Ages is sometimes divided into the Early (450–500 to 1100–1200 C.E.) and the Late Middle Ages (1200–1300 C.E.). Note: While there is little agreement among historians as to the exact dates and terminologies used for these periods, there is general agreement as to the importance of many events that transpired during these historical times.

As mentioned, the Early Middle Ages is another term often used to describe the Dark Ages. It is the period of European history between antiquity and the Renaissance. Some historians consider both the Dark Ages and Late Middle Ages as preludes to the literary and scientific reawakening of the Renaissance. Others consider the Middle Ages as part of the Medieval Period, which is also not well delineated in history. Some historians believe the Medieval Period, a period of intellectual darkness, extended from the end of the Classical Greco-Roman Period into the early part of the Renaissance.

• The Byzantine Era

This period of history somewhat coincides with the Dark and Middle Ages, as it encompasses several similar events of history. The Byzantine Era began when the Roman Empire was being divided between the western Latin empire and the eastern Arabic branch of the United Roman Empire. The capital of the eastern empire was Constantinople (present-day Istanbul, Turkey). Constantine the Great, for whom Constantinople was named, was the 4th-century Roman emperor who established Christianity as the religion for the entire Roman Empire. The Roman Empire had extended its rule both east (as far as Palestine) and west (to northern England) to such an extent that it could no longer effectively govern people of diverse cultures. Subsequently, after the fall of Rome in ca. 476 C.E., the eastern Byzantine Empire became the successor to the Roman Empire and the Christian Church.

Unlike most other historical periods, the Byzantine Era began and ended on rather definite dates. It began in 395 C.E. with the partitioning of the Roman Empire to form the eastern Roman Empire under the dynasty of Theodosius. During the next 1,000+ years the Byzantine Roman Empire was ruled by approximately 12 dynasties, as it developed its own independent form of Christianity known as the Eastern Orthodox Church. The Byzantine period's ending dates also vary, but it is generally agreed to have ended on Thursday, May 29, 1453, with the capture of Constantinople by the Turks and by Mehemet II of the Ottoman Empire. The term "Byzantine" is best used as an adjective to describe the science, art, literature, architecture, and other cultural activities during these years preceding the Renaissance. The term also refers to a type of bureaucracy that emphasizes formal rules and regulations.

- **The Renaissance**—ca. 1300 to ca. 1600 C.E.:

Many historians dispute the beginning, duration, and ending dates for this historical period, as well as its historical significance. This period of 300 years is sometimes referred to as the Late Middle Ages. Some claim that the Late Middle Ages and Renaissance began in the 11th or 12th century; many contend that it commenced in the 14th century and continued into the 16th century. Some extend its existence into the mid-17th century. Thus, there is little agreement with respect to the years to which the term "Renaissance" applies. What is not in dispute, however, is the fact that something unique occurred during the 14th, 15th, and 16th centuries, and even into the early 17th century. For the purposes of this book, the thousand plus years between 500 and 1600 C.E. will be used to delineate the periods referred to as the Middle Ages and the Renaissance as a unique time in history for both scientific development and the reawakening of classical science leading to new discoveries, inventions, and technologies.

Some historians consider the whole idea of a Renaissance a myth that originated during the latter part of the 15th century. This particular belief is based on the theory that literature, art, philosophy, and science progressed on a more or less continuum from the Classical Greco-Roman days, through the Dark and Middle Ages to the 14th and 15th centuries. Some historians absolutely deny the concept of historical periods, *period*. They claim that breaking the continuum of history into periods or ages is pure fiction. Therefore, it doesn't make sense to discuss the Middle Ages and Renaissance or other "periods" in their own contexts. Others claim that for pedagogical purposes it does makes sense to examine our past by identifying events that more or less exemplify the course in which civilization flowed. A number of historians consider the 5th to the 14th centuries as a relatively stationary period of learning, knowledge, and science. Actually, there was a long, relatively intellectually barren period between the Classical Greco-Roman era and the more modern Europe of the 14th or 15th century. Others contend that many scientific developments and discoveries originated during the Middle Ages and, with the reexamination of Classical Greco-Roman literature and science during the Renaissance, a substantial amount of scientific progress was made between the dates 500 to 1600 C.E.

Events in Italy led to the Humanistic Renaissance, which later spread to northern Europe, becoming known, not surprisingly, as the Northern Renaissance. In this context, the word "humanism" described the educational system that emphasized the classical languages of Latin and, later, Greek. Thus, the Renaissance might be called the rebirth of classical literature that expressed new ideas. The Humanistic Renaissance also resulted in a new method of criticism of the theories and treatises that greatly influenced the progress and development of the sciences. As a result, the Renaissance developed a cult of excellence for both scholarship and scientific thinking. The scientific revolution was a new and unique application of scientific ideas and methods that opened the door to a more realistic understanding of the structure of and the relationship between humans and nature. Humankind's ability to develop an abstract understanding of the environment, coupled with intellectual achievements through the ages, has resulted in an unprecedented understanding of the natural world and human nature, as well as the entire universe.

The historical periods of the Middle Ages and the Renaissance overlap and are often treated separately, even though much of the science, discoveries, inventions, and technologies flow from one century to the next. As mentioned, this book covers the years from 500 to 1600 C.E. For clarification, reference will be made to either the Middle Ages or the Renaissance—from a period of intellectual darkness to one of enlightenment. The Late Middle Ages also relates to the Renaissance, a French term meaning "renewal," or "reawakening." It is often described as the period in history when the European rediscovery of ancient Greek and Eastern knowledge, including science, philosophy, art, and, to a lesser extent, religion led to an awakening of new ideas, ways of thinking, and concepts about nature. Today, the term "Renaissance man" refers to a person (of either gender) whose knowledge is both specialized and all-encompassing. The Renaissance man's understanding of the world, science, and society is eclectic, laced with wisdom tempering unsubstantiated predictions and revelations.

Some historians emphasize the Renaissance as the blooming of the humanistic spirit as Europe emerged from the Dark Ages, a period that began with the fall of Rome in ca. 476 C.E. and lasted almost 1,000 years until the Renaissance. Most of the learning, including theology, philosophy, and science, during the Dark and Middle Ages was the province of a relatively elite cadre of literate religious and noble individuals. This period was characterized by the practice of scholasticism, a term used to describe the dogmatic adherence to dominant theological and philosophical scholarly methodologies enforced by early Church fathers. Persons who questioned dogmatic authority and sought new knowledge and understanding about nature and how things work were dealt with severely. During this period, and even later, many who challenged Church authority lost their freedom or their lives, usually by burning at the stake. Human progress from antiquity to the present can be illustrated by the accumulation of scientific developments. Some historians of science contend that scientific progress is not a steady accretion of scientific discoveries and events, but rather a progression of spurts of creative advancements followed by periods of intellectual and creative apathy. The Dark Ages that followed the fall of Rome and lasted through the early part of the Middle Ages is an example of a period of stagnation in science, creative thinking, and civilization in general. Most historians consider the Renaissance as a revival of the classical systems of philosophy and inquiry. This reawakening first developed in the 14th century when some intellectuals of the Roman Catholic Church began to question the Church's leadership authority, corruption, and its inhumane treatment of nonbelievers. Freedom from Church laws that hindered learning for many centuries was followed by the awakening of discovery, science, and technology during the Renaissance, which continued into the later period of history known as the "Enlightenment."

Humans have always believed, and still do, that the past was better than the present or the future and that the search for *real* meaning and truth was, and is, a search for ancient beliefs, knowledge, and methodologies. This mode of thinking led to the concept known as "humanism," where the ideals of progress, revival, and rebirth led to a return to the knowledge and philosophies of the pure civilizations of the past (Greco-Roman, Arabic-Indian, and, to a lesser extent, the Orient) that were neglected during the Pre-Renaissance. The thousand-year period following the decline of Rome up to the Renaissance is also referred to as the Hermetic Period. *Hermetica* is based on the large collection of Greek literature developed under the name of "Hermes Trismegistus" (meaning Hermes the great, great, great; the Greek god in his Egyptian manifestation as Thoth) from the 1st to the 3rd centuries C.E. This collection of Greek literature covered mystical and scientific writing of the time (but not what we today consider science), as well as theology and philosophy. The Hermes Trismegistus writings are a collection of Egyptian and Greek philosophy with Eastern religious elements, including the writings of Plato, the Stoics, and Pythagoreans. The Muslims who later brought this knowledge westward during the Islamic conquests of north Africa and southern Europe originally translated it to the Arabic language. Hermetic philosophy and its related mystical knowledge were founded in the powers of the stars as astrology, the secrets of stones, and the mysteries of plants (alchemy). Alchemy originated and flourished in the early centuries of the first millennium c.e. in Eastern and Arabic countries. At the same time, many of the original Greek writings found their way eastward, where they were translated into Arabic. Later, as Islam spread westward, many of the Arabic translations of Greek literature were returned to Europe. Some historians question the extent to which Islamic countries contributed new scientific ideas and concepts beyond the preservation of Greek documents through their translations. This was also the period when the European Church was intolerant of new or unorthodox ideas, and any creative thinking or the development of new knowledge was considered blasphemous. The disciples of Islam, acting on the words of the Prophet Muhammad, first conquered India and then turned westward, occupying northern Africa (Egypt and Morocco), several Mediterranean countries, and advanced as far as southern Spain, which was also known as Al-Andalus. As Islam spread westward, so did its alchemy, architecture, art, and religion, and, to a lesser extent, its language. At the same time, Europe lost touch with both the ancient Greek and Roman writings, while facing the dilemma of a language in which the population was not conversant. Although there are a number of words of Arabic origin in European languages, including English, the general population of southern Europe never adopted the Arabic language of the Muslims. The powers of that time did not encourage an educated public, and thus outside the Holy Roman Church ignorance and illiteracy were widespread. It was sometime after the end of the Crusades (early 13th century) that the Arabic texts on science and the Muslim translations of Greek literature were retranslated into Latin. (Note: There were four major, plus several minor, Crusades during the Medieval period: 1st 1096–1099, 2nd 1147–1149, 3rd 1189–1192, and 4th 1202–1204.) Robert of Chester (England) in 1144 c.e. produced the first translation of Arabic science and literature into Latin. Most of his translations related to alchemy, a basically Eastern belief and practice that spread throughout Western Europe. It was not until many years later, after the general use of the printing press, that Arabic and Latin texts were translated into vernacular languages.

Humanism, like Hermeticism, was a philosophy that attempted to revive the purity of classical Greco-Roman learning. The term "humanism" is thought to have originated in Germany in the 19th century as a description for the renaissance of Greek and Roman classical studies and writings. Humanists (or neoplatonists) were Renaissance scholars who studied the language and culture of ancient Greece and Rome.

Humanism might be thought of as the study of the past to obtain wisdom for the direction of knowledge of the present and future. While it did not provide the impetus needed for the future advancement of science and civilization, humanism was an improvement over the repression of the Dark or Early Middle Ages. The terms "humanism" or "humanist" have an entirely different meaning today. The modern word "humanity" is related more to a concern with humans based on concepts of benevolence, mercy, compassion, tolerance, warm-heartedness, and the like. The Renaissance concept of humanism was based more on a philosophy of action, curiosity, and learning than just warm feelings toward fellow humans. Sixteenth-century humanists were not only persons of honor, but also exhibited characteristics of fortitude, judgment, prudence, and eloquence. Their philosophy and concept of education and learning exerted an influence that persisted beyond the Middle Ages and the Renaissance and is still evident in the desire for knowledge and understanding of nature that exists in most current civilizations. In other words, university science departments are more related to the philosophies of the humanists of the Renaissance than the current departments of the humanities are.

A major event that contributed to the advancement of knowledge during the Renaissance was the use of moveable metal type and the invention of an efficient printing press by Johannes Gutenberg in Mainz, Germany, in 1450. Although moveable wooden-block characters had been used for printing in China and elsewhere, Gutenberg's press was the first to efficiently make printed material generally available. His innovations of a previous invention freed the general public from illiteracy by creating a demand for learning. A second major contribution is the reformation of the late medieval Catholic Church in the 16th century. The break with the Church's power, wealth, and spiritual corruption was another development that contributed to the emergence of reformers with new ways of thinking. Church reformers, such as Peter Waldo, Jan Hus, John Wycliffe, Martin Luther, and John Calvin, were partially responsible for the rebirth of learning and knowledge associated with the Renaissance. The Feast of All Saints on October 31, 1517, the day on which Martin Luther posted his 95 Theses on the door of the Castle Church in Wittenberg, is usually considered the beginning of the Reformation and Protestantism, as well as setting the stage for the Age of Enlightenment.

During the period of history from Aristotle to the fall of Rome, the theories of the Greeks and others had advanced about as far as their philosophical investigations would allow. Their hypotheses were stalemated because they never went beyond empirical classifications of nature. The ancient Greeks neither conceived of nor practiced experimentation to verify their observations and concepts or made viable predictions to arrive at universal physical laws. Also, during the period from the fall of Rome to the Renaissance, learning, experimentation, and questioning existing dogma were just not tolerated. Objective observation and experimentation were the major processes of science that became the engines that drove scientific progress through the end of the Renaissance into the present century.

The scientists of the Late Middle Ages and the Renaissance were mostly scholars, physicians, and/or alchemists who led the revival of Greek literature and science. Two English physicians, Thomas Linacre (ca. 1460–1524) and John Caius (1510–1573), spent much of their time translating Greek medical texts for use by European physicians, as they considered Greek medicine superior to that practiced on the European continent. Two German astronomers of the Renaissance, George Peurbach (1423–69) and Johann Regiomontanus (1436–76), translated volumes of Greek literature and science, which proved to be a tremendous contribution to humanism as well as science. Interestingly, they used their new information in popular presentations to the general public rather than as academic professors. (The pay was much greater.) In time, Latin translations of Greek and Arabic science were retranslated into vernacular languages rather than Latin. These translations created a new readership of nonscholars. This was particularly true after the printing presses of the Renaissance became a reality. Although the mystic sciences of astrology and alchemy were brought to Spain before the Renaissance, they did not become popular in Europe until the 13th and 14th centuries. The scientists and philosophers of the Late Middle Ages and the Renaissance did not have an astute understanding of the difference between science (the generating of new knowledge) and technology (the application of knowledge). Even today, many people have difficulty understanding the distinction between science and technology, partially because in some new areas of our understanding of nature, the distinction is becoming immaterial, for example, nanotechnology and quantum mechanics. However, in other areas, it is obvious, such as with space exploration, where there is a distinction, as well as a dependency, between the application of scientific principles in engineering technology (e.g., rockets and satellites) and the seeking of new scientific information through the use of this technology (astronomy, etc.). The concept of experimental science became more refined during the late Middle Ages and the Renaissance, as did the advancement of technology, but neither had their origins during this period of history. The same can be said for many important geographical and other types of discoveries. Science in the Renaissance was known as "natural magic" related to the study of magnetism, optics, chemistry, and medicine. Mathematics was a form of natural magic or numerical mysticism. (Numerology was used primarily to predict the future.) However, not all science and technology of the Middle Ages and the Renaissance was of the occult (secret) or magical (deceptive). Exploration, cartography, manufacturing of tools and instruments, mining and metallurgy, the proliferation of specific crafts, and the use of Hindu-Arabic numerals in simple everyday arithmetic spurred the advancements of both science and technology and, ultimately, civilization.

CHAPTER 1

ASTRONOMY

Background and History

Gazing up at the night sky and wondering about the objects wandering against the bright specks of light in the night sky may have been the basis of early religious and philosophical ideologies. Curiosity about the heavens sparked the study of astronomy as a rudimentary science at least 5,000 years ago. It was a science in the sense that ancient humans recognized patterns in the organized and regular motions of stars and other celestial objects, just as present-day scientists search for patterns and trends in natural phenomena. As far as we know, the sun has always risen in the east and set in the west, and the moon regularly appeared with its waxing and waning, thus establishing easily recognized patterns. These and other events became incorporated into the cultures and lifestyles of humans living in all geographical regions. After dusk, as the stars in the night sky appeared in the east, they slowly rotated westward in unison around a fixed point in the sky (the north celestial pole) while maintaining their relationships to each other. Both the bright and dim luminosities all seemed to follow the same pattern, except for a few moving bodies that appeared to wander eastward, remain stationary for a short time, then change direction heading westward, and again reverse their direction eastward. Therefore, they were known as "planets," which is the Greek word for wanderers. Historically, humans learned to use these patterns and regularities as calendar events for organizing human activities such as the planting of crops and religious ceremonies, and as navigational guides for sailors who used the predictable movements of these celestial bodies. One example of an early astronomical instrument is the construction of the Stonehenge monument that is located in southern England. (See Figure 1.1.) The large,

Stonehenge as seen on the summer solstice - June 21

Figure 1.1 Stonehenge
Stonehenge, located in southern England, is a grouping of large stones dating back to about 3000 B.C.E. Their configuration is thought to be oriented with the solstices and moon cycles.

precisely positioned stones of the monument that date back to possibly 3000 B.C.E. were oriented to align with the summer and winter solstices, as well as the moon's cycles, presumably to predict eclipses. The ancient civilizations of China, India, Mesopotamia, Greece, Egypt, and medieval Europe all built stone structures to study and measure celestial events. Their discoveries were the foundation upon which astronomical knowledge advanced under the scholarship of medieval and Renaissance astronomers.

Early astronomers from many countries recognized that the wandering bodies (planets) were unique in the night sky in that they moved about in relation to the other stars. It was sometime before these nonstationary bodies in the heavens were correctly associated with a solar system of planets, moons, and comets revolving around a central star. The entire heavens were seen as the universe, with specific regions related to the various celestial objects. A vast number of concepts related to astronomy were presented from about the third millennium B.C.E. to 500 B.C. by observers from many countries. As a background to the understanding of the nature of astronomy of the Middle Ages and Renaissance, it is helpful to be aware of the early ideas, theories, and discoveries that influenced these ancient observers of the sky and their concepts explaining the nature of the stars, comets, planets, and Earth's place in the universe.

Following is a summary of some of the astronomical systems dating back 3000 to 5000 years B.C.E. that were established by ancient Mesopotamians, Chinese, Indians, Egyptians, Europeans, and Greek astronomers.

Mesopotamia was a country in southwest Asia between the Tigris and Euphrates Rivers. It consisted of several cities, tribes, and civilizations primarily in the region that is present-day Iraq. Some of these ancient tribes included the Sumerians, Semites, Akkadians, Hittites, and Assyrians. The Babylonian Kingdom was a major commercial city whose culture spread throughout the entire Mesopotamian world. The people of the cities built massive towers called ziggurats for viewing the heavens. The ziggurat located in the ancient Sumerian city of Uruk dates back at least 7,000 years B.P. and was used for viewing celestial cycles for astrological purposes as well as for religious ceremonies. The Babylonian astronomers, more accurately called astrologers, used arithmetic rather than geometry to observe the moon and calculate its regular $29\frac{1}{2}$-day progression as well as its phases. Observing how the constellations moved across the western sky throughout the year, they identified zodiacal constellations, developed star maps, calendars, and recognized the cycles of the moon. They also viewed and recorded the first day of each lunar month and were able to ascertain the seasons by tracking the visible stars in the evening and morning skies. By about 1800 B.C.E. these ancient astrologers used this information plus the sun's movement to develop a calendar that was used for over 1,200 years, maintaining records that proved useful for subsequent astronomers to refer to even in the Middle Ages and Renaissance.

Ancient Chinese astrologers/astronomers were not particularly scientific in the sense that Western science and astronomy developed deductive methods of inquiry. Even so, they were keen observers of the skies, and they developed some of the earliest astronomical (and other) instruments. One accomplishment was their calculation of what is known as "the cycle of nineteen years" by the year 3000 B.C.E. This cycle consists of 235 lunar periods and is the point where both the solar and lunar years are harmonized. They also acquired a vast amount of astrological information from the Babylonians that enabled them to develop a cycle to predict both lunar and solar eclipses and instruments to measure the ecliptic's obliquity—at least 500 years before the west accomplished the same tasks. One of the differences between early Chinese and Western astrology/astronomy was that the Chinese were more interested in and attached more importance to the north celestial pole and the celestial equator stars, while western astronomers were more interested in the celestial horizon stars. The Chinese also related the stars' positions to the sun to develop a $365\frac{1}{4}$-day year, as well as a model of a spherical heaven. During ancient times and well into the Middle Ages the Chinese collected great numbers of star maps and records of their astrological observations and measurements that prove, even today, valuable for astronomers. Ancient Chinese astronomy was more of a practical, everyday astrology than a scientific approach to understanding the principles associated with their observations.

Egyptian astronomy, which dates back thousands of years, used simple astronomical methods to measure time and to develop accurate calendars, as well as to align their buildings with the stars. They constructed elaborate temples and sphinxes to worship the sun and stars and to regulate their civil and religious lives. The ancient Egyptians attained technical skills that enabled them to develop a constellational system different from that of other countries.

Just as with the people of most other continents, the inhabitants of ancient India related their observations of the heavens to their spiritual existence as well as their belief in the supernatural. Mesopotamian astronomy influenced India at least as far back as the 5th or 6th centuries B.C.E. The Persian conquest of India brought the concept of Aristotle's concentric spheres, combined with Ptolemy's epicycles, to the subcontinent. These ideas were troublesome for some Indian astrologers who were primarily interested in astronomy for religious purposes and astrological predictions. The Indians constructed large astronomical and time-determination instruments that were used to develop their calendars for observing religious ceremonies. (See Figure 1.6.)

Ancient Europeans erected megalithic monuments—the word *megalith* is Greek for "large stone." Some very large slabs of stones were used to construct monuments and tombs where the slabs were placed standing individually or grouped together in patterns. Many are located in western Europe, southern England, and Ireland. In the region of the British Isles, Druids and Celts formulated primitive calendars and rituals to coincide with lunar and solar cycles. Archaeological evidence from ancient Germanic and Swedish tribes strongly suggests that these ancient peoples developed rituals around the winter and summer solstices.

Following is a quotation from *History of the World*, by J. M. Roberts (1993, p. 127), describing a famous megalithic monument located in southern England.

The most complete and striking megalithic site is Stonehenge, in southern England, now dated to about 2000 B.C. What such places originally looked like is hard to guess or imagine. Their modern austerity and weathered grandeur may well be misleading; great places of human resort are not like that when in use and it is more likely that the huge stones were daubed in ochres and blood, hung with skins and fetishes. They may well often have looked more like totem-poles than the solemn, brooding shapes we see today. Except for the tombs, it is not easy to say what these works were for, though it has been argued that some were giant clocks or huge solar observatories, aligned to the rising and setting of sun, moon, and stars at the major turning-points of the astronomical year. Careful observation underlay such work, even if it fell far short in detail and precision of what was done by astronomers in Babylon and Egypt. (See Figure 1.1.)

Ancient Greek astronomy was mostly myth-based astrology. However, the Greeks developed geometry that provided a more solid basis for developing the science of astronomy than did the arithmetic base used by most other nations. In addition, the Greeks were, to some degree, empiricists and applied deduction to their studies of the heavens. For instance, even though Aristotle developed some elegantly symmetrical models of the universe based on his limited observations, he was still incorrect in his theory of the Earth as unmoving and thus at the center of all things. Other Greek astronomers, by deduction, considered that the Earth moves around a central sun just as do the other planets. Ancient and later classical Greek astronomy was introduced to the Middle East, India, and Islamic countries from which it returned westward to Europe during the Muslim invasions of the early Middle Ages.

Following are brief summaries of a few ancient Greek astronomers (pre-Middle Ages) who made important contributions to the field. Some of their astronomy concepts and theories were widely disseminated and lasted for over 1,500 years until the Renaissance.

1. **Thales of Miletus** (624–548 B.C.E.) not only explained how eclipses occurred but, by using geometry, accurately measured the sun's diameter. He also explained how Ursa Minor could be used for navigational purposes.

2. **Parmenides of Elea** (ca. 515 B.C.E.) provided one clue to the concept of a revolving system when he proposed that the Earth is a sphere. Later this idea was expanded to include a spherical heaven.

3. **Pythagoras** (572–492 B.C.E.) deduced that the Earth must be a sphere by observing ships disappearing over the horizon. He suggested that all other celestial objects were also perfect spheres.

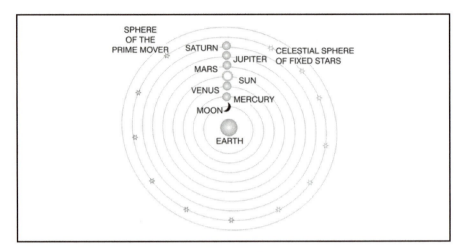

Figure 1.2 Aristotle's Spherical Crystal "Shells" of the Universe
Aristotle's concept of the universe consisted of a series of concentric crystal
spheres with the stationary Earth in the center and a Prime Mover as the power to
run the system.

Pythagoras' study of harmonics led him to his theory called the
Harmony of Spheres. He was the first to explain that the planet
Venus was both the morning and evening star.

4. **Anaxagoras** (ca. 500 B.C.E.) proposed that the light from the
   moon was reflected or false light, and that the moon was Earth-
   like due to its surface features.

5. **Eudoxus of Cnidos** (ca. 400–347 B.C.E.) arrived at a theory of a
   spherical universe based on the circular motions of planets. His
   system required a motionless Earth and 27 crystal-like spheres.
   The sun and moon each had three spheres and each known
   planet had four. His application of mathematics, especially geometry, contributed to the development of the concept of apparent
   motion of the fixed stars and constellations over the period of
   one year.

6. **Heracleides of Pontus** (ca. 388 B.C.E.) used stars to determine that
   the Earth rotates daily on its axis.

7. **Aristotle** (384–322 B.C.E.) developed a geometrical model of the
   universe, first proposed by the Pythagoreans, that included a central fire around which celestial bodies moved in a circular fashion.
   To account for the motion of stars, the sun, the moon, and the
   five moving planets, he developed the concept of uniform circu-

lar motion. His model consisted of 56 spherical "shells," or concentric crystal-like spheres, with the Earth in the center. (See Figure 1.2.)

Aristotle claimed the outer spheres of the heavens did not change but that the spheres near the Earth did change. This inner region consisted of his basic elements: earth, water, air, and fire, plus the transparent, unchanging matter of the universe that he called the *aether.* Aristotle's model and those of other early astronomers were accepted for centuries. Aristotle's physical model of the universe was later used by Ptolemy (ca. 100–170 C.E.) to provide the formal motions to the planets, even though his geocentric solar system required epicycles to describe the apparent backward motion of some planets. Much later, in 1538, Girolamo Fracastoro (ca. 1478–1553) proposed the ultimate elegant model of the universe, consisting of 77 concentric spheres with Earth at the center.

Aristotle conceived of two types of motion: *forced* and *natural. Forced motion* requires that for something to move, it must be pushed. When the push stops so does the object's motion. *Natural motion* requires no push, which he explained as the reason why objects move in the heavens. He believed that the natural motion of celestial bodies was caused by a *prime mover* (supernatural power) that was located in the first (outer) sphere of the heavens. The prime mover was considered the physical explanation for the "ultimate cause," which was adequate to explain the motion of the bodies located in the spheres. He claimed that it is natural for something to fall toward the center of the universe, that is, the Earth. This was one of Aristotle's rationales for placing Earth at the center of his concentric spherical model of the cosmos. Another reason Aristotle placed Earth in the center of his system was that he believed that if the Earth rotated on its axis, we would fly off into space. He did not accept the idea proposed by some astronomers that a spinning Earth could also partially explain the observed motion of heavenly bodies.

8. **Aristarchus of Samos** (ca. 320–ca. 250 B.C.E.) is credited as the first to develop a sun-centered model of the cosmos. His heliocentric treatise was written about two centuries after two 5th-century (B.C.E.) Greek philosophers, Philolaus and Hicates, suggested that the Earth, being a sphere, revolved around a central "fire." Aristarchus proposed his sun-centered cosmos more than 1,600 years before Copernicus explained his heliocentric theory that was published in 1453 C.E. Aristarchus was also the first to use mathematics to determine the distance between the Earth and the sun. He formed a triangle between the moon, Earth, and sun

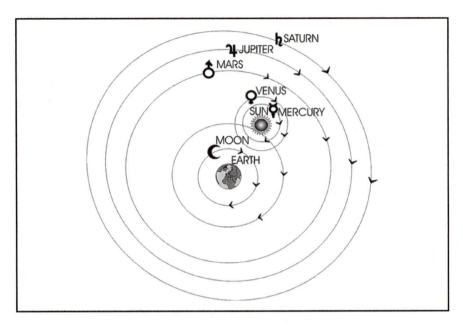

Figure 1.3 Ptolemy's Universe
Ptolemy's concept of the universe required a series of sphere-like epicycles that explained the paths' bodies followed as they revolved around the Earth.

and applied known geometry of right triangles to solve the problems. The result was not very accurate, but it was one of the first uses of mathematics for astronomy.

9. **Hipparchus** (ca. 170–125 B.C.E.) of Rhodes, considered by some historians as the greatest astronomer of ancient times, was one of the first to measure the apparent movement of some stars around a fixed point. He was also the first to realize that the Earth moved—not the stars. This information was later used as the basis for the heliocentric model of the universe. Hipparchus cataloged over 850 stars, including their latitudes and longitudes, and accurately measured the distance of the sun and moon from the Earth. His writings enabled later astronomers to accurately predict lunar and solar eclipses.

10. **Ptolemy** (ca. 90–170 C.E.) is best known for his accurate measurements and predictions that corrected many irregularities in the historical models of the solar system and cosmos. By applying the arithmetic of the Babylonians and the mathematics of Indian

and Persian astronomers, combined with the geometry of the Greeks, Ptolemy employed the concept of the ecliptic, the eccentric, and the epicycle to explain the retrograde motion of the planets. The use of epicycles did not require the concept of a concentric set of spheres. (See Figure 1.3.) Ptolemy maintained that only the spheres required for his concept of 80 epicycles were necessary to explain how other bodies revolved around the Earth. Although he was a competent mathematician for his time, he had difficulty adjusting his epicycle model to actual observations, especially when considering the rate of motion of the planets. Ptolemy's systematic geocentric scheme, including the numerous epicycles used to explain planetary motions, was generally accepted in most of the world well into the Late Middle Ages and Renaissance.

Astronomy of the Middle Ages and Renaissance

The classical Greek period of history was rich with intelligent inquiry into the nature of the universe. Unfortunately, in the Western world this fertile period was followed by a long dearth of science, in general, and astronomy, in particular. After the fall of Rome in ~476 c.e., Europe was a barbaric region inhabited by exceedingly impoverished peoples who eked out a meager subsistence through farming. It was a turbulent time in which scholarly pursuits were limited and contained within the walls of monasteries of the Roman Catholic faith. This changed with the Islamic western invasion that began in the 7th century and which continued to dominate the north African coastal regions and the Spanish peninsula until about the 14th or 15th century. In particular, Spain was an important center for the advancement of Islamic science, as several well-known schools of learning were founded in Andalusia. (Al-Andalus was the Islamic term for Spain.) Scholars from other parts of the European continent traveled to Spain to partake of the repository of Islamic knowledge in the well-known schools of learning located in Córdoba, Toledo, Seville, and Valencia. For centuries medieval European astronomy focused on Aristotle's and other ancient Greeks' contributions to astronomy that had been accepted for generations, both in the West and East, concerning many areas of science. The Ptolemaic beliefs in the geocentric system were also accepted during most of the Middle Ages. In the 12th century Adelard of Bath (ca. 1090–ca. 1150 c.e.) introduced some inaccurate cosmological concepts from India that further confused European astronomers. Not discounting the importance of the Islamic influence on European astronomy, other ancient people living in various regions on the European continent were concerned with celestial mechanics.

In the British Isles, Druids and Celts and their ancestors formulated primitive calendars and rituals to coincide with lunar and solar cycles. Archaeological evidence from ancient Germanic and Swedish tribes strongly suggests that these ancient peoples developed rituals around the winter and summer solstices. As the Church of Rome became more influential and scholars from the East began to take up residence in the centers of learning in Europe, the Hellenistic concepts of astronomy, mathematics, and philosophy were extensively taught. The contributions to astronomy of the ancient Greeks were accepted for generations, both in the West and East. The knowledge of astronomy (and other sciences) of the Eastern countries of China, Egypt, India, and Arabia that was transported to western Europe was fundamental to the development of the science of astronomy in the Middle Ages and Renaissance.

Chinese Astronomy

During the Medieval Period the Chinese were more or less isolated resulting in the development of unique, independent, and original approaches to science and astronomy. Even so, the origins of their astronomical observations and calculations that resulted in a calendar for agricultural purposes were based on magic and religion. By the Middle Ages, Chinese science and astronomy were greatly influenced by the developments in other parts of the world. Their astronomy was enhanced by contact, sometimes through war, with other countries, including the Middle East and Arabia, Korea, and Japan. By the 9th century C.E. Persian astronomers brought Greek methods of computations and designs for instruments to China. One of the main contributions of Chinese astronomers was the maintenance of ancient records over the years that provided observational data not yet available in other countries during that time in history.

Another contribution was the development of instruments to view and measure astronomical bodies and events. There is some evidence that the first astronomical observatories were built in China about 4,000 years B.P. Chinese astronomers had cataloged over 1,400 stars and identified constellations by 500 B.C.E. They also used the gnomon, or shadow stick, to track the lengths of the days before and after the solstices. Su Sung (1020–1101) described an armillary clock sometime between the years 1088 and 1095. This clock was a water-driven astrological instru-

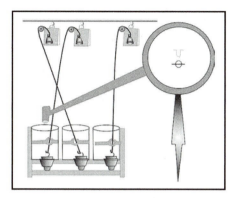

Figure 1.4 Early Weight-Driven Clock
The armillary water weight-driven clock invented in about 1090 C.E. was the first timekeeping device to use an escapement to regulate the clock's movements.

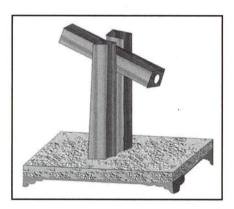

Figure 1.5 Chinese Sighting Tube
This model of a Chinese sighting tube was merely a hollow tube without a lens. It was used to locate faint stars, measure the altitude of stars, and track planets.

ment with an **escapement** consisting of tipping buckets of water onto a chain drive. (See Figure 1.4.)

This device assisted in the production of the first printed book of star maps (using wood-block moveable type). There are a number of ancient astronomical instruments located in the Exhibition Hall of the Observatory in Beijing, China. Many of the following examples were developed, more or less in isolation before and during the Middle Ages, and some of them found their way to the West during the journeys of traders such as Marco Polo in the Middle Ages (13th century). In addition to the gnomon, the Chinese developed the armillary sphere, the celestial globe, both the mural and altazimuth quadrants, and the sextant. They also developed a sighting tube, a hollow instrument (with no lenses) used by astronomers to sight and concentrate on a single faint star or to track a planet and then, using an angle device, determine the object's altitude. (See Figure 1.5.)

A similar hollow tube was also used on the armillary sphere as a protractor-type sighting device to measure the position of stars. Using their simple eyeball instruments, they not only observed lunar and solar eclipses in about 1300 B.C.E., but kept records and made accurate predictions of eclipses 500 years before Western astronomers learned how to predict them. The *equatorial torquetum* by Kuo Shou-ching is a large instrument cast in bronze in 1270 C.E. It was used to observe the pole star from its equatorial mount, thus determining the positions of major stars. The Chinese also observed what we today call *novas,* that is, stars that become very bright and, in a short time, fade out. They reported that a *supernova* (exploding star) could be seen during the daytime. A collapsing supernova is a tremendous explosive event that sends great bursts of electromagnetic radiation (light) into space. This event usually occurs once every century or so. In 1054 C.E. Chinese astronomers observed and documented such an explosion in the constellation Taurus. It reportedly burned as bright as the planet Venus, sending out reddish-white eruptions that could be seen during both day and nighttime hours for 23 days. The Crab Nebula, which is visible today using powerful telescopes, is a remnant of the explosion that took place nearly a millennium ago. (A *nebula* is an immense and diffuse cloud-like mass of gas and interstellar dust particles that is visible due to the illumination of nearby stars.)

Around 400 B.C.E. the Chinese astronomer Kan Te described sunspots that were first reported in the West by Einhard vita Karoli Magni in his *Life of Charlemagne* in 807 C.E. Some Chinese astrologers attributed these blemishes on the surface of the sun to shadows cast by flying birds. Sunspots were not confirmed until Galileo viewed them with his telescope in 1610. Today, we know that these are actually cooler areas of gas on the surface of a hot sun that range from about 3,000 to 62,000 miles in diameter. By the 6th century C.E. the Chinese were familiar with comets and recognized the principle that the comet's tail always pointed away from the sun. But at this time, they were unaware of the effects of solar wind on the gaseous tails.

Over the years the Chinese had several concepts of the universe. For example, at one time they believed that the Earth was flat, floated on water, and was encased in a curved dome containing the fixed stars. Later, they believed that the universe was a giant egg with Earth at the center as the yolk and the shell as the sky. By the early Middle Ages their new theory was that the universe was just an infinite empty space and the sky merely an illusion. Astronomy, as a science, was not as important to the Chinese as astrology that used information about the heavens to predict the future, explain the past, and aid the rulers in making decisions. By the Middle Ages and Renaissance, Western astronomical science and instruments had overshadowed Chinese and Eastern contributions to astronomy.

Egyptian Astronomy

There are no historical archaeological records or writings related to Egyptian astronomy before the first millennium B.C.E. except for some of their buildings. The Great Pyramid of Khufu (also called Cheops) was built in ca. 2600 B.C.E.—or 4,400 years B.P.—and the temple of Amon-Ra in Karnak, Egypt, was constructed about 4,000 years B.P. Both were exceptional engineering feats in addition to being burial tombs oriented according to astronomical events. The pole star of that time, known as Tuban, was aligned with an inner hallway, and by using the star Sirius they were able to determine the length of the solar year. The conquest of Egypt by the Persians by about 500 B.C.E. introduced Near Eastern astronomy methods to Egyptian astronomers. However, prior to this, the Egyptians used simple astronomical instruments and mathematical methods to measure time and to develop accurate calendars, as well as to align their buildings with the stars. The primitive nature of Egyptian mathematics may have contributed to its stagnation. On the other hand, the Egyptians developed a calendar with a solar year of $365\frac{1}{4}$ days, based on three seasons, for farmers to predict flooding of the Nile River. Historically, and perhaps paradoxically, more credit has been given to the Egyptians' contributions to mathematics and astronomy than is deserved. The Egyptians used both Greek geometry and Babylonian mathematics to calculate ovoid epicycles to determine the motion of the moon and planets. Their mathematics, although applied to astronomy, was more advanced than their theories of the solar system and universe during the Middle Ages and Renaissance.

After the fall of Greece and rise of the Roman Empire, many Greek intellectuals, including astronomers, moved to Egypt primarily because of the excellent library established in the new city of Alexandria, Egypt, by Alexander the Great in 332 B.C. Ptolemy was one of the famous Greek citizens who lived and did much of his writing in Alexandria. After the conquest of Egypt by the Muslims on their move westward in the 7th century C.E., Egyptian astronomers made few contributions to the science of astronomy.

Indian Astronomy

Large masonry instruments for observing specific heavenly objects were constructed in India following its conquest by the Persians in the

5th century C.E. Mesopotamian (Babylonian) astronomy influenced Indian astrological omens and religion through translations of Greek astrological writings translated into Sanskrit. Aristotelianism became popular in India sometime in the 6th century C.E. This proved troublesome for most Indian astronomers who accepted unequivocally Aristotle's model of the solar system that consisted of a concentric set of spheres, and then blended it with Hellenistic epicycle astronomy. The Indians were primarily interested in astronomy for religious and astrological purposes and did not contribute significantly to the body of knowledge in this field. They did develop calendars and construct several large observatories and instruments designed for eyeball viewing. One large complex of observatories built at Jantar Mantar in India was constructed during the Middle Ages and includes several large masonry viewing instruments that were used to predict time, measure latitude, and check the positions of celestial bodies. A famous instrument located in the Jantar Mantar observatory that still exists today is known as the Brihat Samrat Yantra, near New Delhi, India. It consists of a large, sloping stonework that is oriented in relation to the sun's solstices and can indicate local time within 2 seconds, as well as determine the declination of the sun. (See Figure 1.6.)

In addition, the observatory contains other masonry instruments, including a large shadow clock and sundials. These massive concrete and brick glass-free instruments were built during the Middle Ages, which raised questions as to why the Indians would build such large structures without optics when Europe was developing much smaller and portable instruments. One claim was that since the European instruments were small, their calculations and measurements were inaccurate, partly due to atmospheric interference. Others claim the Egyptians also built large fixed instruments without optics possibly because they still would be standing in future centuries.

Reportedly, Eratosthenes used a shadow stick (gnomon) to assist him in measuring the Earth's circumference. The Muslims who conquered Northern India in 664 C.E. brought the concept of the gnomon to India. *Gnomon* is the Greek word meaning "one who knows." The original version was a simple vertical stick that cast a shadow when placed perpendicular (upright) to the ground. As the sun progresses from east to west across the sky, a shadow is cast that is longer in the morning and afternoon, and shortest at noon. The shadow's length also changes with the seasons. Smaller versions called "clock sticks" were used in several countries. (See Figure 1.7.)

Figure 1.6 Indian Observatory
Two views of an ancient Indian observatory located near New Delhi, India. The
stone-masonry edifices were designed for viewing various astronomical events.
(Photographs by the author.)

Figure 1.7 Egyptian Clock Sticks
An example of an ancient clock
stick, similar to those used in Egypt
and other countries.

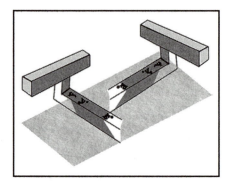

Clock sticks were used as aids in determining the time of day as well
as the time of the year. During the Middle Ages, Indian astronomers
used the gnomon to observe the vernal and autumnal equinoxes and
the summer and winter solstices. They divided the gnomon into twelve
units and developed tables that correlated with the shadow's length. At
this time in history neither India nor most other countries were cognizant that the length of the shadow would be different at different latitudes. Therefore, shadow data for one country might not match that of
other countries because any tables developed would be accurate only
for specific latitudes. There are few records of astronomers in India
prior to the Muslim invasion and the introduction of Greek astronomy.
One famous astronomer who did leave some records was Varahamihira
(fl. 505) who had an astronomical observatory. However, most of the
information that he described came from the West. He and several
other Indian astronomers believed that the Earth was spherical and that
the moon, sun, and other planets, and all the other bodies in the solar
system, had their own proper motion and traveled at the same speed in
circles around the Earth once every 24 hours. They also believed that
the distance of each of the other planets from the Earth was proportional to that planet's periods of revolution. Following are short summaries of a few Indian astronomers.

- **Aryabhata I** (fl. 476 c.e.) advocated astronomy based on revelations
  from Brahma that were included in the *Paitāmahasiddhānta,* which is
  based on a Greek text, in which he explained 4th-century c.e. Greek
  geometric models for precession, the length of a solar year, and calculations for determining an eclipse. He also wrote a number of astronomical texts that were widely used in the Arab countries in the 9th
  century. The Arabs called his text *Zijal-Arjabhar.* It incorporated great

amounts of data gleaned from Greek astronomers, including eccentric tilted epicycles for planetary latitude motion.

- **Brahmagupta** (ca. 589–ca. 665), astronomer and mathematician who directed an early observatory located in central India, made major contributions in these fields. He discovered the solution for quadratic equations that, along with his knowledge of algebra, was used to determine that both Mars and Venus followed an oval epicycle rather than a purely circular one. He also rejected the concept of stationary stars and a moving Earth because he claimed that this would cause structures to collapse. He devised new, complex rules for calculating planetary longitudes and lunar and solar eclipses. Although he was not the first to do so, he described how epicycles pulsated and change in size. This was more evident if the observer was not in the center of the epicycle, which would then appear to change in size at different stages.

After being influenced by the ancient Greeks and Muslims, Indian astronomy did not progress during the Middle Ages and Renaissance much beyond its basic interest related to religion and astrology (calendars). However, after the Renaissance, this began to change slowly.

Mesoamerican Astronomy

Not much is known about pre-Columbian America before its discovery by Europeans. What we consider astronomy was more or less art in the form of rituals and story-telling legends. There were large earthen mounds in the shapes of animals and geometric stone pyramids throughout the region from the southwestern United States to northern South America, a region now known as Mexico and Central America. Four great civilizations flourished in Mesoamerica for about 2,000 years. Their rather rapid demise resulted, in part, from their discovery by the Italian explorer Christopher Columbus (1451–1506) in the late 15th century C.E. along with the many explorers and exploiters, such as Cortez, who followed. The term pre-Columbian generally refers to the Olmec, Zatopec, Aztec, and Mayan civilizations. Purportedly, Europeans destroyed many of the written records of these ancient civilizations in the 15th and 16th centuries C.E. to discourage the native pagan religions and install Christianity. Few Mayan manuscripts describing lunar, solar, and Venus calendars, as well as almanacs that were written on preserved tree bark, survive. As with all other ancient civilizations, the sun, moon, stars, and planets were essential in understanding and divining the events that occurred in their lives. Astrology would be more descriptive of their interests and practices than would the science of astronomy. These were creative but ritualistic civilizations that built pyramids aligned with celestial bodies, and both animal and human sacrifices were integrated into their ceremonial worship of deities related to celestial bodies. The Sun Stone and pyramids in Mesoamerica were great architectural structures as well as temples for religious ceremonies and astronomical observations and calendars. The Mayans were avid astronomers and were aware that at their latitude (about 23.4 degrees north) the sun, at noon twice a year, would be directly overhead because their shadow sticks would show no shadow at this time. The Mayans developed mathematics to study and predict astronomical events, such as equinoxes, solstices, and eclipses. Independent of Western or Islamic astronomy, they developed their own concepts of a layered universe similar to the Greek idea of celestial spheres. Their universe consisted of layers for the stars, comets, planets, sun, moon, and clouds with the Earth at the center, but the 13th layer was occupied by their god. The Aztecs also erected a temple complex near the city of Teotihuacán in Mexico so that it coincided with the constellation Pleiades (seven sisters), which is a grouping of seven stars they used for predicting meteorological and agricultural events.

Few records of individuals involved with the astronomy of these four Mesoamerican civilizations survive. During the late Renaissance period, most information was recounted verbally to the European conquerors and historians who, in turn, recorded the events and stories related to local astronomy and rituals. One of the last mentioned persons associated with astronomy was a ruler in Peru by the name of Atahuallpa who is said to have used astrology to predict the arrival of the Spaniards.

Islamic Astronomy

To understand the character of Islamic astronomy, it is necessary to underscore the influence of Muhammadanism in the 7th century C.E. Muhammad (570–632 C.E.), who was born in Mecca, in present-day Saudi Arabia, received a vision in the year 610 to establish a religion known as Islam. The word *Islam* means "resignation or surrendering" to God and is the religious system of Muhammad. Islamic astronomy can best be understood by explaining Muhammadanism of the 7th century C.E. By this time in history the desert tribes of the Middle East, under Islamic religious rule, began conquering surrounding civilizations. Muhammad's Arab followers believed in the Muslim brotherhood of the faithful and that they were destined to rule the world even if that required eliminating all nonbelievers. By the 8th century C.E. Muslim armies conquered nations to the East as far as northern India and central Asia, and to the West including Egypt, Morocco, and southern Spain. Just 100 years after Muhammad's death Muslims ruled much of the world, from the Atlantic to the Indian oceans. By the 8th century C.E., all the countries of the Arabian peninsula, the north African coast, Armenia, Mesopotamia (Iraq), Persia (Iran), Egypt, Morocco, the northern part of India, and most of Spain had been overtaken by Muslim warriors intent on spreading Islamic teaching of Muhammad. In the year 732 Christian armies defeated the Muslim armies south of Paris, France, thus preventing their advance further north into Europe. Although Islam's march to the north of Europe was halted, the southern Iberian region of Spain was conquered and brought under Islamic control. This region of Spain was known as Al-Andalus, and the rulers showed mutual respect for all religions, including Christianity and Judaism. Islamic governance was shared and intellect was cultivated, especially among Jewish scholars. (It is interesting to note that, at this time in history, the Muslims were much more tolerant and accepting of Jews in Spain than was the Spanish Roman Catholic Church during the Inquisitions that drove the Sephardic Jews out of Al-Andalus.) Over the next few centuries, while under Muslim rule, some of the great cultural and scientific achievements of Islamic civilization, including architecture, philosophy, and astronomy occurred. Several centuries later, when Christians gained control of Al-Andalus through military actions, many Muslims immigrated to Morocco in North Africa while many others remained in southern Spain where their influence is still in evidence.

Muslim leaders believed that it was in their best interest, and also in the best interest of the people whom they now ruled, to promote science, particularly medicine, and astronomy. Since much of medicine was based on alchemy and astrology, it was necessary that one learned more about nature and astronomy in order to become a competent physician. This is the period of the so-called Dark Ages when most of the people of Europe were uneducated serfs as compared to the more enlightened period for the Islamic world. The major contribution of Arabian astronomy (and science in general) was their preservation of ancient science in translations, commentaries, and interpretations. Arabian astronomers of the Middle Ages and Renaissance were faithful followers of the Greek astronomer Ptolemy and the theories contained in his book *Almagest*. This book also contained many diagrams of astronomical instruments that the Muslims copied and, in some cases, improved. They not only used but augmented many Greek astronomical tables. While Islamic astronomers did not advance a body of new knowledge in astronomy, they did validate the work that had been done by their predecessors. After being driven out of southern Europe by the Christian Crusaders, the Muslim Arabic translations of ancient Greek science were retranslated into Latin, thus creating a Western revival of classical Greco-Roman science and astronomy. After the Crusades began, southern Spain continued to be an Arab stronghold, even after their defeat. Spain and Italy during the Late Middle Ages were the birthplaces of the Renaissance of science on the European continent.

Following are some of the people from Arabian countries who contributed to the field of astronomy between the years 500 to 1600 C.E.

- **Abu Abdullah Al-Battānī** (ca. 858–929) conceived a number of astronomical theories based on information gleaned from Ptolemy's writings. He improved on Ptolemy's measurements for the perigee of the sun and Earth (the point at which the sun and Earth are closest to each other) and his calculations of the ecliptic of the Earth's orbit to its equatorial plane. Al-Battānī's figure was 23°35′ for the inclination of the Earth to its plane around the sun, which is very close to the currently accepted figure of 23.5°. He also accurately calculated the length of the year as 365 days, 5 hours, and 24 seconds (the actual length is 365.24220 days). In addition, he made accurate measurements for lunar and solar eclipses and developed the mathematical concepts and tables for sines, cosines, tangents, and cotangents. Copernicus and other astronomers from later periods used Al-Battānī's algebra and trigonometry tables for their own calculations.

- **Abd al-Rahman ibn 'Umar** (fl. 900 C.E.), known as the "wise one," used and improved Ptolemy's calculations of longitudes to prepare a new book listing the fixed stars by their magnitude of brightness.

- **Abu Ali Hassan ibn al-Haitham** (965–1040 C.E.), known as Alhazen in the West, is considered the father of modern optics. He did not discover or invent optical instruments, but as a great physicist, he made contributions to the fields of optics, mathematics, as well as astronomy. He was one of the first to use a scientific method that involved establishing relationships between one's observations, proposing hypotheses, and verifying one's assumptions by experimentation. He studied how light is propagated from one medium to another (diffraction) and developed the mathematics describing the laws of reflection. Understanding both refraction and reflection were important in the development of astronomical optical instruments such as the telescope. Alhazen developed theories of light related to solar and lunar eclipses, sunsets, rainbows, the density and height of the atmosphere (55 miles), and human binocular vision. He was the first to correctly describe why the sun and moon seem larger near the horizon than they do higher in the sky. He also was one of the first to use a pinhole camera (camera obscura) to study light and images. The camera obscura was studied in greater detail during the 16th century. Alhazen's books *Kitab-al Manazir* and *Opticae Thesaurus* were influential in the work of Leonardo da Vinci, Johannes Kepler, and Isaac Newton.

- **Ibn al-Zarqala** (1029–1087), a Spanish Muslim astronomer, published *Toledo Tables*, which not only listed vast astronomical information but also described how he used various instruments, including the astrolabe, to make his observations and measurements.

- **Nasir al-Din al-Tusi** (fl. mid-1250s C.E.), compiled a collection of 400,000 manuscripts and constructed new astronomical instruments, including a large quadrant with a 10-foot radius. Nulagu il Khan, a grandson of Genghis Khan, founded the observatory used by Nasir al-Din for his instruments and studies that resulted in a book called the *Ilkhanic Tables*.

- **Ulugh Beg** (1420–1449), a descendant of a Mongol ruler, founded an astronomical observatory in Samarkand, which is in present-day Uzbekistan in west-central Asia. He built a 60-foot quadrant vertically imbedded in mortar that he used to determine the positions of stars. His was the first original work that did not rely on Ptolemy's data. From about 1430 to 1439 Beg published a new catalog of stars, including a star map that proved to be a great improvement on both those of Ptolemy and Hipparchus. Although it was the first accurate star map to become known in the West, it was not printed in Europe until 1665, and by that time, it had become obsolete.

European Astronomy

There was not much scientific or astronomical progress during the so-called Dark Ages of the Medieval Period in Europe. Any inquiry that attempted to challenge the authority of the Roman Catholic Church could lead to imprisonment, exile, or even death. After the Protestant Reformation and the weakening hold of the Catholic administrators of the Church who controlled most aspects of everyday life, including intellectual activities, the environment for scientific investigations began to change. This is one of the major reasons why the period from about 1400 to 1600 and beyond is known as the Renaissance—a renewal or resurgence of knowledge and inquiry into nature and man's role in the scheme of things.

Scholars from the European continent traveled to Spain to partake of the repository of Islamic knowledge that centered in the libraries of Córdoba and Toledo, and later in Seville and Valencia. For centuries medieval European astronomy focused on Ptolemaic beliefs of a geocentric system. In the 12th century Adelard of Bath (ca. 1090–ca. 1150) introduced some inaccurate cosmological concepts from India that further confused European astronomers. Not discounting the importance of the Islamic influence on European astronomy, other ancient people living in various regions on the European continent were concerned with celestial mechanics. As the Church of Rome became more influential and scholars from the East began to take up residence in the centers of learning in Europe, the Hellenistic concepts of astronomy, mathematics, and philosophy were extensively taught. However, the Church's acceptance of astronomy as a science was limited and only concerned itself with timekeeping as it related to religious rituals. Nevertheless, the thirst for knowledge was pervasive, and a period of translation of the works of Greek and Islamic scholars into the scholarly language of Latin ensued. Opposition existed to the teachings of some of the Greek scholars, particularly Aristotle, because of the astrological influences in his cosmological writing. The study of astronomy was most important at the universities at Oxford and Paris, and a number of scholars in residence advanced some new astronomical theories. Instrumentation such as astrolabes and quadrants would be invented and star catalogs and tables compiled. However, the most revolutionary astronomical theories would have to wait to be developed until the middle of the 16th century when Nicolaus Copernicus (1473–1543) would discover that it was not the Earth at the center of the universe, but rather the sun.

Following is a summary of some of the astronomers and their contributions made during this period of awakening:

- **Henry Langenstein** (1325–1397), also known as Henry of Hesse, was a European astronomer who accepted much of the astronomy that originated in Arabia and other Eastern countries. However, he used physical examples to criticize Ptolemy's epicycle planetary model described in *Almagest*. His main criticism was that Ptolemy's system was not consistent with the real physical aspects of astronomical observations.

- **Nicholas of Cusa** (1401–1464) was born in Cusa, Italy, which is now part of Germany. He used his mathematical and philosophical reasoning to argue against the standard model of the universe bounded by concentric spheres with the Earth in the center. He proposed an "infinite universe" model where nothing is fixed. Therefore, there is no center, no circumference, no infinity, yet everything is in complex motion around everything else. His conclusion: The Earth cannot occupy the center of the universe since there is no fixed center or periphery to the universe. His philosophical concept determined the *place* of the Earth as well as its *motion.* He concluded that the Earth rotates on its axis as it revolves around the sun and that all the stars are also suns. These concepts reflect some ancient ideas of the Hermetic philosophers who followed Hermes Trismegistus, a mythological Greek philosopher/astronomer who believed that a visible, all-seeing God was the light and soul of the universe that guided the family of stars around the sun. (Hermes Trismegistus was the name given by the Greeks to Thoth, the Egyptian god of the sciences.) These ideas influenced both Copernicus and Giordano Bruno of the 16th century.

- **Georg von Purbach** (1423–1461) was an Austrian astronomer and mathematician who attempted to retranslate a 300-year-old Arabic version of Ptolemy's *Almagest.* He died before it was completed. Purbach's student Regiomontanus attempted but failed to complete the translation. Purbach's major contribution was a very accurate and complete table of lunar eclipses published in 1459. His work on the solar system was based on Ptolemy's misconceptions of solid spheres that were accepted for almost 200 years until disproved by Tycho Brahe.

- **Regiomontanus** (Johann Müller) (1436–1476). As a student of Georg von Purbach he completed his teacher's translation of Ptolemy's *Almagest* from the original Greek rather than from less accurate Arabic translations. Regiomontanus' benefactor provided funds that enabled him to build an observatory that contained many astronomical instruments, as well as his own printing press. He made numerous observations of the solar declination that, in 1475, were included in *Tabulae directionum* (Tables of Directions), a useful navigational chart. This volume also provided the first publication of modern trigonometry in Europe. Regiomontanus was the first to make a scientific study of a comet in 1572, later known as Halley's Comet. He printed a book on trigonometry that included a table for sines for minutes and tangents for degrees that was used by Columbus during his voyages of exploration.

- **Girolamo Fracastoro** (ca. 1478–1553), an Italian physician specializing in contagious diseases, introduced the term *syphilis* as the "French disease." He also had an interest in astronomy and in 1538 published *Homocentrica sive de stellis liber* (Homocentricity). His *Book of Stars* attempted to correct the epicycles and eccentrics associated with Ptolemy's geocentric system. He proposed replacing this system with the ancient Eudoxian model consisting of 27 celestial spheres.

- **Nicolaus Copernicus** (Mikolaj Kopernik) (1473–1543), a Polish astronomer, was influenced by the Greek astronomer Aristarchus' theory of a heliocentric solar system. Copernicus destroyed the accepted geocentric physical model of the universe in his famous book, *De Revo-*

lutionibus Orbium Coelstrum (On the Revolution of the Celestial Spheres), written in 1540. However, it was not published until 1543, when he was near death. The major thesis of the book can be summed up in his statement, "All the spheres revolve about the Sun as their midpoint and therefore the Sun is the center of the universe." His concept also involved a spinning Earth and the concept of angular momentum that he described as the planet furthest from the sun was also the slowest revolving around the sun. The Roman Church criticized his book as conflicting with the Christian Bible. His thesis was also not accepted by the Danish astronomer Tycho Brahe, who stated that if the Earth moved, there should be movement of the fixed stars that no one, so far, was able to measure. Copernicus answered this objection by stating that since the stars were so far from the Earth, it was impossible to observe any change in **parallax,** which would indicate motion. The Copernican theory did explain the movement of the moon and planets in a more accurate way than did Ptolemy's geocentric system of epicycles, but Copernicus was incorrect in insisting that the paths of the planets were perfect circles.

- **Thomas Digges** (1546–1595) at the age of 13 became the student of John Dee. His first book, *Pantometria,* described the arrangement of lenses he used for surveying. Since his lens arrangement enabled him view great distances, it is assumed he use this device as a telescope for his astronomical observations long before Galileo used a telescope in 1610. Thomas is known for his application of mathematics to astronomy, particularly trigonometric theorems that he used to determine the parallax and positions of stars. He established the position of a new star that became known as "Tycho Brahe's supernova of 1572." This data challenged the Aristotelian concept of separate celestial spheres for the moon, planets, stars, and heavens. Digges's measurements proved that this new star could not be located between the Earth and moon. Digges also concluded that the Copernican concept of a heliocentric world is correct, and that the theory of the universe contained within an outer sphere was incorrect. In other words, Digges extended the astronomical limits of the universe to include and embrace the theoretical heavens, somewhat similar to the theory Nicholas of Cusa proposed in the mid-1400s. Although Digges expressed his theories in mystical terms, he was one of the first to use mathematics when making his observations to expand the known universe to infinity. Both Digges and Dee influenced future scientists, particularly astronomers and mathematicians, including the mystic philosopher Giordano Bruno.
- **Giordano Bruno** (1548–1600), although basically a philosopher, produced concepts and writings that influenced the astronomy of his day. He wrote a trilogy that explained his ideas in three areas. The first vol-

ume addressed his rethinking of ancient Greek atomism and his acceptance of Euclidean geometry. The second investigated the meaning of Pythagorean number symbolism. The third, *De immense,* related his speculations on the nature of the universe. In this volume, Bruno connected his understanding of Copernicanism with his new concept of the infinite—in particular, his theory of the infinite dimensions of the physical universe that he called the "infinite infinite." Early in his youth he believed that the Earth does not change its inclination with respect to the poles, and it revolves around the poles, and during the course of a year the moon and the Earth do complete an annual revolution around the sun. Bruno not only believed in the Copernican heliocentric universe, but he taught it during a very conservative period of religious history. For his efforts he was brought before the Roman Inquisition in 1593. Refusing to recant at least some of his eight heretical propositions, Bruno was burned at the stake in the year 1600.

- **Tycho Brahe** (1546–1601), a Danish astronomer, reasoned that since comets move, the universe is not static. Even so, he accepted Ptolemy's concept of the geocentric universe rather than the controversial Copernican heliocentric system. In 1560 at the age of 14 he observed a solar eclipse and decided then and there that he wanted to be an astronomer. As a young man Tycho challenged a colleague to a duel to determine who was the best mathematician. The duel with swords resulted in his nose being cut off, which he then replaced with a fake one constructed of gold, silver, and wax. One of Tycho's major contributions to astronomy was the discovery of a new star visible to the unaided eye, known today as a supernova or "exploding star." This nova observed in 1572 became known as the "Tycho Star." One of the most prominent craters on the moon is also named after him. Since this new supernova showed no parallax, which was the same for all other fixed stars, he concluded that his new bright star belonged in the sphere of fixed stars. This disputed Aristotle's theory that change does not take place in the cosmological heavens since the sphere of fixed stars is eternal, incorruptible, and unchanging. And that change can only take place in sub-spheres of less importance than the entire universe. In addition to discovering this new star in 1572, he also discovered a comet in 1577, both of which he used to support his theory that change takes place in the heavens. Tycho constructed several new large astronomical instruments. One was a huge sextant; another was a quadrant that he used to directly sight objects in the sky. (See Figure 7.6.) He made some excellent observations and tables for his day, considering that the telescope had not yet been invented. For some unknown reason he supported, until his death in 1601, Ptolemy's idea that the Earth did not move and was the center of the known universe. He spent 20 years making accurate measurements for the positions of almost 800 stars, which proved invaluable for the work of his assistant, Johannes Kepler.

- **David Fabricius** (1564–1617) was a German amateur astronomer credited with discovering the first variable star, which he named "Mira" (marvelous), known today as the star Omicron Ceti. His son, Johannes, was also an astronomer who, along with Galileo, is given credit for discovering sunspots.

- **Galileo Galilei** (1564–1642). Although some of Galileo's contributions to astronomy were made after the year 1600, he is still considered a Renaissance scientist. In the late 1500s he was the first astronomer known to use a telescope to view and study celestial objects. Either Hans Lippershey (1570–1619) or Zacharias Janssen (1580–1638) invented the telescope in the late 1500s or early 1600s. Galileo learned about this invention and built three models of telescopes of his own designs. His best instrument was a 30-power telescope, which is about the power of a good pair of modern binoculars. Galileo made four important astronomy-related discoveries.

(1) Although large sunspots were observed by the naked eye at least a thousand years before Galileo's time, he was the first to use a telescope to view the surface of the sun in detail. (Aristotle discounted these spots on the sun since they must be impossible in a perfect universe.) Galileo's observations confirmed the movement of darker areas or spots on the sun's surface, and he used this knowledge as the basis for his conjecture that the sun rotates on its axis. He used this observation along with other evidence, to conclude that the Earth and other planets are not only spinning on their axes, but are also revolving around the sun in a circular path. This was the first scientific observational proof for the Copernican heliocentric theory of the planets moving around the sun. Unfortunately, Galileo disagreed with Kepler's laws that stated, in part, that planets move in ellipses, not perfect circles.

(2) Galileo carefully observed the planet Jupiter with his telescope and noticed two small objects that appeared to move around the planet. At first he thought perhaps Jupiter moved, but later, he noticed that two additional moons orbited Jupiter. He claimed that from this evidence it proved that new bodies do exist and that they also move around other objects. This was another rationale for his acceptance of the heliocentric concept of the solar system, which resulted in an inquisition by the Catholic Church and his house arrest for the remainder of his life. His records of the eclipses of Jupiter's moons were used to assist in navigation.

(3) While viewing Saturn, he noticed a slight bulge on each side of the planet. Over time this bulge became smaller and then again larger. His telescope was not powerful enough to distinguish this bulge as the rings of Saturn with which we are now familiar.

(4) Galileo was always fascinated with the number of stars that could be seen with his telescope. Purportedly, he said, "So numerous as to be almost beyond belief." When he pointed his telescope toward the Milky Way, he saw many more stars than in other portions of the sky. Some were very faint and blended together to form what he thought was a cloud. Today we know this as the Milky Way galaxy.

- **Johannes Kepler** (1571–1630) is another astronomer whose major contributions were made after the year 1600. Even so, he is also considered a Renaissance scientist who, as a mathematician, viewed his first comet in 1577. This observation led him to study the mathematics related to planetary motion. In his book *Mysterium cosmographicum* he expressed his belief in the existence of a mathematical harmony that drove the universe. In 1599 Kepler became Tycho Brahe's assistant and was assigned to observe and measure the orbit of Mars. The young Kepler thought it would not take very long, but the task required eight years. During this time Tycho died and left many records later used by Kepler to determine that Mars revolved around the sun in an elliptical orbit, not in a perfect circle as formerly believed. (This information later became important for Newton's calculations.) Tycho conclude that all bodies that revolve around other bodies also have elliptical orbits, and that they follow precise mathematical patterns. Using the data from his observations Kepler developed his three laws of planetary motion: (1) As all planets revolve around the sun in elliptical orbits, the sun forms one of two focal points, while the other focal point is imaginary. (2) An imaginary straight line joining the sun and a planet sweeps over equal areas in equal intervals of time during the period of one orbit. (3) The square of a planet's orbital periods is proportional to the cubes of the semimajor axes of their orbits. (See Figure 1.8.)

There were numerous ancient astronomical instruments developed in many different countries before the Middle Ages. The designs for these crude instruments were not only exchanged between countries, but they were also greatly improved during the Late Middle Ages and Renaissance (1300 to 1600). The astrolabe is an example of an instru-

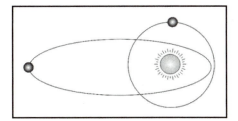

Figure 1.8 Kepler's Law of Areas Kepler's Law of Areas consists of three parts. This figure depicts his first law that explains that all bodies revolving around another body do so in an elliptical fashion, not in perfect circles.

ment developed in many forms over many years in different regions. (See Chapter 7 for details on the astrolabe.)

Astronomy was the scientific cousin of astrology, but the distinction between them was not sharp during the Middle Ages. Astronomy was studied for a variety of purposes, including religion, calendar-making, navigation, astrology, as well as a science that was linked to other systems of knowledge.

There is a long history of sailors using the stars in the celestial sphere as well as the sun and moon to assist in navigating the seas. A question often asked is: Was the development of astronomy advanced by navigation, or was navigation advanced by astronomy? It does seem that many astronomical instruments were developed for land use and later adapted for navigation. For instance, the simple astrolabe used to determine latitudes by determining the sun's altitude was first employed on land, but it was soon adapted for use at sea. Both the early land versions and the marine astrolabe were crude. The mariner's astrolabe, used in conjunction with a magnetic compass, was inadequate on a rolling and pitching ship. The observational calculations that were made at sea were adjusted once the crew made landfall and were able to make observations from a firm land base.

There were a number of astrolabe-like navigational instruments for measuring angles developed during and after the Renaissance. A major one was the sextant. Others were the cross-staff, quadrant, and the back-staff. Although improved instruments assisted sailors in determining their latitude, none of these were of much assistance in determining longitude, which meant that if sailors could not see familiar landmarks, they often became lost. (Note: Measuring latitude can determine your north or south position, while longitude is used to measure how far one has traveled westward or eastward, which involves elapsed time.) One reason that longitude was difficult to determine at sea was the need for a timepiece that could keep accurate time on board a pitching ship. This problem was solved in the mid-1700s when John Harrison built the first chronometers that could keep accurate time aboard a ship. (See Chapter 7 for more on timekeeping inventions.)

Optical instruments for astronomical and navigational purposes have a long history of other uses. The ancient Egyptians used polished metal plates as mirrors, but there is no record of using glass as optical devices. Ptolemy conducted the first study of the basics of reflection and refraction in his book titled *Optics* in the mid-200s C.E. He described refraction as light rays bending when they passed from one medium to another in lenses. He used this concept to explain how the atmosphere affects the appearance of stars that seem to be located at higher altitudes than their actual positions.

Calendars

Background and History

The word "calendar" is derived from the Latin word *calendarium*, meaning an account book or register. It also refers to the Latin *calendae*, which was the first day of the Roman month when ritual and market days were officially announced.

From the earliest days of human civilization the positions and movements of the sun, moon, planets, and stars were studied and utilized as a means of organizing human events. The five wandering bodies (planets), along with the moon and sun, moved around the sphere of stars within a narrow region referred to as the zodiac. Ancient astronomers attempted to arrange the years, months, and days into a consistent time system using the regularities observed in the heavens. Since a year on Earth does not contain an integral number of days, calendar makers had to assign a different number of days to some of the months as well as to years. Later, calendars were used as a means of societal control through the development of religious dogma based more upon superstition and astrology than on astronomy. Ancient astronomers sought out patterns in the heavens, which they then recorded, interpreted, and translated into everyday experiences. Thus, the invention of calendars was a means of organizing astronomical units of time to serve practical, agricultural, commercial, social, and religious purposes.

Unlike timekeeping devices that measure units of a single day, calendars are designed to apportion time over extended periods, that is, longer than one day. (See Chapter 7 for detailed information on ancient clocks.) Hundreds of different types of calendars have been developed over the span of human existence. Some are based on observable astronomical cycles, others on seemingly abstract cycles with no obvious relationship to astronomy (e.g., old Chinese calendars), even though they are designed to repeat themselves. Most were, and still are, inaccurate due to erroneous astronomical data or unrelated religious interpretations. The earliest known calendar dates back to 4236 B.C.E. in Egypt. It is believed to be the first one using a 365-day year and 30-day months divided into 10-day weeks.

Middle Ages and Renaissance

During the Middle Ages and Renaissance, refinements and adjustments based on improved celestial observations were made to the calendar. These changes are still in widespread use in much of Europe and the Americas, as well as other parts of the world. Although this particular calendar is known as the Christian calendar, its early evolution had little to do with Christianity. The old Roman calendar, before the birth of Christ, was inaccurate and thus dates for specific events became confused. In the year 45 B.C.E. Julius Caesar introduced a new calendrical version instituted after his romantic affair with the Egyptian queen, Cleopatra (69–30 B.C.E.). It is referred to as the Julian calendar and is based on 365 days per year divided into 12 months (with no relationship to the moon), with 7-day weeks. It was not until the 6th century that year 1 was assigned to the accepted date of the birth of Christ, although Christ's actual birth date has always been in dispute. Most ancient calendars started with year 1, not 0. The concept for zero was invented by ancient Greek mathematicians, but it was first used as a place-value symbol by Indian mathematicians and was not well known when European calendars were developed. Our modern version of tracking historical time continues to ignore the first year of 0, which sometimes makes it difficult to determine when to celebrate the end of a century or millennium. The Julian calendar also provided an extra day once every four years—thus the leap year. Over the centuries the Julian calendar became inaccurate for determining the solstices and religious holidays, thus necessitating serious revisions.

A Muslim, Abu Abdullah Al-Battānī (858–929 C.E.), better known as Albategnius, was the first astronomer to accurately determine that the length of the year was 365 days, 5 hours, 46 minutes, and 24 seconds. (This is very close to current figures of 365.24211896698 days.) He also accurately determined the Earth's ecliptic (~23.5°), which affects the seasons, as well as establishing the mean orbit of the sun, the angular diameter of the sun, and accurate orbits for the moon and planets. His recorded discoveries later became useful to astronomers of the Late Middle Ages and Renaissance.

In 1582 Pope Gregory XIII reformed the Julian calendar on the advice of the astronomer Christoph Clavius. This new Gregorian calendar had century years that are not divisible by 400, which means there was no longer a leap year included as there was in the old Julian calendar. The old calendar had overestimated the length of each year by approximately 10 minutes and 48 seconds. This new calendar corrected this error from the old Julian calendar by eliminating 11 days between October 4th and October 15th. This resulted in 1582 being the shortest year in history with only 354 days. By 1584 most Catholic countries accepted the new Gregorian calendar. However, the European Protestants rigidly adhered to the Julian calendar until the early 1700s when they finally accepted the Catholic or Gregorian version. The Gregorian calendar was adopted by the new American colonies in 1752, and in 1784 Benjamin Franklin proposed the adoption of daylight saving time. Eventually, but much later, other countries accepted the modern version of the Gregorian calendar: Japan (1873), Russia (1917 and 1940), and China (1949). Somewhat surprising was the decision by the Eastern Orthodox Church in 1971 to reject the Gregorian calendar in favor of the Julian version.

Calendars have always provided a link between humans and the observable universe, with the illusion that in some way individuals can control time itself. Calendars also served a social and religious purpose providing cultures with a document to organize the lives of their citizens. Ancient astronomers and those of the Middle Ages and Renaissance, who astutely observed the heavens, recorded what they saw, and made mathematical calculations, contributed greatly to the science of astronomy, but, in essence, calendars were always social tools, not scientific documents.

CHAPTER 2

GEOGRAPHY AND EXPLORATION

Background and History

Prehistoric humans were hunter-gatherers who were quite familiar with their local geography. As the weather changed, the food supply migrated and thus so did the people, which provided them an opportunity to explore new territory. No doubt as they moved from place to place, they memorized landforms and trails so that they could retrace their progress. As populations increased and tribal interactions became more frequent, it is not difficult to surmise that information was exchanged in the form of legends and mythological tales concerning hunting and the geography of regions with which one group or another was familiar. It is likely they even scratched out simple maps that indicated mountains, rivers, caves, and other landmarks, in order for others to find their way when traveling to new regions. These crude sketches on dirt, bark, or leather may have been the first geographic maps.

Recently, archaeologists have found skulls and other evidence that three humanlike species coexisted in south-central Asia at about the same time 1.7 million years ago. It was formerly believed that *Homo habilis, H. ergaster,* and *H. erectus* arrived out of Africa at very different times. Now the evidence indicates that these different protohuman types may very well have all originated in Africa, migrated northward, and lived together at about the same time in ancient history. Also, most anthropologists concede that over a period of millions of years *H. sapiens* did migrate out of Africa to Australia and the Americas as well as to Europe and Asia. Thus, it is obvious that our most ancient ancestors were, indeed, explorers.

Prehistoric economics, in the form of trade, was one incentive for ancient people to travel and learn more about the geography of different regions. Objects from one culture and geographic region have been found in other distant places. Several thousand years ago trade routes were established in regions around the Mediterranean Sea. From about 2000 to 2500 B.C.E. the inhabitants of several ancient countries in the Mediterranean were involved with trade. The first were the Minoans of Crete, followed by the Mycenaeans of Greece who reportedly ruled the eastern Mediterranean Sea. A thousand years later Phoenicia and Carthage, Egypt, and other countries dominated the Mediterranean. By 1000 B.C.E. the Greeks controlled the trade on the Black Sea and colonized the region now known as Turkey.

This early travel, trade, and exploration resulted in the practice of record keeping that included diagrams and sketches of the routes taken as well as the major landmarks. These first maps were crude and often inaccurate, but they did serve as guides for those who followed the same paths. (See Cartography.)

The Phoenicians from city-states in present day Lebanon were early Mediterranean traders, most likely sometime before 1000 B.C.E. An interesting story that illustrates how economics and trade played an important geographic role in the year 500 B.C. considers the ability of Phoenician sailors, who were accomplished sea navigators and possibly the first to circumnavigate Africa. At this time in history, the metal tin was mixed with copper to form the alloy bronze, thus making harder tools and weapons than those made with soft copper. The tin mines in the eastern Mediterranean region were depleted. (Supposedly this was the first time in history when humans mined a mineral to the extent that they depleted the source.) The Phoenicians sailed out of the Strait of Gibraltar into the North Atlantic Ocean and discovered what they called the "tin islands." Since this new source of tin was an important discovery, its location was kept secret. It is believed that they landed on the southwest coast of England now known as Cornwall, which still produces some tin.

Many people made contributions to the fields of geography, navigation, and cartography before the Middle Ages and Renaissance. Interestingly, more knowledge was gained and progress made in these fields (as well as other areas of science) during the thousand years from 500 B.C.E. to about 500 C.E. than in all of the Dark Ages of the Medieval Period (from 500 C.E. to about 1200 or 1300 C.E.). Following is a summary of some of the important contributions to geography made by these ancient geographers.

- **Eratosthenes** (ca. 276–ca. 194 B.C.E.), a Greek living in Egypt, was a poet, historian, mathematician, and geographer. His most famous contribution to geography was his method of determining the circumference (and thus size) of the Earth. His figure of 25,000 miles (or 250,000 strata, which was the unit of distance used in those days) is very close to the current equatorial circumference of 24,902 miles.

- **Aristotle** (384–322 B.C.E.) can be described as a proto-Renaissance person because of his many and varied talents and accomplishments. Although many of his ideas were derived from previous scholars, his theories and writings influenced others for centuries and his philosophy continues to influence how modern people think. He was knowledgeable of maps of his time and thus was aware of the regions around the Mediterranean Sea, such as the "Pillars of Hercules" (Strait of Gibraltar), and the western region of the Atlantic Ocean, including the British Isles, which at that time was the western edge of the known world. He was also familiar with the Indian Ocean, the Red Sea, the Caspian Sea, and northern Africa as forming the other boundaries of what was then considered the entire world.

  In his book *Meteorologica* Aristotle divided the Earth into zones and distribution regions for flora and fauna based on their latitude of habitation. This was the first time that actual lines of latitude were used to designate zones on the surface of the Earth. His conclusion was that the polar and equatorial zones were the least desirable for habitation, while the mid-latitude regions were best suited for plants and animals as well as humans.

- **Pliny the Elder** (ca. 23–79 C.E.) is best known for his 37-volume *Natural History* (published in Latin in 77 C.E.), which was an encyclopedic work covering astronomy, geography, geology, botany, zoology, agriculture, pharmacology, and mining. Much of what he wrote was based on past books from earlier philosophers. A keen observer who was interested in facts, he engaged in little theorizing, which was unusual for his time. Pliny accepted the concept of a spherical Earth as well as celestial spheres. He described much of the geographic area and cities of present-day Arab nations, as well as coastlines and seas of the Middle East.

- **Ptolemy** (Claudius Ptolemaeus of Alexandria). There were actually 14 Ptolemys who ruled Egypt throughout antiquity. Three became particularly famous. Ptolemy I (ca. 366–ca. 283 B.C.E.) ruled Egypt after the death of Alexander the Great and established a museum of learning in this Mediterranean coastal city. Ptolemy II (ca. 308–246 B.C.E.) was the son of Ptolemy I. He expanded the museum of learning established by his father. Finally, Claudius Ptolemaeus (ca. 90–170 C.E.) was not related or associated in any way with the other two Egyptian Ptolemys, even though they share the same name. Most likely he was born in Egypt but became Hellenized and subsequently very knowledgeable about Greek science. He is most famous for his books that compiled over 500 years of Hellenistic science—namely, astronomy (*Almagest*), astrology (*Tetrabiblios*), geography (*Geography*), and optics (*Optics*).

In his eight volumes of *Geography*, Ptolemy described how the astrolabe along with mathematics could be use to determine the latitude and longitude of 8,000 different locations on the Earth, thus creating maps of the then known world. He introduced the use of the grid system for latitude and longitude in mapmaking, which is still used by cartographers. He was also the first to orient maps with north at the top and east at the right, possibly because most of the then known world was in the northern hemisphere. Ptolemy's maps reflected his concept of a spherical world, contrary to most flat maps of his time. His estimation for the size (circumference) of the global Earth was about 18,000 miles, much smaller than the 25,000-mile equatorial circumference determined by Eratosthenes about two centuries earlier. One reason for this underestimation was that he calculated just 50 miles for each of the 360° for the Earth, while Eratosthenes correctly used 70 miles per degree. His map exaggerated the eastward extension of the coast of Asia at 180° rather than the correct figure of 130°. He also placed the equator much further north than it really is, and he did not include the Americas in his maps. Ptolemy's maps, both geographic and astronomical, and his calculations and measurements (particularly of the planets) were the best available for many centuries. His *Geography* volumes were translated into Arabic in the 9th century C.E. and later brought westward to Europe where they were retranslated into Latin. Exploration of new lands between the years 500 and 1400 C.E. by travelers and later discoveries by sea in the 1400s and 1500s opened up new worlds. Ptolemy is rightfully known as the "Father of Geography," while Aristotle might be considered the "Grandfather."

Exploration and Discoveries

Economics in the form of trade, coupled with the human desire to explore the unknown and spread the word of a particular religious faith, has been responsible for most of the geographic discoveries since the beginning of recorded history. The period following the fall of the Roman Empire is known as the Dark Ages because of the dearth of exploration and discoveries and the decline in knowledge in general. This was particularly true on the European continent where trade and the expansion of knowledge between regions were greatly restricted. During this period, most travel was restricted to religious pilgrimages to the Holy Land, the site of the Crusades that began in the 11th century.

By the time of the early Renaissance following these dark medieval times, economics and trade again were the impetus for exploration and discovery, as well as a reawakening of curiosity about distant places. Trade, particularly the spice trade, was established throughout the Mediterranean region, India, and Southeast Asia, and later between Europe and the Orient. Once overland routes were established between China and the West, a desire by Europeans for more exotic products from the Orient and India increased. New routes between the East and West were proposed. As these land routes became more dangerous due to thieves and bandits, new and safer sea routes were considered. This necessitated the construction of larger ships by many nations to withstand the rigors of the open seas. These ships were expected to reduce the risk of attacks from bandits, as well as reduce the prices of imported goods from the East. Keen trading competition developed, leading to explorations for new sea passages between the East and West. The ambition to establish new sea routes that would enhance trade, increase wealth and power, and ultimately produce a better way of life led to the discovery of new lands. Trade expeditions by seagoing ships were the defining hallmarks that distinguished exploration during the Renaissance (1300 to 1600 c.e.).

Following are short synopses of selected explorers, their travels, excursions, exploits, and discoveries from the 5th to the 14th centuries, followed by major explorations and discoveries during the Renaissance years (1300s to the year 1600):

- **Fa-Hsien** (fl. 5th century c.e.) was a 75-year-old Chinese traveler who, between the years 399 and 413 c.e., made a Buddhist pilgrimage west of China on foot. After following the Great Wall of China, he crossed the inhospitable Taklimakan Desert, then through Tibet to the Middle East and Afghanistan. After traversing the length of India he sailed by boat to Sumatra and then returned to China. The main purpose of Fa-Hsien's mission was to record the history and customs of Buddhist countries and collect religious books and images of Buddhist gods. Because he followed much of what was known as the East-West "Silk Road," his records provided new geographic knowledge of these regions.

- **Hsuan-Tsang** (fl. early 7th century c.e.) was another Chinese traveler who made a Buddhist pilgrimage using established land routes. His records of his travels between the years 627 and 643 c.e. provided excellent information about the terrain, customs, and religious practices of many distant countries. From western China he followed the Silk Road to India. He walked south following the west coast of India, north following the east coast, and then east as he returned to China.

- **Leif Eriksson** (fl. 1000 C.E.) was the son of Eric the Red, the king of the Vikings of Scandinavia. The Vikings were early raiders, explorers, and traders of the North Atlantic who traveled as far south as Italy. Iceland was discovered by accident in the year 860 by the Viking Gardar Svarsson who was blown off course. Greenland was also discovered by accident in 930 and later colonized by Eric the Red in 982. North America was also most likely discovered by accident in the year 986 when the Viking explorer Bjami Herjulfsson missed Greenland. It is said Herjulfsson also explored the northeastern coastal regions of North America but did not land. The best-known early exploration and discovery of North America was accomplished by Leif Eriksson in the year 1000 C.E. when, after several attempts, he landed on what he called "Vinland." He established a colony on present-day Newfoundland (or possibly New England's upper coast) on a place called L'Anse aux Meadows. There is no doubt that Vikings explored Iceland, Greenland, and northeastern North America, but most of the original Viking inhabitants perished or the colonies were abandoned by the 15th century.

Trade between the Orient, Asia, Africa, and western Europe soon overshadowed the explorations, discoveries, and trade between the Vikings and some of their neighbors. One reason was economics—the goods from the Far East were more exotic than were those from Greenland and northern Europe, and thus they provided greater profits.

The Silk Road

Travel and minimal trade between the West and East existed for over 1,000 years and is considered to be an important factor in the exchange of cultures as well as ideas and marketable goods before and during the Middle Ages. From the 3rd century to the early 1300s caravans used the Silk Road as a trade route. A few ancient travelers from Europe visited Asia and China, bringing back tales of great wealth. In addition, a few Chinese traveled throughout the Near and Middle East and southern Eurasia. Since these early explorers followed several established trails, the early Silk Road was not really a road in the sense that we think of a modern road. Rather, it consisted of a variety of trails and routes through mountain passes, over plains, valleys, rivers, and deserts, joining villages and hamlets. It was merely a system of paths that silk traders followed, with the routes changing over the years as local politics changed. As the traffic increased, new prosperous towns and cities developed along the road. Also, it was not a one-way trade route. Chinese silk, porcelains, lacquered boxes and dishes, mirrors, and many other items that seemed exotic to Westerners were transported west—

mostly on camels (in the East) and horses (in the West). At the same time the Romans and several Arab countries shipped eastward glass, coins, gems, jewelry, diamonds, and coral, along with many exotic items such as myrrh, frankincense, ambergris, cardamom, carpets, dyes, minerals, and ivory. The West also sent gold and silver as well as weapons, including bows, swords, spears, and later firearms. Historically, the Silk Road's most important contribution was the role it played as a highway for the exchange of cultural, technologies, and religious ideas, such as Buddhism from India and China, Christianity from central Asia and the West, and Islam from the Arabian countries to both the East and West.

The Spice Trade, Exploration, and Conquest

Hieroglyphics on the walls of Egyptian pyramids and tombs indicate that spice trading between the Orient, India, Arabia, and Egypt dates back at least several thousand years. In the 4th century B.C.E. ships plied the coastal waters around China, India, Sri Lanka, the east coast of Africa, the Easter Islands, the Malaysian peninsula, and Sumatra. This early spice trade was somewhat limited to the countries in the East and the Orient, although the Greeks did some overland spice trading with the Arabs. According to Christian legends, three Gentile sages, who were representatives of the kings of Arabia, Persia, and India (commonly known as the three wise men, or the Magi), brought gifts of gold and the spices frankincense and myrrh to the Christ child in the city of Bethlehem, Palestine. This is said to have occurred in what is now known as year one of the Common or Christian Era (C.E.).

From the time of the Old Testament of the Bible there were accounts of the Arabs as masters of the spice trade. The Arabs carefully guarded from the Europeans the Eastern sources of the spices in order to maintain a monopoly. For example, the Prophet Muhammad reportedly married the widow of a rich spice trader, and as his power, influence, and religious dogma spread throughout the region, so did his Eastern sources for spices and wealth. By the Middle Ages the Europeans made demands for enormous amounts of spices, which tended to expand both the trade and trade routes.

Both the Silk Road and the spice trade between the East and West were mostly conducted overland, with sea transport used only for short distances between countries or to augment the land travel. Because of the small vessels, poor navigation, and notoriously bad weather in these regions, sea traders avoided open seas, staying close to coastlines. Whether overland or sea, the spices came through Alexandria on the Mediterranean coast of Egypt and then to Italy (modern-day Venice) and southern Europe by the Mediterranean Sea. Spices were extremely expensive, and many fortunes were made in the silk and spice trade. For instance, during the Middle Ages the cost of a pound of ginger from China and the cost of one sheep were the same. A pound of mace cost three sheep or one-half of a cow, and cloves were equal to $20 a pound (present-day value). The most prized spice, pepper, was sold by each individual peppercorn and was worth more, by weight, than gold. During the 11th and 12th centuries local city taxes and rents could be paid in spices, particularly pepper, when a small bag of this spice was worth more than a person's life. Overland caravans of thousands of camels carried spices and other merchandise from the Orient and cities in India to Babylon, Carthage, Alexandria, and Rome. Another famous route was the "golden road to Samarkand" that extended from Jordan (northwest of Arabia and southwest of Asia) to Egypt and beyond. This route existed for over a thousand years transporting cloves and pepper from India, cinnamon and nutmeg from the Spice Islands, and ginger and other products from China. Europeans had no way to preserve food (refrigeration was centuries away) and most of their diet was bland. A few types of spices could be used to preserve some foods as well as greatly improving their taste, and thus Europeans desired the exotic spices and other products imported from Egypt and Rome from the East. Traders were only too happy to supply the increased demand at inflated prices. One typical sea route of the spice trade was based on the seasonal monsoon winds of the Indian Ocean blowing eastward in the summer and westward in the winter. Traders used these natural air currents to their advantage. With increased sea traffic many ships were lost to storms and to pirates that plundered the traders' ships. The Greeks and Romans were willing to pay high prices for their spices, seeming to make the benefits outweigh the risks for the traders. However, this arrangement resulted in primarily the rich being able to afford the spices. Then as their use expanded beyond the enhancement of food, so did the demand for cosmetics, soaps, medicine, wines, incense, and so on, most of which came from the East. This clamor for the exotic products of the Orient and India was the driving force behind the development of more efficient and safer channels of trade between the East and West. Thus, economics was a major driving force for the improvement of navigation and shipbuilding to bypass these ancient overland passages that were proving to be very expensive and dangerous avenues for goods. Larger, more seaworthy ships and improved maps and navigational instruments made it possible to sail in open seas around the

Horn of Africa to the Middle East, India, the islands of Southeast Asia, and the Orient. The new trade routes serendipitously resulted in increased explorations and discoveries of new lands and cultures. (Serendib is the old name for Ceylon, known today as Sri Lanka, the island off the southern tip of India. The word "serendipity" originates from a story in which the heroes of Serendib were making unexpected discoveries by accident. Accidental discoveries sometimes occur in modern science, for example, the serendipitous discovery of penicillin.) It might be said that the trade in silk and spices, as well as the evangelical and proselytizing of both Christianity to the East and Islam to the West, were a portent of the Age of Exploration and Discovery.

Following are some of the most important early geographers and explorers of the Middle Ages and Renaissance:

- **Abul Hasan Ali al-Masu'di** (ca. 895–957 C.E.) was an expert geographer and historian who traveled to many eastern countries, including China, Madagascar, India, Ceylon (Sri Lanka), as well as Arabian countries newly converted to Muhammadanism. In his book *Muruj-al-Thahab*, Ali al-Masu'di described much of the geography, climate, and local religions. His second book consisted of 30 volumes chronicling in detail the countries he visited. He is noted for his analytical style in which he relates the sociology, ecology, topography, and geography to the history of different countries. Because of his "scientific style" of writing, he is credited with influencing for many years the fields of geography, geology, biology, and earth science.

- **Al-Idrisi** (1099–1166 C.E.), a Muslim born in Spain, is considered by some as the greatest geographer and mapmaker of the Middle Ages. He not only traveled extensively, but he collected great quantities of geographic and biologic information (medicinal plants) and made accurate measurements of land features that he incorporated into his maps. Al-Idrisi calculated the circumference of the Earth as 23,000 miles, but it has not been determined how he arrived at that figure. (The Earth's polar circumference is 24,857 miles, while the circumference at the equator is greater at 24,900 miles due to centrifugal force created by the spinning Earth.) He also constructed an ~850-pound silver spherical globe for the King of Sicily, the surface of which included the outlines of the seven known continents, numerous trade routes, cities, rivers, lakes, and mountains. The globe contained distances, the heights of mountains, and other measurements. His major book, *Nuzhat al-Mushtaq fi Ikhtiraq al-Afaq* (The Delight of Him Who Desires to Journey through the Climates) is an encyclopedia of the geography and botany of the countries and regions to which he traveled. In the early 17th century his books were translated into Latin. Columbus reportedly used copies of al-Idrisi's maps when he sailed westward.

- **Marko (Marco) Polo** (1254–1324 C.E.) and his brother, Niccolo Polo, were Venetian traders who were the most famous of all the travelers of the Silk Road. Marco recorded most of his travels while in prison, but his legacy, producing descriptive geographic documentaries, captivates to this day. During one of his trips he was forced to join a caravan to the Mongol capital of China where he met the great leader Kublai Khan and received a friendly reception. When he headed back west, he was asked to petition the Pope to send 100 "wise men" to educate the Khan's people. When Marco, his brother, uncle, and father later returned to China, they had just two scholars with them, plus many gifts from the Pope. They stayed with the Khan for 17 years, acting as administrators and travel advisers. Supposedly, Marco claimed these were the best and most exciting years of his life. He was involved in trading, business, and diplomatic affairs. In 1295, they returned by sea to Persia and then on to their home in Venice. During this voyage home, Marco saw for the first time the large, Chinese oceangoing ships. One of them had 60 cabins and was divided into sections by watertight bulkheads designed to prevent the ship from sinking if rammed by a whale.

  Marco described many exotic places, people, and customs in his writings, and much of it was, and still is, discounted as fantasies and interesting myths. Marco was not given credit for opening new trade routes for the Venetians. The Doge of the Venetian Republic ordered him to be silent about his discoveries, while others benefited greatly from his exploits. Regardless of the veracity of his chronicles, he brought different cultures and civilizations closer together and opened up travel, exploration, and trade for future generations.

- **Ibn-Batuta** (1304–1369 C.E.) is known as the "Muslim Marco Polo." After visiting Mecca in 1325, he decided to dedicate his life to travel. In his lifetime he traveled 75,000 miles, which was more than any other person had traveled in his day, including Marco Polo. He maintained records of the regions he visited, which included China, India, Russia, Africa, as well as most of Arabia. The encyclopedia of his travels stressed the religion of the region, particularly Islam, rather than the geography.

- **Zheng He** (Cheng Ho) (fl. 1400s), a Chinese Muslim, was the most acclaimed admiral of the Ming Dynasty. In 1414 he commanded a fleet of 62 large Chinese trading vessels. His ships were much larger than those built by the Dutch, Portuguese, and Spaniards. For instance, the largest of his ships was 400 feet long and 140 feet wide and could carry 1,500 tons. By contrast, the largest of Columbus's ships was only 90 feet long and 30 feet wide and could carry only 400 tons. After Genghis Khan conquered China in the 1200s, private trade was prohibited with the West, as was sailing in the Indian Ocean. By 1368 the Chinese expelled the Mongols and established the Ming Dynasty, which ruled for the next several centuries. By the 1400s China attempted to reestablish itself in the western seas, with the goal to become the "Middle Kingdom" and rule land to the west of China. Chinese explorers visited the islands of Southeast Asia, India, Ceylon, and Africa, reporting many strange animals and human customs. By 1432 China had seven large oceangoing fleets that numbered in the hundreds of boats and some 20,000 men for crews. Thus, China was quite capable of ruling the seas. But for some unknown reason Zheng He's fleet did not go westward beyond the African Cape of Good Hope and into the Atlantic Ocean. One theory is that the Chinese Navy and its conquest goals were suspended when a new emperor objected to spending great sums of money on navigational pursuits. This decision resulted in the isolation of China for many centuries. By the early 1500s and beyond, the waters east of Africa were controlled by the Dutch, British, Portuguese, Arabs, Spaniards, Indians, Japanese, and, later, Americans.

By the 14th and 15th centuries a combination of numerous land and sea routes through and between many countries were used for trade. As a consequence, these routes became increasingly unsafe for the richly packed caravans, as bandits became more numerous. The Arabs controlled much of the land routes in the Middle East as well as most of the coastal sea-lanes of trade between the Orient, India, Egypt and the West. The various nations and cities along the routes imposed taxes on the spices and other goods that passed through their territories. A number of factors acted as incentives for the development of safer and more economical trade routes. These factors included disasters at sea (storms and pirates) and on land (lack of vegetation along desert trails for camels and horses), and, of course, bandits and taxes. These negative elements of trade meant that the cost of goods from the East was extremely high and unreachable for many. Thus, economics was a major impetus for the great explorations and discoveries that occurred across the Atlantic Ocean to the west of Europe during the Late Middle Ages and Renaissance.

- **Prince Henry the Navigator** (1394–1460), a Portuguese royal, did as much as anyone during the Renaissance to establish the countries of Europe as one large trading block. At 19 years old, he was charged with building a fleet of ships that he subsequently and successfully used to lead a crusade to oust the Muslims from Morocco in northwest Africa. As a result, he believed that great wealth could be generated through trade in Africa. In addition to jewels, gold, silver, salt, pepper, ginger, and other spices arriving from the Far East, the African trade also included slaves. Prince Henry, supported by the King of Portugal, commanded a series of expeditions to explore the west coast of Africa. A shipbuilder, cartographer, navigator, as well as an explorer, Prince Henry oversaw the building of new types of ships called *caravelae*. These were much smaller vessels than the large square-riggers, but they were much faster. One reason for the new design was economics. These new, faster ships assured that the crew and cargo could get home in fewer days even when sailing into the wind. The *caravelae* combined the designs of both the cargo ship for trade and the maneuverability of river sailboats using **lateen** sails instead of square rigged sails. These newly designed boats could sail into the wind at a 55° angle, while square-riggers required a 67° approach to the wind. This meant that the *caravelae* could take shorter tacks into the wind and thus reduce the distance about one-third for the return voyage. By the mid-1400s most of the west coast of Africa had been explored by Prince Henry and the captains of his ships. After Prince Henry's death, Bartolomeo Diaz voyaged further south in 1487 along the west coast of Africa. During a storm he was driven eastward and passed the Cape of Good Hope without seeing or realizing that he had sailed around the southern tip of Africa. On his return west in 1488 he did discover and explore the region of the Cape. Several other Portuguese explorers also sailed around the Cape and up the east coast of Africa as far as Ethiopia, but they never made a return voyage. This discovery of a sea route around the tip of Africa directly to India bypassed the land routes and opened up a new era of sea trade between the East and West.

- **Christopher Columbus** (1451–1506) came from a poor family of wool weavers in Genoa, Italy. He received little formal education and learned navigation mostly by experience at sea. One story tells how he arrived in Portugal. While he was sailing in a Genoese ship to England, it was attacked and then sunk. Columbus grabbed some driftwood or an oar and swam to the coast of Portugal where he stayed for a number of years. Later he sailed along the coasts of Europe, England, Iceland, the Azores, and Africa. In 1479 his marriage into Spanish nobility opened up contacts in the royal court. Based on the maps and writings of Ptolemy and other navigational documents, he believed it would be a relatively easy westward voyage from Europe to the east coast of Asia. He calculated that it would be faster to sail westward from the north coast of Africa than from the coastal regions of northern Europe. He finally convinced Spanish royalty to finance his trip westward to establish a new route to the riches of the Far East. One reason the Portuguese did not sponsor his trip is that they accepted Eratosthenes' calculation for a larger circumference of the Earth than Ptolemy's estimation. Thus, the

Portuguese believed Columbus would be unsuccessful. Also, they were more interested in establishing and controlling a trade route to the Orient around the southern tip of Africa. Columbus's first trip included three small ships: the *Santa Maria* carried 40 men, the *Pinta* carried 26, and the *Nina* carried 24. Each of the ships could fit crosswise (gunwale to gunwale) inside one of the large ships of the former Chinese Navy. The *Nina* and *Pinta* returned to Europe, while the *Santa Maria* shipwrecked, leaving 39 volunteers to stay behind and establish a settlement in Haiti. (The entire crew of the *Santa Maria* was killed by the natives.)

There is some controversy as to Columbus's skills as a navigator. Mostly he used **dead reckoning,** but he also had some experience with maps and tools for celestial navigation. While at sea, he made several attempts to measure his latitude. Latitude, both on land and sea, can be determined by measuring the celestial latitude (declination) of a particular star at the same time each night. By measuring the angle between the star and the point of the overhead zenith, latitude can be calculated. He used a quadrant and, at least once, an astrolabe, for celestial navigations. Most of his measurements were inaccurate. Also, his claims that he was able to determine his longitude were somewhat fraudulent. He claimed that he determined his longitude by timing eclipses during his voyages. It has also been alleged that Columbus entered different figures for each day's travel (leagues) than his pilot did. Thus, there may have been a difference in distance equal to a full day's travel for the total trip. However, in all fairness, during the 15th century no one had a clock accurate enough to be used to determine longitude while at sea. Since Columbus used Ptolemy's figures for the circumference of the Earth, rather than the more accurate calculation of 25,000 miles made by Eratosthenes centuries earlier, he underestimated the distance (and thus travel time) from the west coast of Africa to the east coast of the Far East. Also, Marco Polo's map indicated that China and the Asian continent extended much further eastward than they really do. Columbus thought that it was just 3,000 miles from the coast of Europe to Asia—thus his trip was longer than expected. After going south from Spain to the Canary Islands (off the north African coast), he headed west by what turned out to be the shortest route, as well as the fastest, by picking up the west-blowing winds to the New World. It is still remarkable that in just three weeks he landed on one of the islands in the Bahamas. It is unclear if he actually landed on Watlings Island of San Salvador or the Samana Cay, an island just south of San Salvador. After exploring the other islands and collecting some spices, he returned to Europe by a more northerly route to pick up the trade winds blowing eastward. Columbus never landed on the continent of North America, but later he did make landfall on the northeastern tip of South America. He made a total of four trips to the Americas. His second voyage in 1493 consisted of 17 ships and 1,200 men. For his third trip he recruited one woman for every ten emigrants. Columbus introduced tobacco to Europe after he learned of its use by natives. He also was the first to bring horses to the New World. Contrary to some stories, Columbus did not set out to prove the world was round, nor did he ever say that he discovered a new continent, as he always considered the islands he visited as part of Asia. Also, his crews consisted of experienced sailors, not criminals as is often reported. Accounts that maintain that his first trip was in peril, that it experienced foul weather, and that several sailors died for lack of food are all inaccurate. None of this is true—the weather for his first trip was fine, there were no storms, and no one died during the trip west. However, they did lose two ships to a hurricane on the return trip to Europe.

As a navigator, Columbus was extremely lucky. He arrived at his theory of sailing west to India and Asia based on sailing experiences down the west coast of Africa. He decided that sailing west from north Africa would be faster then going east around the Cape of Good Hope. Although he had some of the best maps and navigational equipment of the time, he had to reinvestigate the geography through the writings of Aristotle, Ptolemy, Strabo, Pliny, and others. He depended on Marco Polo's description of Japan that claimed the islands were 1,500 miles off the east cost of Asia, and thus he thought it was possible to sail west and reach Asia or India. Columbus had no way to determine his longitude. This means that he was unable to determine how far he had traveled each day or, for that matter, from the time he left Spain. Columbus did instruct a cabin boy to turn over the sandglass clock every half-hour. However, the clock sometimes ran slow, sometimes fast. Using the simple quadrant and astrolabe to establish the positions of the moon, sun, and stars, Columbus had some idea of his latitude; however, he had no success in determining his longitude.

• **Vasco da Gama** (1460–1524). With four ships out of Lisbon, Vasco sailed around the African Cape of Good Hope in 1497 to explore a passage to India and the East. (This was several years after Columbus's westward trip to find a passage to the East going the other direction.) Vasco da Gama sailed up the east coast of India to Calcutta in 1498, which, in a sense, completed the goal of Prince Henry to find a sea route to compete with the overland trade. This was the first trip that lasted long enough to cause death to most of the crew as a result of scurvy.

• **Ferdinand Magellan** (1480–1521) was a Portuguese navigator whose fleet of five ships set out on a round-the-world ocean exploration in 1519. Because Spain funded his expedition, Portugal tried to prevent his trip. He eventually set sail westward to find a southern passage to the riches of the Spice Islands, a goal that had evaded Columbus about 25 years earlier. Magellan knew that he must pass the unexplored southern tip of the South American continent, but this was more of struggle than he had anticipated. It required several weeks during storms and high seas to slowly and cautiously pass through these treacherous straits, now called the Strait of Magellan. After passing the tip of South America he entered the open, calm, great new body of water that he named Pacific, which means "peaceful" or "calm." It was much larger than he expected, and his fleet sailed for four weeks without sighting land. Many of the crew were in the process of dying of hunger when they finally landed on an island, present-day Guam. After securing food, they continued to the Philippine Islands where natives attacked the party. Magellan died from wounds received in this battle in 1521. The exploration continued westward. Only one of the original five ships, the *Victoria,* and only 18 men under the command of Juan Sebastian de Elcano made the return trip to Spain in 1522. They managed to arrive with enough spices to pay for the trip and to make a profit. Spain soon gave up the Spice Islands for the more lucrative adventures of collecting Inca gold in the New World. Portugal dominated the Far East spice trade until the end of the 16th century, and then Dutch expeditions controlled the trade until the end of the 18th century. After the Dutch were defeated by Great Britain, England ruled the Spice Islands as well as India and the trade routes in the Far East until the early 20th century.

Magellan's circumnavigation of the Earth proved that Eratosthenes' calculation of 25,000 miles for the Earth's circumference was correct while Ptolemy's was not. Geographically, his trip indicated that huge oceans cover the face of the Earth, with large island-like continents covering only a small area of the Earth's surface. The southern route from Europe—down the east coast of South America, through the Strait of Magellan, across the broad Pacific Ocean, through the waters of Southeast Asia and India, and around the Cape of Good Hope at the tip of Africa, then north along the west coast of Africa to return to western Europe—proved to be a long, treacherous, and costly journey. By the end of the 1400s and throughout the 1500s a number of expeditions departed from various European countries seeking a shorter, more direct route to the treasures of the East. It was commonly assumed that by sailing west on a more northerly route, the northern regions of the New World could be bypassed, and it would be possible to sail north of Cathay (China), then south to India. When looking at a globe, it is obvious that the more north the latitude, the shorter the distance from one meridian to the next. Meridians are great circle markings from pole to pole. Longitude is one's location between curved polar meridians. Greenwich, England, is the location on the globe where meridians start at 0 and end at 360 degrees. Thus, distances between meridians are greatest at the equator and least near the poles. This idea soon became known as the Northern Route to Asia or the Northwest Passage. Another reason the British, Dutch, and French all decided to seek a northern passage to the East was that the route around southern Africa (Cape of Good Hope) was controlled by Portugal, and by 1521 the route around the southern tip of South America (Strait of Magellan) was controlled by Spain.

Following are short summaries of the explorations and discoveries of a few of the many who attempted to find a northern passage to the East—both westward by the polar route north of North America and eastward by the polar route north of Eurasia. In 1553 England paid explorers to find the northern route to Asia, and the Dutch offered an award of 25,000 guilders to locate a route to Asia.

- **John Cabot** (Giovanni Caboto) (1450–1498) was born in the seafaring city of Genoa but grew up in Venice, which at that time was vying with Genoa for primacy of the seas. Cabot knew that Columbus's voyages to seek a passage to the Indies of the East sailed by the southern route. The distances between longitude lines (meridians) are shorter the further north one proceeds, and all meridians converge at the poles (zero degrees). Thus, Cabot decided it would be a shorter route to sail to Asia by a route at latitudes closer to the North Pole. Cabot also believed that the spices being shipped to Europe originated in northern China, not southern India, which, if so, would make the trip even more profitable. He received a grant from King Henry VII of England and embarked in 1496 with 18 men on a single, small (50-ton), fast ship, the *Matthew*. He only got as far as Iceland on his first voyage in 1496. In 1497 he tried again and landed somewhere along the North American shoreline. Although there is no evidence, it is assumed that he landed in the present-day Canadian province of Newfoundland, or Labrador, or possibly present-day Maine. Even so, Henry VII gave him credit for discovering and claiming Newfoundland for England. The king gave John Cabot and his three sons "letters of patents," which are somewhat like lifetime grants for land and wealth they discovered. 1498 he set sail again with five ships. All but one met disaster, and in distress, it reached Ireland, but Cabot was lost and never heard from again.

  His son, Sebastian Cabot, who accompanied his father on his voyages, also led another expedition to the New World in 1508, seeking a northwest passage. Although the accuracy of his report of repeating Magellan's voyage around the world has been questioned, Sebastian has been given credit for doing so.

- **Giovanni da Verrazano** (ca. 1485–1528). In 1520 Verrazano explored the mid-east coast of North America. He knew that the Spanish had explored the Florida region with no westward passage found, and others had explored Newfoundland and the northern coast of the continent, also without success. Therefore, he believed a passage along the broad unexplored middle between these two regions might exist. What he did discover was New York Harbor. The present-day world's longest suspension bridge (Verrazano Narrows), connecting Brooklyn and Staten Island in New York, was named in his honor.

In 1609 another English explorer, **Henry Hudson,** entered the river harbor discovered by Verrazano. He sailed as far north in the river as the present-day city of Albany. This river that flows into New York Bay is now called the Hudson River. He was also seeking the Northwest Passage and later entered what is now known as the Hudson Bay in northern Canada. His men mutinied and left him to die in 1611 in present-day James Bay, located in the southern region of the Hudson Bay.

- **Jacques Cartier** (1491–1557) was a French navigator who explored the northeastern regions of North America for a passage that might lead from the Atlantic Ocean to the Pacific Ocean and then to India. In 1534 Cartier discovered a large bay-type opening between Labrador and Newfoundland (now called the Strait of Belle Isle) that he followed westward. He encountered friendly native Indians along this river that he believed to be the passage he had sought. On his second voyage he followed the river until he saw the rapids upriver that indicated this was not the passage to the Pacific. Cartier and his men suffered from scurvy and were close to death when Donnacona, a Huron chief, told him about a cure that involved the needles and bark of the white cedar tree. He and his crew stripped the trees, ate the material, soon recovered, and returned to France with rumors of gold to the north of the river. The river and adjacent land to the north was claimed as French territory (Canada), which the French ruled for over two hundred years. The river and its mouth were discovered on August 10, the day of St. Lawrence; therefore, he named this river the St. Lawrence River.

  Later in 1603, **Samuel de Champlain** was commissioned by the French to explore the coastal area south of Nova Scotia and then proceeded up the St. Lawrence River. He established the first major French settlement in 1608. It was then, and still is, called Quebec, one of the provinces of Canada. Later, he discovered a large lake to the south that was named after him (Lake Champlain).

- **Hernando de Soto** (1500–1542) was a Spaniard who explored Florida and other regions of the present-day southern United States. He is credited with discovering and exploring the Mississippi River in 1541. Hernando died while he camped by the Mississippi River in 1542.

- **Sir Hugh Willoughby** and **Richard Chancellor** (dates unknown). Their expedition was one of the first of what later became known as the

British "Company of Merchant Adventurers." Sir Willoughby was not a navigator, but he was a natural leader. Chancellor acted as the pilot for their three ships (*Bona Esparanza, Edward Bonaventure,* and *Bona Confidentia*). After Cabot's return to England from his unsuccessful search for a northwest passage, and with the knowledge that both the southern passages around South Africa to India and the southern passage around South America were controlled by other countries, Willoughby and Chancellor considered several alternative routes. One route, over the North Pole itself, was impractical. The one to the northwest of the New World was tried, but no passage was found. However, the route to the northeast past Scandinavia and Russia, then possibly continuing to northern China and down to India, had not yet been explored. Their three ships left England in 1553. During a storm Willoughby's ship became separated from the other two and later became stuck in the ice at Murmansk near the Arizina River. A year later Russian fishermen found the ship with all of the crew frozen to death. Chancellor reached the White Sea and landed at the harbor now known as Arkhangel'sk in northern Russia. The Russian czar Ivan IV (the Terrible) invited Chancellor to visit him in Moscow where the explorer was royally received. On his return trip to England in 1554 Chancellor brought documents that resulted in an extensive trade agreement between Russia and England.

A number of explorers from many countries attempted to find a northwest passage to the Orient. Just to mention a few: **Willem Barentsz** (Barents), an explorer from the Netherlands lost his life trapped in polar ice. The Barents Sea, an arm of the Arctic Ocean, is named after him. **Olivier Brunel** was an early Flemish navigator who, in 1556, sailed beyond Lapland searching for the Northeast Passage to China following the route established by the Company of Merchant Adventurers, now known as the English Muscovy Company. Brunel had difficulty breaking into the established trade between Russian, England, and Denmark. He did manage to land a small boat with goods for trade, but when returning to his ship the landing boat capsized, drowning Brunel and some of his men. His ship made its way back to Holland loaded with semiprecious stones that were inadequate to pay for the voyage. **Jens Munk** (1579–1628) was one of Scandinavia's great polar explorers. After serving in the Danish Navy he was sent to discover the Northwest Passage that was then believed to connect the Hudson Bay north of Canada to the Pacific Ocean. The goal of the Danish-Norwegian king Christian IV was to control the Arctic Straits that had already been discovered, and thus dominate the Northwest Passage and the riches of the East. In 1620 Jens Munk's ship was trapped in the ice, his men had scurvy, all but two of his crew died, but somehow Jens Munk and the two men returned to Norway.

Unlike Antarctica, which is a land continent covered with thick snow and ice, there is no land base for the northern Arctic. It is basically thick sheets of ice. Technically, there is a sea route north of Canada that could be called the Northwest Passage that extends from Baffin Bay west of northern Greenland extending westward to the Beaufort Sea located at the entrance of the Arctic Ocean. But it is virtually impassable by surface ships due to year-round ice. Today, nuclear submarines are the only types of vessels that regularly travel beneath this frozen body of water. Nuclear submarines have also passed under the polar ice sheet from west to east following the Great Circle Route (the shortest route) from North America to Russia. These subs have explored and mapped the bottom of the Arctic Ocean and discovered that it is one of the shallowest of the major seas even though it has a thick layer of ice on its surface. It might be said that nuclear subs have explored and found the famous Northwest Passage. At about the same time this North Atlantic seagoing traffic was searching for the short route to China and India, others were seeking riches in the southern parts of the New World. This included Spain, a country that explored and exploited much of the Caribbean and the coasts of the Americas.

- **Amerigo Vespucci** (Americus Vespucius) (1454–1512), an Italian navigator, led several expeditions along the eastern coast of South America. He was the first to realize that the North and South American continents were separate continents and not part of the Asia described by Marco Polo. Many explorers, including Columbus, also assumed incorrectly that the North and South American coastlines were the coastal areas of the Asian continent. In 1502 Vespucci published his theory explaining that the New World was an entirely new continent. He explained that Asia must be far west, across the Pacific Ocean, beyond the New World just being explored. In 1507 a mapmaker decided that this new land should be shown as a separate continent to be named America after Amerigo Vespucci.

- **Francisco Fernandez de Cordoba** (1457–1525) was one of many Spanish explorers who sailed up and down the coasts of Mexico, Central America, and South America, and was the first European to land on the Yucatán peninsula of Mexico. He found only traces of the Mayan civilization that existed near the coastal area of the peninsula several centuries earlier.

- **Hernando Cortés** (1485–1547) led a large expedition of sailors and soldiers, including 17 horses and 10 cannons. He pressed into central Mexico where his forces overwhelmed the Aztecs. The local populace had never seen horses or cannons or other types of Spanish arms. Their king, Montezuma II (1466–1520), considered the Spaniards as gods and thus did not oppose them until he realized that the Spaniards were destroying the Aztec civilization. Cortés explored western and northern Mexico and discovered Baja California.

- **Pedro de Alvarado** (1486–1541) was a Spanish conquistador and a major officer for Hernando Cortés. Alvarado ruled the local populations while Cortés explored northern Mexico. Alvarado exerted absolute control and, as such, created several rebellions. Cortés sent him to explore Central America where he conquered Guatemala and Salvador. He founded several cities and colonies and ruled as governor of this region. Alvarado led a force against a revolt in 1541 and was killed in retreat. His wife, Doña Beatriz de la Cueva, succeeded him as governor of Guatemala.

- **Francisco Vásquez de Coronado** (1510–1554) was governor of the area now known as Sinaloa and Nayarit, Mexico. He is an example of the many Spanish explorers who had great interest in the stories of gold and set out to find these civilizations of wealth. He led a large expedition into what is now New Mexico, west Texas, and as far as Kansas tracking down the rumors of gold supposedly held in local villages. What he found were poor Native American tribal villages. Only 100 of his original expedition of 340 Spanish, 300 loyal Indians, and 1,000 slaves returned to Mexico with him. He died in Mexico City in disgrace in 1554.

Not all Spanish explorers were "gold hungry and bloodthirsty." There is a story of a priest, **Jeronimo de Aguilar,** and an explorer, **Gonzalo Guerrero,** who were shipwrecked off the Yucatán peninsula and survived along with several others in the party. The natives were cannibals and killed most of the men, but these two escaped into central Mexico where they met with friendly natives. Both Aguilar and Guerrero assimilated with the Mayan civilization and learned their language. Guerrero was presented a young wife by the king and raised two children, while the priest is said to have kept his vows as he was appointed chief of the king's harem. After Cortés arrived in central Mexico in 1519, the Mayans resisted the Spaniards. Guerrero, now more Mayan than Spanish, refused to join the Spaniards. He chose to fight alongside the Mayans until he was killed. After the battle, Aguilar joined his countrymen on a return trip to Spain. Another example of a humane explorer is the priest **Bartolome de las Casas** (1484–1566), who was known as the 16th-century human-rights advocate for fair treatment of the natives. He helped establish laws that prevented natives from being used as slave labor and denounced the Spaniards' treatment of the Aztecs, Mayans, and other tribes. After objections from local colony governments, de las Casas established a farm colony in what is now present-day Venezuela that treated both the Spanish and native inhabitants as equals. However, after much resistance, the colony failed and de las Casas spent the rest of his days writing about his experiences.

Cartography

Introduction

Cartography is the science and art of mapmaking. Maps can be classified in a variety of ways, for example, specific types: road, aerial, weather, political, relief, physical, orthophoto, and grid (latitude/longitude). They can also be classed as abstractions, such as topographical (world or local land-feature contour lines), plats (surveys of property boundaries), and charts (nautical and aeronautical navigation). Mapmaking is as old as humans' ability to abstract the reality of their earthly surroundings into crude sketches, figures, or picture graphs. Evidence indicates that maps evolved independently throughout most regions of the world many thousands of years ago. One might consider the trails formed by prehistoric humans to guide others as preliminary maps. Once they sketched out the trail's path in the dirt for a new person to follow, an abstract representation of the territory was created, that is, a map. Some anthropologists consider the simple diagram types of mapmaking older than the art of writing. The science of semantics is the study of the meaning of abstractions, such as words, but it might just as well be thought of as the science of the meaning of abstract map-type diagrams. For many millennia people have confused the written word, symbol, icon, or image for the real object it represents (the map is confused with the territory, or a flag for the country), or the spoken word for an act (the epithet "Go to hell" does not mean one is Hades-bound). It often seems that only children are aware of the distinction between object and image. When taunted, they may say, "Sticks and stones may break my bones, but words will never hurt me." The inability to read an abstract map is something akin to not understanding the abstractness of language, either written or oral, which has resulted in irrational behavior. Some humans can be, and are, map illiterate as well as word illiterate.

History of Cartography

No doubt there were Stone Age maps, but since ancient maps were most likely drawn on perishable material, such as bark, wood, and bone, there is not much evidence of prehistoric cartography. However, a mammoth tusk with a sketch illustrating a stream and houses made some thirteen or fourteen thousand years ago was found in the Ukraine. Manifestations of Old Stone Age maps are skimpy, while New Stone Age cave drawings (ca. 6200 B.C.E.) recently found in Turkey clearly indicate the outlines of villages, streams, rivers, and even volcanoes. Chinese historians referred to maps as far back as the 7th century B.C.E. There is also some evidence that crude footprint trails served as maps during the pre-Columbian period of the New World.

It is believed that the first real map was the one engraved on a silver vase buried in a tomb in the Ukraine dating from about 3000 B.C.E. It clearly depicts rivers flowing from mountains to the sea, with animal figures along the river's bank. More practical maps appeared in Mesopotamia after ca. 2000 B.C.E. A clay map found in northern Iraq and dating back to about 2300 B.C.E. illustrates a distinct area of hills, canals, and boundaries, along with cuneiform symbols denoting landowners' holdings and names. This is the first clay-tablet map known to display the cardinal points of north, south, east, and west. Another clay-tablet map dating from about 1500 B.C.E. depicts the Babylonian city of Nippur. This particular clay tablet was the first city map that included the city's major buildings drawn to scale. The first known world map was also designed in Babylonia in 600 B.C.E. It indicates the countries surrounding Babylonia, including Assyria (present-day Iraq), Urartu (present-day Armenia), Persia (present-day Iran), and a large unknown ocean. The map was designed to show Babylonia as the center of the world with the four regions that surrounded it.

A number of ancient Greek philosophers and scientists also attempted to construct maps of various regions of the known world. Among them were the following:

- **Thales of Miletus** (ca. 624–546 B.C.E.) proposed using a disk-shaped map to depict oceans as well as land.

- **Anaximander of Miletus** (611–554 B.C.E.) was the inventor of cylindrical maps that depicted land on a curved surface. Although no copies of his maps have been found, his successor, **Hecataeus of Miletus,** drew a map in ca. 500 B.C.E. based on Anaximander's originals. Hecataeus's map was circular and illustrated the world as land surrounded by the oceans.

- **Marinus of Tyre** (ca. 70–130 C.E.). Tyre is a city in ancient Phoenicia (now Lebanon). Marinus was a geographer and mathematician and an early mapmaker who is sometimes known as the father of mathematical geography. Although his works have not survived, Ptolemy used Marinus's ideas and maps, which were the first to utilize latitudes and longitudes, in developing his own maps.

- **Ptolemy of Alexandria** (ca. 90–168 C.E.). The easiest way to represent a three-dimensional spherical Earth was to construct a two-dimensional circular map with the land in the center and oceans around the edges. Although this depiction was not accurate, it sufficed until the idea of **projection** and a system of coordinates were developed. Ptolemy understood the difficulty in depicting a spherical surface on a flat surface. As far back as the time of the Babylonians, maps of the Earth were constructed as flat circles. It was Ptolemy who, in about 150 C.E., first suggested the use of projection and coordinates to draw maps. On his early maps Ptolemy placed east at the top, with Jerusalem as the central feature and the three major continents—Europe, Africa, and Asia—separated by the Mediterranean Sea. Ptolemy used the ideas and maps of other cartographers just as cartographers of the Middle Ages used some of Ptolemy's concepts. Even though Ptolemy underestimated the size of the Earth, this error, along with the orientation of his maps, was corrected. Today, all modern maps are oriented with north at the top.

An early and detailed Chinese map dating back to 221 B.C.E. recently found in a tomb was quite detailed, with grooves for rivers and raised mountains. No doubt, this was the first relief map that indicated the elevation of topographical surface features.

Greek cartography influenced the art of mapmaking into the Middle Ages and some of the older engraving techniques were used until the era of the printing press during the Renaissance.

Cartography of the Middle Ages and Renaissance

The concepts and methods of constructing maps that were developed in Europe and Asia preceding the fall of Rome greatly influenced cartography during the Late Middle Ages and Renaissance. Following the fall of Rome in the mid-5th century C.E., western Europe experienced what is known as the "Dark Ages" under a theological dictatorship that obstructed all types of learning except within the Church. This suppression of learning and knowledge resulted in a long period of social, political, and economic deprivation that stymied the development of classical intellectual activity for several centuries. The only types of maps made by Europeans during this medieval period of the Early Middle Ages were elaborate, religious-oriented maps of the Mediterranean regions. These crude, stylized Church-approved maps were known as "T-O" maps because they consisted of a *T* within a circle, with the three points of the *T* touching the inside of the circle. Asia was placed above

Figure 2.1 T-O Maps
Two examples of T-O maps. The bottom one is translated into English. The name is derived from seeming to have the letter *T* contained within an *O*.

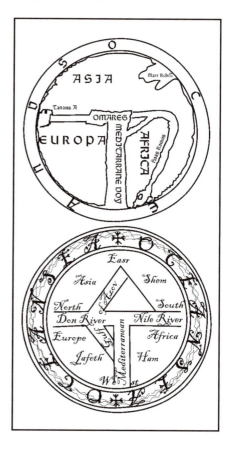

the crossbar of the *T*, with Europe situated on the left of the vertical bar of the *T* and Africa on the right, while Jerusalem was located just above the junctions of the crossbar and vertical of the *T*. (See Figure 2.1.)

In part, the backwardness of Europe provided an opportunity for the invading Muslims, Arab, and Mongol traders in the 7th to 9th centuries c.e. to explore new lands eastward as far as China and Russia and west to present-day Spain. The Muslim navigators designed new maps based on Ptolemy's world maps. By the early Middle Ages elaborate maps were being block printed in China, and Muslim and Arab geographers were producing large world maps.

While Western Europe was in the intellectual doldrums, Chinese cartographers constructed elaborate maps, an example of which is the map carved in stone that was produced in the Shensi province in northern China in 1137. Designed on a rectangular grid system, it used a scale of 100 units to 1 mile. This stone was a very accurate representation of the geographic region it depicted. By the 1400s the Koreans produced an elaborate world map that accurately indicated the shapes of Africa, Europe, and islands off the west coast of Africa and Europe. Oriental maps were the first to attempt to depict topological features in three dimensions. In the late 5th century c.e., Hsieh Chuang, a Chinese ruler, instructed his geographers to build a 10-foot square relief jigsaw map from wood. Topological features such as rivers and mountains were shown, and each province was a separate piece of wood so the map could be disassembled and then reassembled similar to a modern jigsaw puzzle.

The following men are considered to be the most prominent and influential navigational mapmakers of the Middle Ages and Renaissance:

- **Abdullah Mohammed ibn al-Sharif al-Idrisi** (1154–1192) was considered the greatest mapmaker of the Middle Ages. As with most Muslim cartographers, al-Idrisi used Ptolemy's maps as well as Ptolemy's instructions of how to make maps utilizing a set of points, lines, and areas. The Arab traders and explorers combined this knowledge with their own experiences to produce very stylized maps that sometimes distorted the coastlines of continents. His maps included details of northern Egypt, the Nile River, and the Mediterranean Sea, but as with all Arabs of this period, al-Idrisi did not indicate much detail beyond the Strait of Gibraltar since he feared the wild Atlantic Ocean beyond the Strait. Al-Idrisi is most famous for his world map, which is still considered an excellent representation of his time in history.

- **Ishaq Ibrahim ibn Muhammad al-Farisi al-Istakhri** (ca. 1193). His books and maps are only known through his biographers. Al-Istakhri's texts and maps, along with some other Arab geographers comprise about 21 maps, sometimes referred to as the "Islamic Atlas." These range from: (a) a world map, (b) a map of the three known seas (the Mediterranean, Indian Ocean, and Caspian Sea), (c) a map of the 17 provinces, (d) a map of geographic regions within the Islamic Empire including Egypt, Syria, Arabia, Armenia, and the Persian Desert, and (e) a map of other geographic regions under Islamic control. One purpose of his maps was to designate the caravan routes across the Khurasan and Persian deserts.

Following are short descriptions of some of the more important mapmakers of the Late Middle Ages and Renaissance, which was a period of extensive exploration by ships. Portuguese ships sailed farther out into the unknown Atlantic Ocean, exploring down the western coast of Africa by 1487 and later in the early 1500s sailing west as far as Brazil and Canada. Although the Portuguese depended on Ptolemy's techniques for mapmaking, they continually updated and improved their maps, producing large master charts that indicated all the hazards of the coastlines they explored. Since these maps were state secrets, most of these navigational charts have been lost. However, sections that depict the coasts around major Atlantic harbors and ports, as well as their dangers, have been preserved.

- **Christopher Columbus** (1451–1506), the Italian navigator, used Ptolemy's maps as well as the updated Portuguese versions of these maps. Columbus was not a mapmaker, but he was experienced enough to be able to combine features from several different maps to arrive at a route for his westward venture. He used Ptolemy's 3,000-mile estimate for the westward voyage from Europe to the east coast of Asia. This error required more time than he expected to reach what he thought were the islands off the west coast of Asia. Columbus made several maps of the islands he visited in the New World. Navigators in the late 1400s and early 1500s used Columbus's maps while exploring the coasts of North America and northern South America.

- **Piri ibn Haji Memmed** (Piri Re'is) (dates unknown), an admiral in the Turkish navy, drew a map, using the equidistant-projection method and mathematics, that included distorted coastlines of North and South America. He claimed he used a number of different sources, including Columbus's maps of the Americas, to construct the map that he completed in 1513.

- **Abraham Ortelius** (1527–1598) was a Flemish map illustrator who, with no navigational experience or science background, published a collection of 70 maps in 1570. He called his popular book the *Theatrum Orbis Terrarum* (Theater of the Whole Word). Considered the first modern world atlas, it went through several printings in several languages. Another Flemish geographer, Gerardus Mercator, published his atlas of the world in 1587. These two atlases were among the first major map collections to be printed on the new printing presses of Europe. They were well received by the public, as well as geographers, because these atlases were works of art that also provided detailed information. They are still highly valued by present-day libraries.

During the late Middle Ages and into the Renaissance, interest in ocean travel increased. Maps of all kinds were developed in Europe as well as in other countries. A number of Europeans contributed to the science of cartography during this period. Among them were the following:

- In 1341 **Petrarch and Robert of Anjou** of Italy produced the first accurate land map of Italy.

- In the early 1500s **Peter Apianus** of France was the first to use mathematics and measurements to establish that the Earth is not a perfect sphere.

- **Thomas Hariot** and **Edward Wright** were colleagues of John Dee, the English creative mathematician who was interested in problems of mapmaking. They developed new mathematics that could be used to produce accurate loxodromic charts, that is, charts based on sailing on a rhumb line. This simply means that a ship's path maintains a fixed compass direction, which is shown on a map as a line crossing all meridians (longitudes) at the same angle. Hariot and Wright also demonstrated how to use these rhumb lines, as well as new charts that indicated the longitude and latitude lines to plot a true course. However, they kept secret the mathematics involved.

- In the later 1500s **Reinhardus Gensfelder** published the first land map of central Europe.

Over 200 maps have been collected from the period 400 to 1600. Many ancient maps, prepared by cartographers from many countries, are located in various museums around the world. Over the years many mapmakers contributed to the science of cartography that is based on projecting the three-dimensional surface of the Earth onto paper as a two-dimensional image of a particular geographic region or the entire world.

History of Map Projections

Since the time when many geographers, explorers, and astronomers realized that the surface of the Earth is curved, they pondered how to depict the features of the Earth's three-dimensional surface onto a flat two-dimensional surface. The earliest and easiest procedure used was to form a circle and then sketch the land and sea markings within the circle. Over the centuries this proved ineffective, and other techniques were explored, which led to the idea of projecting the parallels (latitude lines) and meridians (longitude lines) of a globe's surface (i.e., the Earth's surface) onto a plane surface. Depending on the shape of the plane surface, the projected image can assume many different shapes and characteristics. Many of the following different types of projections were originally proposed hundreds of year ago.

- **Ptolemy's** contributions to mapmaking have already been discussed. Although he made no reference to what is known as conic projections, he did introduce the concept of depicting parallel latitudes on a map as circular arcs while his meridians were more or less straight lines.

Ptolemy acknowledged Marinus of Tyre (ca. 70–130 C.E.), a geographer and mathematician, who established correct latitudes and longitudes for particular land areas represented by his maps.

- **Giovanni Matteo Contarini** (ca. 1445–ca. 1507), an Italian mapmaker, not only modified Ptolemy's world map but also improved it by increasing Ptolemy's meridians (lines of longitude) from 180° to 360°, which represents the current concept of meridians. He also extended Ptolemy's latitude markings to include the North Pole in a more accurate conic projection.

- **Johannes Ruysch** (ca. 1471–ca. 1533), a Dutch cartographer, was the first to use a projection that placed the North Pole at the center of a circular arc of latitudes along with parallels that were equally spaced from the pole. It depicted a view as if looking down on the North Pole. The map extended to about 38° south.

- **Gerardus Mercator** (Gerhard Kremer) (1512–1594) was born in Flanders (present-day Belgium), educated in the Netherlands, and died in Cleve (present-day Germany). He learned engraving from Gaspar à Myrica and mathematics from Gemma Frisius while in the Netherlands. The three men constructed a terrestrial globe in 1536 and also a globe of the heavens that illustrated the major stars. Mercator's first map projection was of the area of Palestine (present-day Israel) in 1537. In 1538 he used a type of cylindrical projection to construct a map of the world. Mercator is credited with inventing a crude cylindrical projection process for making maps, at least in Europe. Several hundred years earlier the Chinese used the concept of a celestial sphere to project a distorted map of the heavens onto a cylindrical surface. The pole was at the center, with successive positions of the moon representing its motion. The Mercator type of map projection is the most common type of map of the world found in homes, schools, and so on. However, it distorts the sizes of continents near the polar regions. On a Mercator map the meridians (lines of longitude) and parallels (lines of latitude) do form rectangles with right angles, but the areas of the rectangles become greater as one proceeds from the equator to the poles. (See Figure 2.3.)

Types of Map Projections

Envisioning and executing a map projection required a fair amount of imagination as well as knowledge of mathematics. For instance, for the spherical surface of the Earth (a globe) to be projected onto a flat surface, one has to imagine a point source of light at the center of the globe that casts a shadow of the land/ocean features onto the flat surface. In addition, the plane surface can itself assume different orientations to the globe's curved surface that, in turn, produces various types of projected images, including varying degrees of distortion. The plane

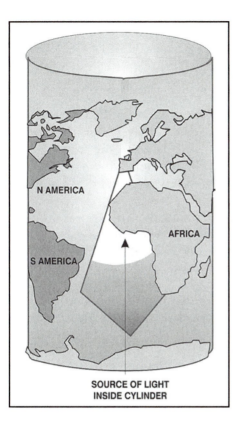

Figure 2.2 Cylindrical Projection
An example of how a cylindrical projection map is formed. The lighted globe inside the cylinder projects its surface image of the continents onto the inside surface of the cylinder. This type of projection exaggerates the sizes of those landmasses closer to the polar regions.

N AMERICA

AFRICA

S AMERICA

SOURCE OF LIGHT
INSIDE CYLINDER

surface can lie flat at a point (tangent) on the globe's surface, it can form a tube or cylinder with the globe inside the cylinder, or it can form a cone shape with the globe situated inside the cone.

Cylindrical Projection:

An example: Take a large flat plane, such as a large sheet of paper, and wrap it around a globe to form a cylinder with the globe centered inside the cylinder and the globe's equator touching the cylinder's inner surface. Then project the grid system of the globe's surface onto the inner surface of the cylinder. If the paper cylinder were unfolded and laid flat, a projection of the globe's surface would be transferred to the flat paper. In other words, a cylindrical projection can be envisioned by inserting a clear globe with meridian lines (longitudes) and parallel lines (latitudes), as well as the outlines of continents on the globe's surface, inside a cylinder. When a light bulb is turned on inside the globe, a projected (but somewhat distorted) image of the globe's surface features will be projected onto the inside surface of the cylinder. (See Figure 2.2.)

The meridian and parallel lines that intersect on a Mercator projection are at right angles to each other. This type of projection can be a true representation of the size of areas for only a few parallels on each side of the equator. However, at the projected higher latitudes north and south of the equator, the land and sea areas become increasingly distorted and appear larger than land areas closer to the equator. Nevertheless, the cylindrical type map is useful for navigation, particularly in the mid-latitudes because all lines are related to the azimuth (see Azimuthal Projection for description). In addition, when using a compass with a Mercator map the compass points are related to the map's rectangular orientation. (See Figure 2.3.)

A number of other cylindrical projection techniques are employed to reduce or eliminate distortions of areas near the polar regions. These are referred to as Cylindrical Equal-Area Projections. Other types of cylindrical projections are Oblique Mercator, Transverse Mercator, and the Universal Transverse Mercator. These variations are different orientations for Mercator maps.

Pseudo-cylindrical Projections:

The cylinders in pseudo-cylindrical projections are *not* regular rectangular plane surfaces wrapped to form a symmetrical cylinder. Rather they are cylinders in which the cylinder's ends are curved inward toward the poles. The result is a grid projection where the parallel lines are straight, while all meridian lines, except the central meridian, are curved inward toward the poles due to the inward curving of the ends of the cylinder. Several examples of these types of projections are the Cylindrical Equal Area Projections, and the Sinusoidal Projections.

Conic Projections:

When placing a similar lighted clear globe with parallels, meridians, and land markings inside a cone instead of a cylinder, a different type of image is formed. The amount of distortion varies with the orientation and placement of the globe inside the cone. Normally, the parallels are projected as concentric arcs of circles, while meridians are projected as straight lines radiating at intervals from the apex of the inside, but flattened, cone surface. Because of the excessive distortion, conic projections are limited to maps of mid-latitude regions such as the United States. There are several versions of the standard conic projections. They are the Albers Equal Area Conic Projection, the Peter's Conic Equal Area Projection, the Mollweide Equal Area Projection, the Lam-

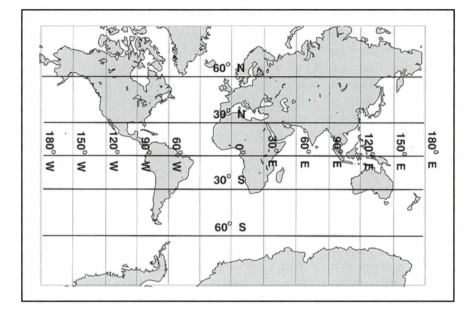

Figure 2.3 Mercator Projection
Gerardus Mercator used a cylindrical projection to create a map in which the meridians of longitude were parallel (instead of curving toward the poles). This made navigation easier since a compass bearing could be drawn in straight lines. Note: One advantage with this type of projection map is that the lines of longitude and latitude form 90°-angle rectangles.

bert Conformal Conic Projection, and the Equidistant and Poly-conic Conic Projections.

Azimuthal Projection:

An azimuthal projection of the Earth is designed so that straight lines from a given point to any other given point will be the shortest distance between the two points. The most common is the Azimuthal Equidistant Projection type of map, which indicates polar coordinates for the Great Circle Route. Airlines and ocean vessels use azimuthal maps to follow the shortest distance between two faraway points. Any point on the Earth's surface can be used as the central point to make an azimuthal map. The further the distance from the central point on the map, the greater the distortion. The most common type of azimuthal projection is a view looking straight down upon either the north or south polar regions.

Following are several other types of azimuthal projections where the imaginary point source of light is placed at different positions on the globe in relation to a plane surface. The Lambert Azimuthal Equal Area Projection is used for large ocean areas. This projection has a straight line for the central meridian, but all other meridians are curved. It is useful for following the Great Circle Route. Orthographic Projections exhibit distorted shapes but have true distances on parallel lines. The Sterographic Azimuthal Projection maps are oriented from above the view of the North or South Pole downward, somewhat similar to the standard azimuthal projection. Although the size of scale is distorted as one moves from the central pole, the directions are true which makes it ideal for polar navigation.

Other Types of Map Projections:

Other types of maps that are *not* constructed from a projection simply use longitude and latitude as a rectangular-grid coordinate system. City, state, and road maps are often non-projected, because they show only limited geographical areas. While non-projected maps depict countries around the world, large areas are increasingly distorted in size and shape toward the poles. Following are some examples of non-projected types of maps: (a) The Planar Surface Projection map, in which a flat two-dimensional plane is placed tangent to a particular spot on the surface of the clear globe. The surface image is projected by the globe's internal light onto a flat plane surface. The image becomes more distorted as the distance increases from the tangential point of contact between the plane surface and the globe; (b) Robinson Projection maps, a compromise between cylindrical and conic projections, have equally spaced and parallel lines, along with curved but equidistant meridian lines; (c) The Texas Statewide Projection is a cartographic standard developed to provide standards for techniques used for mapmaking; and (d) The Space Oblique Mercator is a projection map used for the Landsat 7 satellite images. The Landsat 7 is an Earth-circling U.S. satellite designed to acquire images of the Earth's land surface and coastal areas. It provides images of land cultivation, erosion, lake levels, and disaster areas. Landsat 7 is a project of NASA (National Aeronautics and Space Administration) maintained out of the Goddard Space Flight Center in Greenbelt, Maryland.

Another interesting concept that depicts the surface of a globe onto a flat paper surface is sometimes referred to as the "orange peel map." Using a marking pen on the surface of a grapefruit or orange, sketch some vertical and horizontal lines to represent meridians and parallels, plus adding some continents. Then slice through the skin on opposite sides from each pole almost to the equator. Leave about $^3/_4$ inch uncut at the center or equator of the skin. Try to make the vertical cuts through the oceans, not on the continents, and make the cut only as deep as the skin so that it can be peeled. As carefully as you can, without breaking the skin into separate pieces, separate the intact skin of the grapefruit or orange from the fruit part. Lay the skin on a flat surface with the curved surface of the peel raised up. Then gently push down on the raised peel until it flattens out. The result is similar to a type of map that is sometimes used to reduce distortion.

Conclusion

Some historians and anthropologists consider the crude maps that represent territories as old as the written symbols that represent words. As humans expanded their horizons, their ability to draw maps depicting their enlarged universe also increased. Ancient humans who understood that the Earth was a sphere used circles with known land areas placed near the center of the circle, while placing oceans around the edges of the circles. As travelers, traders, and even terrorists roamed from one country to another, and sailors ventured along continental seashores, the need for more accurate maps soon became imperative. Maps were drawn showing the Silk Road and spice trade routes that indicated the land and coastal passages, as well as the dangers, for caravans and ships traveling between the East and West. Regardless of the improvements in the quality and accuracy of navigational maps that were drawn during the Middle Ages and Renaissance, the exploration of new sea routes, lands, and the transportation of bounty over the Earth's great oceans was a dangerous enterprise for these explorers. Many died at sea, either from scurvy or other diseases, or on land at the mercy of hostile natives and cannibals. In addition to storms, explorers faced other major hazards during their voyages. The determination of longitude was a serious navigational problem because it was difficult to ascertain how far a ship sailed each day or determine the ship's total distance from the homeport. An accurate clock that could be used aboard ships was not invented until the mid-18th century. (Due to the natural motion of all ships, existing clocks could not keep accurate time.) John Harrison invented the first clocks, called chronometers, which were able to keep accurate time aboard ships. Harrison invented several versions of his chronometers in England between 1737 and 1770. They soon became essential for navigators to track distances over time and, thus, longitudes.

Hundreds of maps produced during the Middle Ages and Renaissance were used to search for new sea routes to the Orient and other unexplored lands. Today, cartography has progressed to an exact science. A Global Positioning System (GPS) operating from a system of Earth-orbiting satellites can accurately determine Earth's geographic positions and topological features. In addition to their use for map-making, handheld GPS systems can determine the position of the user within a few feet anywhere on the Earth's surface. It is obvious that humans still possess the urge to roam, explore, and map the solar system and universe.

CHAPTER 3

THE BIOLOGICAL
SCIENCES: BOTANY
AND ZOOLOGY

Background and History

Early prehistoric humans depended on plants and animals to supply their food, shelter, clothing, and medicine. Through experience, observation, and trial and error, they learned which plants to select for specific uses. In a sense, making these early distinctions between useful vs. useless, harmless vs. harmful, and edible vs. inedible plants was a form of early plant classification (taxonomy). Likewise, early humans soon learned the habits of animals, the dangers associated with hunting certain kinds of beasts, and which animals' meat was edible. Knowledge of this primitive taxonomy and stereotyping from generalizations of types (species) of plants and animals often made the difference between life and death. Early people considered themselves as part of nature. However, at the same time they were cognizant that they had a sense of self and could exert some control over their environment. It is generally accepted that around 9000 B.C.E. primitive hunter-gatherer tribes began to cultivate the most desirable plants and domesticate some animals for a more dependable food supply. Evidence has been found that by 8500 B.C.E. Mesopotamians living in the fertile valley between the Tigris and Euphrates rivers in present-day Iraq raised sheep and goats and cultivated several different types of grain crops.

By the Greek Classical Period (fl. 7th to 1st centuries B.C.E.) a number of philosophers and proto-biologists raised questions about the universe and concluded that the natural world, including the creation of plants and animals, was not the product of supernatural forces. By the 1st century of the Christian Era the Romans were aware of at least 1,300 to 1,400 different types of plants. They could distinguish between species, but their system of naming them was arbitrary and not related to modern taxonomy. There was no real distinction between medicine and biology during the classical Greek/Roman period.

The term *biology* was first introduced in 1802 by Jean Baptiste Antoine de Monet, who called himself Chevalier de Lamarck (1744–1829). Biology did not become a scientific body of knowledge until the late 19th century. Biology is the branch of the natural sciences that studies living organisms and systems. It encompasses several specialized areas, such as cytology (the study of cells), zoology (the study of animals), botany (the study of plants), as wells as ecology, genetics, biochemistry, as well as other subspecialties.

The early physicians and philosophers who were interested in the natural world made the most advances in biology and, in particular, zoology. The early Greeks accepted the concept of **spontaneous generation** and the differences between types or classes of living things, but they did not understand either **micro-** or **macroevolution.** Although the Greeks were not scientists as we think of the term, they were observant and categorized (hierarchical listings) by characteristics of both plants and animals. Following are some of the proto-biologists of the Classical Period:

- **Anaximander** (ca. 610–546 B.C.E.), the father of cosmology, developed a somewhat mystical philosophy of the world where change between substances is possible, that is, animals could be changed from one type into another. His concept that life evolved from the sea and that land animals are descendants of sea animals predated Darwin's evidence for organic evolution by more than 2,000 years.

- **Xenophanes** (ca. 560 B.C.E.) was the first to recognize that fossil-type shells were at one time animals that were buried in mud by floods.

- **Empedocles** (ca. 492–ca. 430 B.C.E.), a Pythagorean physician and philosopher, believed in the four main eternal and immortal elements that make up the universe: water, fire, earth, and air. This theory of matter was accepted by Aristotle and others and continued to persist until the age of modern chemistry in the 18th century. Empedocles' belief that plants, like animals, had a soul, emotions, and common sense, and could reason was based partially on the fact that plants will point their drooping leaves toward the sun (heliotropism). This seemed to him a rational, common-sense reaction of plants.

- **Aristotle** (384–322 B.C.E.) expanded Empedocles' theory of "four elements" by compartmentalizing the universe into four main categories: inanimate, sea and air, plants, and, at the top, animals. He also theo-

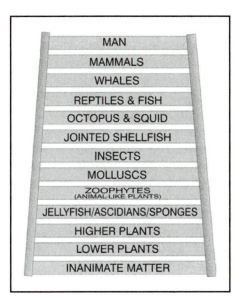

Figure 3.1 Aristotle's Ladder of Life
Aristotle's "Ladder of Life" is based on his concept of the lowest species of life progressing in complexity in steps to man, the highest.

rized that animals had mobile and sensitive souls, plants had vegetative and nutritive souls, and humans had obtained rational and intellectual souls. Aristotle was a careful observer of nature who prepared a list over 500 animals based on his observations of their appearances and habits, which he organized into a reasonable classification system of living things called the *Scala naturae,* or "Ladder of Nature" or "Ladder of Life." He placed the simplest organisms on the bottom rung of his ladder, with humans occupying the top rung. (See Figure 3.1.)

Aristotle recognized that animals are different, sometimes in subtle ways that might escape the observer. One example: He did not categorize all animals in the sea as "fish" because he recognized that animals like dolphins and whales, even though they live in the sea, are warm-blooded (unlike cold-blooded fish), breathe air, and give birth to live young that are nourished by milk from mammary glands. Therefore, he classed them with land beasts rather than fish. Aristotle also classified plants as well as animals but with less detailed distinctions between species. Thus, Aristotle is often referred to as the "founder of zoology" although he gave no hint that he believed one type of animal on a lower rung of his ladder could evolve and occupy a higher rung. He is often credited with introducing the science of taxonomy, a form of classification. It is the science of identifying, naming, and grouping living organisms into a system based on both external and internal anatomical structures and physiological functions, as well as the organism's genetic code. Aristotle, like most Greek philosophers, believed in vitalism, or the idea that life is based on a vital force (a spirit force different from physical and chemical forces) and is not materialistic. Accepting the concept of vitalism means one cannot accept the mechanistic aspects of nature.

- **Theophrastus** (ca. 380–287 B.C.E.), a Greek philosopher and disciple of Plato and Aristotle, wrote more than 200 papers describing the characteristics of over 500 plants. He also developed a classification system for plants based on their morphology (study of the form, structure, and parts of organisms), types of reproduction, and general natural history. He described in detail pepper, cinnamon, bananas, asparagus, and cotton. Two of his best-known works, *Enquiry into Plants* and *The Causes of Plants,* have survived for many centuries and were translated into Latin. He is sometimes considered the "grandfather of botany."

- **Crateuas,** a 1st-century B.C.E. Greek artist and physician, produced the first illustrated herbal pharmacological book for medicinal plants. His book influenced medicine for many centuries.

- **Pedanius Dioscorides,** another 1st-century B.C.E. Greek physician, described over 600 different kinds of plants, including their growth patterns and useful qualities for herbal medicine. His illustrated publication was used for pharmacology and medicine as late as the Renaissance years.

- **Titus Lucretius Carus** (ca. 99–ca. 55 B.C.E.) wrote a book titled *On the Nature of Things* that was presented in typical verse style. He accepted the atomic theories of the physical world of Democritus and Epicurus and the classification of living things by Aristotle. He disagreed with most ancient philosophers about the spirit nature of the soul. Lucretius believed the soul was not immortal, but rather that it was material and disappeared at death. He also professed that the idea of an immortal soul, along with the belief in gods, is the cause of all human misery. His primary contribution to biology was his explanation of the different kinds of sensations experienced in life, including the detailed sensations of sexual life. Due to his hostility to the church, he was ignored during the Renaissance, but his writings strongly influenced Giordano Bruno, who himself wrote many freethinking scientific verses, which resulted in his being burned at the stake in 1600.

- **Pliny the Elder** (23–79 C.E.) was more a compiler of past Greek works than an original writer. He based his 37-volume encyclopedia *Historia Naturalis,* 16 volumes of which were devoted to plants, on the works of Theophrastus and Dioscorides. Volumes VII to XI were dedicated to zoology, while volumes XII through XIX dealt with botany. All of Pliny's other volumes were haphazard translations of the works of other Greek writers on physics, geography of the ancient word, medicine and drugs, and minerals and geology. During the Middle Ages, Pliny's *Natural His-*

tory encyclopedia was accepted as the final word on all the plants that existed in the world. Pliny's work was the main source of knowledge of natural history for over 1,500 years. It was not until the 15th century during the Renaissance that Otto Brunfels challenged Pliny's statements and misconceptions.

- **Claudius Galen** (ca. 130–200 C.E.), while practicing medicine in Greece, described human anatomy based on surgery that he performed on wounded gladiators, as well as his dissections of pigs, dogs, and sheep. His dissections of monkeys provided the basis for his assumption that human anatomy was similar to animal anatomy. Although many of his conclusions related to zoology in general and human anatomy in particular were misconceptions, they persisted for over 1,000 years.

Botany

From the period of the Middle Ages until the Renaissance there was not much in the way of the development of original science in Europe. Some scholarship, mostly in monasteries, consisted of copying the ancient works of Greeks and Romans and later translating Greek science from Arabic into Latin. During the Dark and Middle Ages of Europe science was dormant for about 1,000 years, and botany was still a descriptive discipline, not much advanced from Classical Greek/Roman botany. Botanists of the Greek and Roman eras, the Dark Ages, the Late Middle Ages, and into the Renaissance were known as *herbalists* because they mostly collected, grew, dried, stored, and sketched plants. Some became expert in identifying and describing plants according to their morphology and habitats, as well as their usefulness. After the development of the printing press, the most famous herbalists published their findings. Some of their books included beautiful drawings and paintings of their plants, many of which are highly valued today. Following the fall of Rome (ca. 476 C.E.) science experienced a revival in the Far East and in the Arab world, while the Roman Christian Church suppressed European science until the Renaissance. Neither the physical nor biological sciences of this medieval period advanced to any great extent. Botany seemed to be a favorite pastime for the few Western intellectuals of that time, but it did not become a true science until many centuries later. As the Muslim world expanded both east and west during the Early Middle Ages, so did their versions of science, which were based on Arabic translations of famous Greek writers.

A short summary of the careers of a few early botanists of the Middle Ages follows:

- **Avicenna** (Ibn Sina) (980–1037), a Persian philosopher/scientist, interpreted much of Aristotle's work, in particular in the field of drugs and medicine. He is best known for two comprehensive textbooks on all then known medicine, drugs, and diseases: *Canon* (Al Qanun) and the *Book of Healing*. Both were used over the next 500 years in both the East and West. His account of the use of herbs for curing diseases was translated from Greek to Arabic and later into Latin in western Europe. Hence, his books were instrumental in reintroducing Aristotle into Europe during the Late Middle Ages and the Renaissance.

- **Hildegard von Bingen** (1099–1179) was a German abbess (the mother superior of a convent) who collected and assigned Germanic names to over 300 different plants. She was the first person to use original observations to describe in Latin the plants of Europe that were included in her collection.

- **Averroës** (Ibn-Rushd) (1126–1198), a Muslim physician and philosopher born in Spain, is best known for his interpretations of Aristotle's writings. He advanced Aristotle's "Ladder of Nature" (see Figure 3.1) one step further toward the modern concept of evolution, as well as expounding on his own theory that the potential for all living things exists at early, undeveloped stages, which predates the concept of inheritance of characteristics (i.e., genetics). Thus, the mature plant exists in the seed, just as the adult animal exists in the sperm or embryo, similar to the ancient Greek Anaximander's philosophy. Averroës' concept that the seed and embryo are potential plants and animals was based on Aristotle's idea that a block of granite is a potentiality of the statue that is carved from it. (All a sculptor needs to do is remove the granite not needed to reveal the statue.) Averroës was known as the "Aristotle of the Middle Ages."

- **Albertus Magnus** (Saint Albert the Great) (1200–1280) was a German scholar as well as a clergyman who traveled extensively over the European continent. His writings were the first to describe the plants of Europe in detail, and many later botanists copied his style of writing. He wrote many books on numerous subjects, ranging from theology to physics to biology. His book on botany, *De vegetabilibus,* consisted of seven volumes. Although he included many of Aristotle's ideas, including spontaneous generation, he also reported his own observations and theories. For example, in his study of the anatomy of leaves he described in detail the venation (patterns of veins) of plants, as well as their sexual anatomy and reproductive physiology. He also was the first to describe the difference between **monocots** and **dicots** (see Figure 3.2) and compared vascular- with nonvascular-type plants. One of his

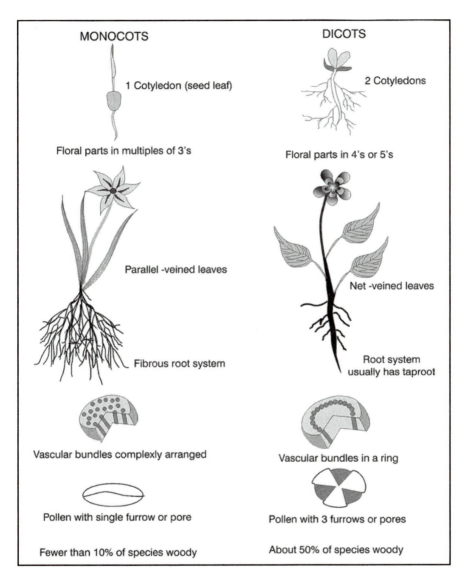

Figure 3.2 Monocots vs. Dicots

Monocot plants have seeds that sprout with only one cotyledon (seed coat) and are mostly related to grasses, bamboo, and such, while dicot plants have two seed coats and form various types of plants, mostly woody shrubs and trees.

students was Thomas Aquinas, who tried to relate learning within Christian doctrine.

During the Renaissance the study of botany began to resemble a discipline based on critical observations. Rather than having physicians and philosophers act as herbalists, academic professors actually studied plants in greater depth. Herbalists of the Renaissance also recognized a more practical form of botany, that is, exploring the use of plants for both medicinal purposes and agricultural uses. Several early European universities established studies in botany, not only as a science, but also as a utility for medicine. During the middle of the 16th century universities in northern Italy developed the first botanical gardens. Herbalists of this period were still not scientists in the conventional sense. While they were critical observers and documenters, they often were more concerned with the philosophical rather than the physical and objective nature of plants. They were collectors who raised local and foreign plants in gardens, often sketching and painting the various species—especially flowers—and describing the plants' structure and physiology in detail. After the printing press was perfected, their works were widely distributed, many of which still exist in modern libraries.

A short resume of several well-known herbalists of the Renaissance follows:

- **Otto Brunfels** (1488–1534) was a Roman Catholic monk who later became a reformed Protestant Lutheran, a teacher, and a physician. He was influenced by the writings of the ancient botanists Dioscorides and Pliny the Elder. In 1530 Brunfels published in Latin *Herbarum vivae eicones,* a collection of over 800 species of native plants that was well illustrated with woodcut prints. It is said that this book inspired Carolus Linnaeus to bestow the title of "father of botany" on Brunfels. Actually, Brunfels did not divert greatly from older descriptions and listings of plants. Rather, he assigned different names in different languages, which made his book much more popular in many countries, and thus better known. His main interest was the medicinal qualities of plants.

- **Luca Ghini** (1490–1556) established the famous garden of Pisa that included not only local flora but also foreign plants. He is the first botanist to press and dry plants in order to preserve them in herbariums for further research and study.

- **Hieronymus Bock** (1498–1554), also a German teacher and physician, traveled extensively to collect plants that he then cultivated in his own gardens. Bock included over 240 species of plants in his herbarium, describing their structure and the development of their vegetation, as well as locations at which they were found throughout Europe.

- **Leonhard Fuchs** (1501–1566) was a German teacher of medicine whose goal was to replace the Arab language and Islamic rule of medicine and pharmacology in Europe with Latin, which was the language of science in Europe during the Renaissance. His main work, *De historia stirpium commentarii*, published in 1542, did not describe plants, their growth habits, or physical characteristics, but rather just listed the species in alphabetical order. While he used the terms *form, habitat, season for collecting*, and *temperament* to describe his plants, he excluded types of flowers in his descriptions. The South American plant fuchsia is named after him.

- **A. Cesalpino** (Caesalpinus) (1519–1603) was an Italian professor at Pisa who ignored other scholarly works in botany and arrived at philosophical and theoretical conclusions based on his own empirical observations of plants. Although he described flowering plants and their fruits in detail, he concluded that plants had only one, not two souls, which was located where the stem meets the soil and becomes a root. His system of classifying plants was one of the first to emphasize the forms of plants as the basis of distinguishing types (species). Later botanists, including Linnaeus, claimed that Cesalpino was the first to develop a definitive classification for plants.

- **John Gerard** (1545–1612) was a well-known English herbalist who established a famous garden at his home near London. His book *A General History of Plants*, published in 1597, describes over 1,800 known plants. He used Latin botanical names to identify his plants, their blooming habits, practical uses, as well as some plant folklore.

- **Caspar Bauhin** (1550–1624) was known for his system of classifying plants by their natural and common similarities and differences in their actual forms, rather than some of the older artificial classifications. It is speculated that he may have perceived the distinction between genus and species, but it is doubtful that he fully comprehended this future concept.

- **Prospero Alpini** (1553–1617), an Italian botanist and physician, served as a physician in Cairo, Egypt. While in Egypt, he observed and documented the existence of male and female sex organs in date palms. Although Theophrastus had recorded the existence of the two sexes of palm trees about 1,000 years earlier, this was the first recognition during the Renaissance that plants were not asexual. Alpini also was the first to describe the coffee plant in detail.

- **Jan Baptista van Helmont** (1579–1644) was one of the first botanists to conduct a well-planned experiment with plants. His hypothesis was that plants grow, increase in weight and size, but do not reduce the weight of the soil in which they grow. He first noted the weight of the soil in which his test plants were to grow. He also measured the weight of the water provided for the plants, and after a period of five years, he measured the weight of the plants and the original soil in which the plants grew. He found that the mature plants weighed 165 pounds but the soil lost only about $\frac{1}{4}$ pound in weight. Unfortunately, his conclusion that the water was responsible for the plants' weight increase was incorrect. This was years before the 1961 Nobel Prize winner in chemistry, Melvin Calvin, traced the path of carbon dioxide through the physical reaction of photosynthesis, which only occurs in the presence of chloroplasts in living green plants exposed to sunlight. Van Helmont also conducted another experiment that involved the burning of wood. Sixty-two pounds of wood ended up as one pound of ash. He attributed the lost weight to water vapor and four gases he named (1) *carbonum*, (2 & 3) both called *sylvester*, and (4) *pingue*. He assumed all four gases were dispersed since he could not collect and weigh them. Although van Helmont was the first to use the term *gas*, which means "chaos" in Greek, it was some years before his four gases were correctly identified and given their current names, as carbon dioxide, carbon monoxide, nitrous oxide, and methane.

There were several other lesser-known Renaissance scholars who produced works that mainly listed and/or sketched plants they collected. Botany as a true science had to wait several centuries for the period of enlightenment.

Zoology

Just as with plants, animals have been observed and studied as far back in history as when humans were hunters and gatherers. Humans are notorious compulsive classifiers. Today, we classify everything in minute detail, for example, political systems, cultures, physical characteristics, languages, income levels, races, clothing styles, and music. In this sense, prehistoric humans were no different than their modern heirs. Today, we justify the classification of everyday things as a way of handling our complex society. The science of naming and classifying plants and animals is called *taxonomy*. Regardless of the names used by ancient humans for various animals, by observation and experience they knew which ones to fear and which ones were available for food.

Aristotle is credited with developing one of the first formal systems of plant and animal classification. Known as *Scala naturae*, or the "Ladder of Life" or "Ladder of Nature" (see Figure 3.1), it was organized from the simplest to most complex life forms. This model was used for centuries. Pliny the Elder wrote a series of books called *Natural History* that recorded all the known scientific discoveries up to that time in history. Volumes 7 to 11 were a summary of what was known about zoology and, as with the volumes related to other sciences, his zoological text was accepted as an authoritative source into the Middle Ages. Muslims translated these and other works from Greek into Arabic. Christian Europe considered the relatively newer Arab Islamic invasion of northern Africa and southern Europe a serious threat to established Christianity. Instead, in a few centuries, the invasion proved of some benefit as the Muslim import of Greek and Eastern science, once retranslated into Latin, led to the Western Renaissance. As the Muslims added their science to this classical Greek heritage, it was spread by Muslim conquests of other countries both to the east and west during the 8th and 9th centuries. Much of the early knowledge about biology and other sciences during the European Middle Ages and Renaissance was actually a reawakening of Greek science retranslated from Arabic into Latin.

Also, as with botany, a major emphasis for the study of zoology in the Middle Ages (and later) was related to philosophy and medicine—anatomy in particular. Zoology in the Middle Ages was a combination of superstition, myth, folklore, and misconception consisting of simple descriptions of known and exotic animals and their seemingly odd behaviors. Many of the ancient Greek myths were passed on for centuries, including some of Aristotle's misconceptions and, later, Galen's erroneous writings on anatomy. During medieval times in Europe the general population accepted folktales of **bestiaries** and fictional creatures. To their credit, zoologists and physicians of the Renaissance were more interested in the morphological (external forms and parts of organisms), embryological (development of organisms from fertilization of egg to birth), and anatomical (internal forms and organs of organisms) differences between animals and humans.

It might be noted that during most of the Middle Ages biology was, in essence, a Muslim science. The Arabs were not great innovators nor did they advance many new biological concepts, but they did discover and preserve a great deal of ancient Greek astronomy, mathematics, and medicine by translating these works into Arabic and in some cases adding to them. They also spend copious amounts of time studying the translations. In both the Muslim and European worlds zoology did not progress much beyond the collections of animals (both native and exotic) in zoos by a few wealthy people. For instance, the wealthy Islamic Abbasid prince created the famous zoological gardens in Samarra (present-day Iraq), famous for its many exotic animals. Descriptive zoology based on observations was not an area of study until the Renaissance. Those who had an interest in zoology (and biology or science in general) during the Middle Ages were usually connected with the Roman Church or the newly formed church-dominated universities. Almost all scholars of this period interpreted or reinterpreted Aristotle and other Greek writers. One of the few extensive zoological writings of the Middle Ages is the *Physiologus,* which is a collection of stories by a number of authors. It relates and interweaves the animal world into religious sermons. Following are a few scholars from the Middle Ages and Renaissance who demonstrated some interest in zoological studies:

- **Al-Asmai** (740–828 c.e.) made a major contribution to biology, and to zoology in particular, with a book on animal husbandry that described the breeding of camels and horses.

- **Al-Jahiz** (776–868) wrote a famous Arab encyclopedic book titled *Kitab al-Hayawan* (The Book of Animals), consisting of seven volumes that covered many fields of science. A keen observer of nature, he studied the organization of ant colonies, the communications and diet of animals, and the relationship between people and animals. However, he also accepted the Greek concept that life emerged from water and mud (spontaneous generation).

- **Albertus Magnus** (ca. 1200–1280) was a Dominican monk whose main contributions were in chemistry. As a zoologist, he was enthralled by the rediscovery of Aristotle's ideas, including his misconceptions. One of Magnus's pupils was Thomas Aquinas.

- **Frederick II of Hohenstaufen** (1194–1250) was an emperor of the Middle Ages who enjoyed hunting. He described the migratory habits of birds and dissected them to study their anatomy. After his death in the 13th century, the church destroyed much of the medical and zoological information he had compiled.

- **Cantimpratensis** (dates unknown) was a Dominican friar who wrote *De naturis rerum,* a collection of the many theories of nature presented by Greeks of the classical period, including Aristotle's. His writings included moral lessons along with the biological content. Another brother, Vincentius Bellovacensis, also relying on ancient texts, wrote a similar volume, titled *Speculum naturae* (Nature's Mirror).

By the end of the Middle Ages biological scholarship abandoned the literary method of interpreting historical writings of others and merely classifying nature. Replacing this nonscientific convention was a system wherein personal observations and deduction formed the basis for conclusions. Roger Bacon was a key person in this break from medieval practices to the use of the modern scientific methods during the Renaissance. Following are short descriptions of the most important zoologists of the Renaissance:

- **Roger Bacon** (1220–1292), an Englishman and Franciscan friar, is considered the scholar who, along with several others, led the transition of biology from a regurgitation of ancient writings during the Middle Ages to the critical observations of the Renaissance. His liberal views and insistence that science must be based on personal experience gained from direct observation of nature distinguishes him from earlier sages. This break with the traditional, approved manner in which medieval scholars studied and compared biological texts was more than the Roman Church would tolerate. As a result, Bacon spent many years in prison as punishment for his views.

- **Conrad Gesner** (1516–1565) is considered the greatest naturalist of the Renaissance even though he was somewhat gullible for accepting without question the conclusions of Greek naturalists. In the late 1500s he wrote the classic *Historia animalium* (Natural History of Animals) in which he included the works of classical scholars as well as studies of his contemporaries. His history of animals consisted of four large volumes containing over 3,500 pages. In addition to organizing it according to Aristotle's principles, he added a great deal of personal information, particularly when he disagreed with heretofore accepted experts. His inclusion of woodcuts made the book more explicit and popular than other more simple and basic zoological texts. Gesner's natural history book became the standard natural history authority for over 200 years. Near the end of his life (he died of the plague in 1565) he published the first book on the subject of fossils, titled *De omni rerum fossilium genere* (A Book on Fossil Objects, Stones, and Gems). This was significant, since the ancient Greeks considered fossils to be prehistoric creatures, while Aristotle believed they were freaks of nature that were formed by mud. Many early Europeans presumed that fossils had been placed on Earth by God to fool men, while others thought fossils were the results of God's early experiments with life. Fossils were not explored or explained in detail until the new science of paleontology was developed in the 19th century.

- **Ulisse Aldrovandi** (1522–1605) first studied to be a lawyer, later changing to philosophy and medicine. He taught pharmacology for many years and grew medicinal plants in his garden. This rankled the local apothecaries, who considered growing and collecting medicinal plants to be their personal prerogative. This controversy was finally resolved through papal intervention. The four volumes of Aldrovandi's *Natural History* dealt primarily with birds and other groups of animals, but also included plants and stones. With the technology improvements in printing presses and artistic reproductions that had been made since the publication of Gesner's volumes, Aldrovandi's *Natural History* was well received when it was published. His work on the anatomy of birds continued to be the best science available on that subject for many years.

- **Guillaume Rondelet** (1507–1556) is best known for his book *De piscebus marines,* which describes and illustrates marine (aquatic) animals. He made no distinction between marine mammals (seals and whales), crustaceans (mollusks), fish, and other marine invertebrates, such as worms. He dissected whales and many types of fish and compared their organs. Rondelet had difficulty in comparing the anatomy of invertebrates with vertebrates, mainly because he had no concept of species and their distinguishing anatomical differences.

- **Pierre Belon** (1517–1564) traveled extensively throughout the European continent and the Middle East collecting specimens and recording his observations. He published two books, *L'Histoire naturelle des estranges poissons marins* (Natural History of Strange Fish) and *La Nature et diversites des poissons* (The Nature and Differences of Fish). His concept of fish was broader than that of Rondelet. Belon included other animals, such as beavers, otters, and hippopotamuses (since they lived at least part of their lives in water), along with whales, crustaceans, and fish in general. His reasons had little to do with the science of comparative anatomy. He justified this odd classification because the Catholic Church said it was permissible to eat the flesh of these animals during fasts since they were "fish," and thus not "meat." His classifications of animals did not always make sense since he included some desert lizards as fish. Belon did distinguish between those with bony skeletons (vertebrates) and those without skeletons (invertebrates). He also classified oviparous animals (development of the embryo within the mother's body as an egg, e.g., birds, insects, fish, and snakes) and viviparous animals (developing of the embryo within the mother's body where it receives nourishment through a placenta, e.g., mammals, including humans). This distinction is still valid. A respected zoologist in his prime, muggers murdered him while he traveled along a highway.

Comparative anatomy was basically descriptive until the late Renaissance, when it was practiced as a science, primarily by physicians and artists, not zoologists. Examples of men involved in the renaissance of anatomy as related to medicine will be addressed in Chapter 4, Medicine, Disease, and Health.

Concepts of Evolution

Although humans have always recognized change, the concept of change has not always been understood when applied to living organisms. It was obvious that after birth, both plants and animals grew larger over a period of time, aged, and later died. These facts were unquestioned and accepted as part of life. But how did life in general originate? And why are there so many different types of plants and animals? These are questions that scientists and philosophers still ponder, although there are many more realistic answers available at present. Evolution is the keystone of the arch of life sciences, and it is also one of the most radical ideas proposed by humans. Organic evolution is terrifying and unattractive, as well as symmetrical and satisfying, as a way to view and explain nature. The concept of organic evolution challenges the history of religion as well as some philosophical beliefs. The well-established and scientifically accepted concept of organic evolution based on change and adaptation continues to be challenged in today's enlightened world.

The ancient Greeks had many gods, one or more for almost all occasions of unexplained natural phenomena. Even so, they were more or less practical people who knew their gods were not responsible for all things that occurred in their world. They assumed "irrational necessity," randomness (chance), and even chaos were responsible for many naturally occurring events that did not require any help from a deity. They accepted that the universe was *not* created by a supernatural force, also believing that the world did not change much over time. The ancient Greek philosophers, Thales of Miletus (ca. 650–580) and Anaximander, also of Miletus (611–547 B.C.E.), among others, explained that living beings evolved from a primordial form of procreation, most likely from a mud-covered Earth.

Anaximander's theory stated that plants were created first, then animals, followed by humans. Humans evolved from fish that lived in water, having altered their fishlike nature and moved to dry land to become early human beings. Some historians credit Anaximander with predating Darwin's theory of organic evolution. Anaximander and his followers also theorized that the present universe evolved from a former primordial universe that will, in time, transform into a new universe in a continually evolving system. (This is a very up-to-date cosmological concept.) This idea led to the concept of a mechanistic or clockwork universe where every event has a natural cause. His theory was similar to Sir Isaac Newton's deterministic mechanical laws of motion. Anaximander's theory of cause and effect is not based on supernatural powers and may have led to later rationales for evolution. Diogenes of Apollonia, a follower of Anaximander, believed that life originated out of the earth (soil, not water) with the assistance of solar energy, which is a reasonable concept for the origin of life. Plato taught that every living being had its own essence that was unique to each organism and to know its essence was to know that organism. This concept did not allow for change or evolution and was taught for many generations. One of Aristotle's main contributions was his thought system related to life. His classification (organization) of life from the simple to the complex was related to his hierarchical concept of nature. Aristotle's theories of evolution and his methods of analysis influenced the development of science through the Middle Ages and into the Renaissance, resulting in the conceptualization of organic evolution by Wallace and Darwin in the 19th century. Aristotle was also interested in the processes of reproduction and how seeds and embryos developed. Some of his predecessors believe that the sperm (seed) was in essence a small invisible person or animal that grew larger once it joined with a female egg, and that the embryo contained all of the adult body parts (known as preformationism). Aristotle's studies of fertile chicken eggs at different stages of embryonic development led him to reject these theories and realize that the embryo only develops when an egg is fertilized by a sperm, and that growth and development occur from the time of fertilization to birth, growth, and death.

It was not until the late Renaissance that comparative embryology became a rudimentary science as related to evolution. Girolamo Fabrizio (Fabricius ab Aquapendente) (1537–1619), a surgeon who was a student of Gabriele Fallopio, studied the developmental stages of eggs of many vertebrates, including mammals, reptiles, birds, and sharks. He not only illustrated the stages of embryonic growth and development but also the placenta and embryonic tissues. He also advanced the field of comparative anatomy by studying and illustrating the eyes, ears, throats, and vascular systems of many different animals.

There was not much intellectual discussion involving evolution during the Middle Ages and early Renaissance. One exception, however, was the debate involving the age of the Earth. Church scholars used what was referred to as the "begat" system, which traces backwards all the generations of kings recorded in the Judeo-Christian Bible. Using this system, estimates to determine the age of the Earth vary from 6,000 years ago by King Saul; 3,184 years ago by Constantine; 7,500 years ago by St. Augustine; 5,993 years ago by Johannes Kepler; and 6,000 years ago by Isaac Newton. In the 16th century James Ussher proposed a more exact date for creation as nightfall on October 23rd, 4004 B.C.E. And a few years later John Lightfoot proposed an even more accurate time for the creation of the Earth as exactly 9:00 A.M. on October 25th, 4004 B.C.E. Since the Renaissance, numerous estimates for the age of the Earth based on religious writings vary from 6,000 to 10,000 years ago. More modern estimates of approximately 3.5 to 4.5 billion years ago are founded on the analysis of radioactive elements with long half-lifes and trace fossils. Geologists use several methods to date rocks and arrive at an accurate estimate of the Earth's age based on an absolute or chrono-metric time scale. Radiometric dating makes use of the natural decay of radioactive elements and their isotopes. The constant half-life for the decay of fissionable elements, when used for dating techniques, is also referred to as an atomic clock. Uranium was one of the first radioactive element identified. Uranium's isotope U-238 has a half-life of 4.5 billion years, which means that half of its atoms will be transmutated into lead every 4.5 billion years; thus, that half is now lead instead of U-238 and will no longer be radioactive, while the other half of U-238 will slowly continue its radioactive decay. This provides a means for comparing the age of rocks containing a proportion of U-238 with the proportion of its fission products (e.g., lead), thus determining when the rocks were formed. Other radioactive elements have different half-lifes, for exam-ple, Samarium-147's half-life is 106 billion years, Rubidium-87's is 48.8 billion, and Thorium-232's is 14 billion years. Carbon-14 has a half-life of only 5,730 years, but since carbon is found in all living things, past and present, it can be used to measure the age of any reasonably old organic matter.

Vitalism and spontaneous generation were two antievolution theories held by many from the Middle Ages into the Renaissance and beyond. Belief in these and other nonscientific concepts delayed the acceptance of Darwin's theory of organic evolution based on natural selection.

Vitalism is a metaphysical concept that has existed for millennia, and it continues to be espoused even today. It is a conviction that life is more complex than deterministic biology, chemistry, and physics. Rather, life is driven by spiritual or vital forces that control all living things. Animals and humans have their own vital forces that determine their birth, growth, development, and death. Vitalism has never been isolated, observed, or experimentally identified; therefore, there is no scientific basis for this concept. Spontaneous generation, or spontaneous cre-ation of life, proposes the theory that life can form from nonliving mat-ter. During the Medieval Period many believed that flies and worms were spontaneously generated from rotting fruit, meat, and manure, and that mice and rats just appeared from piles of dirty rags. This seemed logical since this is where these vermin and their offspring were found. The concept of spontaneous generation was disproved in the 17th century after Francesco Redi (1626–1697) conducted an experiment in which he placed rotten meat in both open jars and covered jars. No maggots or flies appeared in the covered jars. On the other hand, there are modern biologists who theorize that organic molecules could have formed from inorganic matter through a process of **autopoiesis.** Some biologists consider Darwinism too mechanistic and have proposed autopoiesis as an alternative. It is based on a self-organizing and self-maintaining physical/chemical system of life. The system relies on metabolism for its energy to maintain the physiological processes of life. Autopoiesis does not negate organic evolution; it just adds physiology (life's processes) to Darwin's evolution of the anatomic (structural) aspects of life.

In a sense, modern creationism theory is a holdover from the religious theologies of the Middle Ages and Renaissance period. To update the theory, proponents have recast creationism as intelligent design, which implies a more scientific orientation. Intelligent design is basically the same as creationism, but uses some scientific/technical terms to describe its belief—something like medieval scholasticism. These religion-based theories attempt to counter modern science, in general, and biological evolution, in particular, by identifying perceived weaknesses in the concept of organic evolution, such as the gaps in fossil evidence. Although not all the evidence is in, nor accepted by all biologists, the current understanding of modern evolution *is* the foundation of modern biology.

CHAPTER 4

MEDICINE, DISEASE, AND HEALTH

Background and History

Diseases are as old as humankind. They respect no culture—past or present. Some historians claim that many known diseases actually pre-date humans and have existed from the time that that the earliest life forms appeared on Earth—for example, bacteria. Also, modern research has indicated that diseases have not modified greatly over the past several million years. Bacteria fossils that are over 500 million years old and shells that are 200 million years old have been found with indications of parasitic infections. Reptile and dinosaur remains show evidence of osteomyelitis, meningitis of the brain, and other infectious diseases. No doubt, prehistoric humans were also afflicted with a great variety of diseases, as evidenced by arthritic joints found in ancient skeletons and other telltale signs of degenerative and infectious diseases. Paleontologists have examined teeth, fossil bones, mummies, and ancient artwork to study the history of human diseases. At the same time, these artifacts do not provide much evidence of soft-tissue disease, with the exception of some tissue that has been preserved in mummies. Bones and fossils provide the best evidence for diseases and prehistoric practice of medicine. One ancient treatment to relieve a person of evil spirits, and possibly to reduce cranial pressure, is called *trephining* (also known as *trepanning*). A hole is bored into a person's skull to provide an exit for the spirits, and to treat a migraine headache or epilepsy. (See Figure 4.1.)

Numerous trephined skulls have been discovered in Neolithic sites on the European and South American continents. There are reports that the practice still exists within some South American tribes, and there

Figure 4.1 A Trephined Skull
The practice of trepanning, that is, boring holes in human skulls, was, and still is, practiced with the belief that it will relieve pain and/or provide an exit for evil spirits.

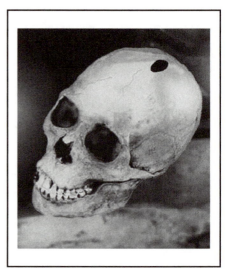

have been reports of recent trephining the southern United States. Early humans were infected and infested with lice, worms, trichinosis, tetanus, **schistosomiasis,** salmonella, yaws, and many other parasitic diseases transmitted to them from the wild animals they captured, butchered, and ate. Prior to the civilization of Mesopotamia, little is known about how ancient people diagnosed and/or treated various afflictions and diseases. It is assumed that by trial and error they discovered ways of grinding and mixing plants and other materials to prepare teas, purgatives, laxatives, poultices, fermented beverages, gruels, and other dietary substances to treat common ailments. At sometime in our history, humans applied what they had learned from butchering animals to the anatomy of their own bodies and thus were able to set broken bones and perform simple surgery. In other words, they gathered, processed, and applied natural products and experiences to practice a primitive form of medicine that progressed over the ages into our modern health-care system. Primitive preventive medicine was basically supernaturalistic, as was the treatment of diseases, since the accepted causes of all diseases were also the whims of supernatural forces. Ancients tried to prevent evil maladies from occurring by performing a variety of rituals that included human sacrifices, amulets, painting and tattoos, mutilations of the body (circumcision and surgical insertion of objects into body parts), the special disposal of human excrement, and the cutting of hair and nails in order to prevent spells and illnesses. About 10,000 years ago human tribes initiated farming and animal husbandry that provided a more settled way of life. Consequently, during the early days of civilization humans developed a ritualized system of medicine and improved health practices. The major difference between primitive medicine and modern medicine is that the practice of modern medicine, although still an art, is based on natural laws of science rather than supernatural forces.

The history of medicine and disease can be thought of as a study of the development and the progress of cultures from prehistoric to ancient Far Eastern and Middle Eastern cultures, followed by the classical Greek/Roman civilizations, the Middle Ages and Renaissance, the Age of Enlightenment (which led to medicine based more on science, although still an art), and currently into the highly technical system of modern medicine. Observations of symptoms, diagnosis, pharmacology, and treatment have always been a complex of knowns, unknowns, and speculations. In antiquity, one person was designated as the healer or medicine man or woman who gathered medicinal plants and attempted cures. The complexity of factors affecting the health of humans has led to the current era of specialization. However, this complexity is also the reason why superstition, spirituality, pseudoscience, and just plain quackery have been a part of ancient as well as modern medicine—for example, unproven alternative or complementary medical practices. Fortunately, the human body has the ability, to some extent, to cure itself of a variety of ailments, even if violated by questionable treatments.

About 5,500 years ago the Sumerians settled in what is known as the Fertile Crescent of land that became the "cradle of civilization" in Mesopotamia located between the Tigris and Euphrates rivers in present-day Iraq. Their medical documents inscribed on clay tablets have survived much better than those of the papyri of ancient Egypt. One reason why Mesopotamia is referred to as the site of civilization is related to its development of cuneiform writing, which used a stylus to make wedge-shaped indentations in wet clay tablets that were subsequently dried or baked in the hot sun. This preserved the tablets and writings for many years, and they could be moved easily from place to place. Eventually, these tablets were dispersed over many lands, and thousands of them still exist today, providing information about the Mesopotamian culture and medical practices. While the information on these tablets is not as detailed as that contained on Egyptian papyri, they do indicate that much of their medicine was related to religion and the supernatural. Mesopotamian healers (physicians) were priests who related the signs of disease to various spirits; thus, they considered disease a punishment for some transgression or impure act, that is, sin. One concept stated that if a person committed a taboo that displeased the gods, the person would lose the protection of the gods, resulting in exposure to the many devils or sorcerers that existed in Mesopotamia. Still, many of these tablets describe how medicine during this period of time began to evolve beyond superstition. Diseases were described, diagnostic techniques explained, and prescriptions for a great variety of drugs, including mandrake and opium, were recorded. In addition, the Sumerians understood the concept of contagion and public health. They developed sewage systems, water systems, and quarantined persons with communicable diseases. While ancient Mesopotamian medicine did not influence the Greeks to the same extent as Egyptian medicine, the tablets did provide a significant sampling of the medical history of this ancient and extinct civilization.

Some of the earliest written documented medical history is found in a series of Egyptian papyrus records that date back from three to four thousand years ago. The medical topics range from skin diseases (irritations and ulcerations), stomach diseases (including parasites), digestive diseases (including those of the colon and anus), diseases of the head (including migraines), kidney/bladder diseases (including kidney stones and flow of urine, possibly prostate problems), and numerous references to diseases and injuries of the toes, fingers, muscles, and bones. Tuberculosis of the spine resulting in a humped-over deformed skeleton, poliomyelitis resulting in leg bone deformities, dwarfism, clubfeet, pneumonia, gallstones, appendicitis, and possibly smallpox were also described and prevalent in ancient Egypt. These ancient writings prescribed surgery and treatments, including herbal prescriptions for specific ailments, as well as prophylactic treatments. For example, a concoction of crocodile dung mixed with honey and milk was prescribed as a female contraception.

Despite the magical slant to medicine, the Egyptian physicians developed an effective method of examining patients similar to the procedures used by current physicians. First, they tendered a preliminary and tentative diagnosis of the patient's medical problem. Second, they examined the patient and identified symptoms. Third, a diagnosis and prognosis for the particular disease was provided. And fourth, they prescribed appropriate therapeutic treatments that included surgery and drugs as well as magical incantations.

Hebrews (Israelites) from about 3,000 years ago are credited with the development of the practice of preventive medicine. Laws were passed that prohibited the contamination of wells and the eating of pork to prevent parasitic infections. Ritual baths were prescribed for both men and women to cleanse them of both guilt and dirt after contact with someone who was considered unclean. Isolation of persons with certain diseases was practiced, including those with venereal disease and leprosy. The religious ritual of circumcision is believed to have started in order to show that a male gave part of himself to God, but later it was possibly performed as a prophylactic against diseases. Hebrew medicine was based on common sense, and thus was the beginning of a more empirical and practical form of medicine.

Chinese medicine dates back to almost 5,000 years ago and was unchanged for many centuries. This medicine had a philosophical base, in particular the concept of *yin* and *yang*. These were the two opposite poles affecting life as well as the universe. Yin was negative, dark, passive, female, weakness, cold, and the right side of the body. Yang was positive, light, active, strength, warmth, male, and the left side of the body. Yin and yang principles apply to maintaining the harmony within the five organs of one's body as well as harmony among the five elements, five planets, five seasons, five colors, five sounds, and five directions. Acupuncture was an outgrowth of yin and yang and is still considered an alternative to modern medical practices, although its effectiveness is questionable. Chinese herbal medicine was highly developed. One book on herbs listed 365 herbs, with both prescriptions and poisons. Included are opium (pain relief), rhubarb (laxative), rauwolfia (tranquilizer), kaolin (diarrhea), ephedrine (respiratory problems), iron (anemia), sodium sulfate (purgative), and ginseng (vigor).

The origin of Indian medicine is traced back to the Vedic Period about 3,500 years ago. It was based on the supernatural, and disease and illness were believed to be the products of sinful behavior or demonic possessions. The physician/priest identified illnesses such as malaria, cholera, tuberculosis, diabetes, and plague. The Brahmanic Period of medicine existed about 2,800 years ago. After the Aryans invaded northern India, a caste system developed that soon placed the Brahman caste (priests) at the top and other classes were below, with the darker skinned people at the bottom. Indian medicine had three components: diagnosis, surgery, and prescriptions of herbal concoctions as treatments.

Ancient Greek civilization dates back about 5,000 years ago, and its medicine continuously changed over much of this time. Modern medicine has its roots in Greek medicine, not only in the terminology used by physicians but also in the philosophy of the practice of medicine. The Greeks had many gods, and some were responsible for curing diseases and healing. Homer's poems, the *Iliad* and the *Odyssey* (written between 800 and 1000 B.C.E.) are the earliest writings that describe ancient Greek medical practices. The *Iliad* describes various treatments for injuries incurred during the Trojan War. Homer listed detailed descriptions for more than 150 types of wounds. It was not until the 5th century B.C.E. when Aesculapius, a legendary physician god who supplanted Apollo, the god of both disease and healing, developed classical temple medicine. From this time until Greece was conquered and the fall of Rome in the 5th century C.E., many Greek physicians advanced to the profession of medicine. The most famous was Hippocrates, whose influence extended into the Renaissance.

• **Hippocrates** (ca. 460–ca. 377 B.C.E.). The Pythagorean philosophers, including Empedocles, greatly influenced Hippocrates who, in time, became known as the "father of medicine." About 70 of Hippocrates' books were collected and stored in Alexandria during the last centuries before the Christian Era. They are compiled into what is known as the *Corpus Hippocraticum,* although Hippocrates did not author all the volumes. Hippocrates taught at the medical school at Kos, which emphasized diet, hygiene, and service to the patient. The use of drugs and surgery was limited to situations in which other methods failed. Surgeries did include bloodletting, trephinations, gallbladder and kidney surgery, and a few other last-resort measures. Hippocrates' books stressed natural causation of disease, practical diagnosis, critical observation leading to prognosis, treatment of the whole patient rather than just a single ailment, and withholding treatment from those who were incurable. In other words, Hippocratic medicine was more practical than philosophical and theoretical. For example, two famous sayings attributed to Hippocratic writings are "Treat the patient, not the disease" and "First, do no harm." Even today, graduates from accredited medical schools in the United States recite a modified version of the 2,500-year-old Hippocratic Oath that has been approved by the American Medical Association.

From about the third century B.C.E., when medical knowledge and physicians moved from Greece to Alexandria, Egypt, and later after Rome conquered Greece during the Punic Wars in 148 C.E., the practice of medicine was fairly static. The most important physician of this period was Galen (ca. 130–200 C.E.). Known for documenting his many ideas and theories about the human body, diseases, and treatments, Galen based much of his knowledge on the writings of earlier physicians and philosophers. Galen's writings and authority as a physician were accepted as the last word in medicine for the next 1,500 years. This revered authority was interpreted by many as the "tyranny of Galen" since it impeded the advancement of medicine throughout the Middle Ages and into the Renaissance.

Medicine of the Medieval and Renaissance Period

Modern medicine is not only built on the contributions and discoveries made during the Middle Ages and Renaissance, but modern medical practices are often not far removed from those of our more distant ancestors. One of the most important ancient practices is alchemy, the roots of which date back about 3,000 years, when a philosophical view of medicine involved drugs, herbs, chemicals, and magical incantations.

Medieval medicine encompasses the period from the fall of Rome (about 476 c.e.) through the period of the Renaissance (about 1400 to 1600). Medieval also means the Middle Ages. As far as medicine is concerned, this time in history can be divided into two periods. The earliest years of the Middle Ages are often referred to as the Dark Ages, while pre-Renaissance medicine is known as *monastic* medicine, since monks not only acted as physicians but were also compilers of ancient Greek medical texts. During this period in history, medicine was a smorgasbord of classical Greek/Roman, pagan/barbarian, and Christian/Islamic religious conceptions and misconceptions of human anatomy, physiology, disease, drugs, and cures. The practice of medicine in the Western countries was stratified more by status than by knowledge or skill. Following are short descriptions of the various types of medical practitioners that were prominent during this period in Europe.

(1) *Physicians* were at the top rung of the status ladder of medical practitioners and were usually connected to universities. Physicians were required to lecture as well as debate their own points of view about diseases, drugs, and treatments. Since physicians were supported by the government, the church, and/or universities, and since they had membership in professional guilds, they were the only practitioners called "doctors." During this time in history most European cities had only a few university-educated physicians in residence. For example, in the year 1296 only six university-trained physicians practiced in Paris, France. These physicians only tended to the wealthy and elite, while the masses, if treated at all, were treated by other types of practitioners.

(2) *Barbers,* who had experience with sharp instruments, expanded their services to include surgery. It was considered beneath a physician's station to perform surgery, partly because of the Roman Church's beliefs relating to blood as somehow sacred and

partly because surgery was associated with dissection, which was an inferior trade. The most common surgeries performed by barbers were hernia repair, removal of gallstones, and cesarean sections.

(3) *Monks* contributed to medicine primarily by copying ancient medical manuscripts from Greek into Latin, thus enabling them to be read by the educated elite of this period. Monks also ministered to the poor and the aged members of their monasteries. Commoners often sought the medical services of monks and, in time, some of the their infirmaries developed into early hospitals.

(4) *Dentatores* were the dentists of the Medieval Period. Only the wealthy could afford their services since their fees were excessive for the service provided. The common belief of the time was that worms caused tooth decay. The primary material for fillings was ground bone. By the Late Middle Ages, gold was also used for fillings, and crude dentures were constructed from animal bone.

(5) *Folk Healers* were either male or female, and their practice varied depending upon their geographic location. Folk healers were sometimes called "leeches" because they had no formal education. Even so, they provided practical medical services based on observations and applications of folk remedies. Herbalists traded the secrets of their profession between male and female healers. Most herbalists were knowledgeable women familiar with the health benefits of specific plants, flowers, and trees. Some of these remedies were as follows: wearing a bag of buttercups around your neck cures insanity; taking St. John's wort to cure fever; swallowing an egg white mixed with nettles cures insomnia; heather, when boiled in water and applied to the head, cures a headache; a spider encased in a raisin, then swallowed, cures ague; rubbing goose droppings over a bald head grows hair; and applying the skin of an eel relieves leg cramps.

(6) *Midwives and Nurses* were medical professions open to women. Midwives were generally trained by physicians or other experienced midwives. Nurses from the medieval era were often, but not always, members of a monastic order. They were responsible for the basic care of patients in hospitals and were in great demand during epidemics.

Medicine was at a low point in the Western world during the early Medieval Period. In contrast, during this time in the Arab world a new religion grew around the prophet Muhammad, who was born in the Arabian city of Mecca in 570 C.E. Based on his visions in 610 Muhammad founded Islam (*Islam* means submission to God). After Muhammad's death in 632, his followers consolidated illiterate Arab tribes, formed large armies, defeated both Persia and the Byzantine Empire, and extended the Islamic Empire eastward to the western border of China and westward along northern Africa to Morocco and across the Mediterranean Sea to southern Spain and the shores of the Atlantic Ocean. For the next several centuries Muslims translated Greek scientific and medical texts into Arabic, while adding innovations of their own. Muslim science, and medicine in particular, reached its height during the Medieval Period partly because the prophet Muhammad is quoted as saying "Make use of medical treatment, for Allah [God] has not made a disease without appointing a remedy for it, with the exception of one disease, namely old age." Medical knowledge followed the Islamic religious path through a very backward western Europe. At this same time in history, the Muslims preserved the cultures of the countries they conquered by preventing the destruction of homes and property and by recognizing the Christian and Judaic beliefs of their new subjects. During their reign, the Muslims established clean and efficient hospitals that were open to all citizens. These facilities were also available for educating medical students. This was also the period of history when, for the first time, health and hygiene were taught and followed in Europe. This was also a time when only qualified physicians were allowed to practice medicine. It was also the time when the art of preparing and dispensing drugs, that is, pharmacy, was first recognized as a profession.

Physicians of the Middle Ages

During the Medieval Period surgery was separated from medicine, thus retarding both professions. The Roman Church was the dominant religious and civil influence during this time in history, and since Church policy disdained the shedding of blood, physicians who were, for the most part, clergymen, no longer practiced surgery. Barbers conducted all bloodletting and surgery while physicians often observed. This was also the period in history when physicians and medicine moved westward during the Muslim invasions. Following are a few of the physicians who contributed to the advancement of medicine, health, and hygiene during the Middle Ages:

- **Claudius Galen** (ca. 131–ca. 200 C.E.). Even though he lived before the Middle Ages, he is mentioned because of his influence on physicians and medicine in Europe until the 17th century. Since the dissections of human bodies were outlawed in his day, he performed surgery and experiments on both live and dead animals, mainly pigs. This limitation resulted in many misconceptions about the human body, and thus his writings were rife with errors—at least two hundred. Despite this failing, Galen made a number of important discoveries. He was the first to become aware that the organs and tissues of the body are formed as they are for a specific purpose. (This idea is now known as "form follows function," or one might say, "anatomy follows physiology.") He was also imbued with the ancient concept of the Aristotelian "threes": the three types of living things are vegetable, animal, and rational intelligence, and each has its own soul. Galen believed the human body has three basic forms of actions, referred to as *pneuma*: (1) The *pneuma psychicon,* or animal spirit that controlled sensation and movement, was centered in the brain; (2) The *pneuma zoticon* was centered in the heart and controlled the blood and body temperature; and (3) The *pneuma physicon,* located in the liver, controlled nutrition and metabolism. All kinds of triads were concocted during medieval times. For example, the status of humans was ranked: ordinary humans came first, then angels above humans, topped by the Holy Christian Trinity (itself another triad). Animals were also classed in a triad hierarchy of birds, fish, and mammals. Galen's explanations of how the blood flows from the heart to the lungs and body were basically backwards, but they persisted for hundreds of years. He was the first to prove that veins carried blood, not air. However, he also incorrectly taught that the nerves were filled with fluids.

One reason for Galen's influence on physicians through the Middle Ages and into the Renaissance is that his works were translated into many languages (Latin, Arabic, Syrian, Hebrew, and later into several European vernacular languages). Another reason for his long-lasting influence was his belief that everything is made by God to serve a particular end as determined by God. This theology fit very well with Christian, Jewish, and Islamic monotheistic faiths based on the concept that humans were formed by the implementation of an intelligent plan by a single supernatural being (although not necessarily the same One God for all religions). Everything was designed and created to fulfill His will, and therefore, to know man one must know and accept God. Acceptance of this philosophy was one reason why scientific inquiry into natural phenomena was absent during the Dark Ages.

- **Jurjis ibn Bakhtishu Jibril Yuhanna ibn Masawayh** (ca. 700–800 C.E.) translated works of Hippocrates, Galen, and Aristotle from Greek into Arabic. His works, as well as other Arabic texts, were later translated into Latin, which contributed to the renaissance of classical Greek learning in Europe.

- **Al-Razi** (841 or 863–926 or 930 C.E.), born Abu-Bakr Mohammed ibn-Zakaria, became a competent physician at an early age. In later years he was a court physician and also the chief physician of the Baghdad hospital. He wrote several books that were translated into at least five Western languages, including Latin, and which stressed three aspects of medicine: (a) preventative medicine; (b) maintenance of personal as well as public health; and (c) specific treatments for specific diseases. He preached moderation and balance in all things, especially eating (diet), drinking, exercise, and ambition. He was an excellent surgeon and was the first to use opium as an anesthetic. Al-Razi was also the first to use mercury as a drug in the treatment of patients. His books on therapeutic drugs enjoyed widespread popularity in Europe for many years.

- **Al-Zahrawi** (930 or 963–1013 c.e.), also known as Abu-Al Quasim Khalaf ibn'Abbas al-Zahrawi, was a court physician. He was known for his practice of surgery as well as medicine. Sometimes referred to as the "father of surgery," al-Zahrawi's encyclopedia of medicine contained about 30 volumes, which illustrated over 200 different types of surgical instruments. He also classified all types of fractures, including his famous descriptions of skull factures caused by blows from swords. His detailed account of how to strip varicose veins is essentially the same procedure used today. He advocated that, before practicing, all physicians should understand basic science, anatomy, and physiology in order to prevent mistakes that could result in the death of patients. His books were used as medical texts for the next 500 years.

- **Ibn Sina** (980–1037 c.e.) is also known as Abu-ali Husayn Ibn-Abdulla Ibn-Sina and, in the West, as Avicenna. He was a child prodigy, and by the age of 18 he not only completed his study of medicine but also was proficient in the Arabic classics, mathematics, anatomy, logic, and philosophy. He wrote many books, including 21 in the field of medicine covering principles of medicine, a listing of the then-known drugs, and descriptions of diseases of specific organs, as well as diseases in general. One of his books contained information on how to make and test drugs. Ibn Sina was the first to recognize and describe tuberculosis as a contagious disease of the lungs, how water and soil can spread disease, and that drugs should be tested on animals before being administered to humans. His encyclopedic medical work influenced physicians for many generations.

There were many Muslim physicians in the early Middle Ages that contributed to the revival of medicine in Europe. Western Muslims imported from the East the preserved medical knowledge that was translated from Greek into Arabic and was stored in the great libraries of Damascus. These volumes furthered the development of the medical profession during Arab occupation of the west. The most important physicians of the Middle Ages were the following:

- **Al-Biruni** (973–1048) was also known as Abu Rayhan al-Biruni. He was a physician who specialized in surgery and published descriptions of medicinal plants.
- **Constantinus Africanus** (1020–1087) resided in the Monte Cassino cloister in Salerno and is known for his translations of medical texts from Arabic into Latin.
- **Ibn Zuhr** (Latin name: Avenzoar) (1091–1161) was an excellent surgeon who was known for his skill in performing dissection. His medical books, including those on anatomy and parasitology, were written so that common people as well as professional physicians could understand them.
- **Ibn-Maimon** (Latin name: Maimonides) (1135–1204) translated Greek medical writings that described poisons, hygiene, and public health into Hebrew and Latin. He is better known as a Jewish philosopher whose medical fame came as a supporter of Galen.
- **Ibn an-Nafis** (ca. 1210–1288) was best known for his discovery and description of the pulmonary system, where blood is exchanged between the lungs and heart.
- **Arnald of Villanova** (1235–1312) was a famous physician of the Middle Ages who spent much of his time on diplomatic missions for royalty. He was famous for his criticisms of Galen.

Physicians of the Renaissance

Although Galen's concepts of medicine were prominent during the early part of the Renaissance, a new approach to surgery and medicine was needed. The reawakening of Greek humanism in science in general and medicine in particular, along with other developments and inventions in Europe (e.g., the printing press) provided the impetus for this revival. (The term *humanism* is a description of the renaissance of Greek and Roman classical studies and writings.) Humanism is the study of the past to obtain wisdom for the direction of knowledge of the present and future.

Following are the most influential scholars and physicians of the Renaissance who contributed to the revival of medicine:

- **Jean Francois Fernel** (Latin name: Ioannes Fernelius) (1497–1588), a physician in the French court, was known as a great clinician. His primary book, *Universal Medicine,* was a trilogy of three volumes, *Physiology, Pathology,* and *Therapeutics.* He subscribed to the Galenic "humoral theories" even though he did not always agree with Galen's other concepts. However, Fernel did correct some of Galen's errors and was the first to describe, in detail, influenza and the transmission of syphilis. He also proposed that while gonorrhea and syphilis are distinct diseases, they are contracted in similar manners through sexual intercourse. An excellent pathologist, he was the first physician to identify, postmortem, tuberculosis as a cause of death. Fernel was one of the few physicians of his time who was versed in philosophy, mathematics, and astronomy, while denying the validity of astrology.

- **Giovanni Filippo Ingrassia** (1510–1580) was the first to isolate and describe the specific diseases chicken pox and scarlet fever. He also conducted studies of the structure of bone.

- **Felix Plater** (1536–1614) was the first to make a reasonable classification of diseases, including both physical and mental diseases.

- **Guillaume de Baillou** (1538–1616) was the first to clinically describe whooping cough and rheumatism.

Diseases and Treatments: Middle Ages and Renaissance

Since the beginning of time, disease has always been a fact of life. However, their causes and cures were seldom known. Prehistoric humans lived in scattered groups of 50 to 100 people, thus contagious diseases were contained or limited to that group. The smaller tribes might be eradicated, but epidemics generally did not spread to people in distant geographic regions. Low-density populations are predominantly resistant to the pathogens of diseases and plagues, and besides, if a small tribe polluted an area, they just moved on. It was only after larger, more permanent settlements were made possible by the practice of settled farming and animal domestication that the transmission of many diseases became problematic for the communities. Among these diseases were trichinosis, sleeping sickness, tularemia, tetanus, schistosomiasis, smallpox, leprosy, typhus, salmonella, yaws, and syphilis, plus a number of other contagious and parasitic diseases.

Little progress was made in understanding the causes, effects, and treatments of diseases from the prehistoric to the Classical Greek/Roman Period, and into the Middle Ages and Renaissance, primarily due to the lack of knowledge of the **germ theory.** However, the most important reason for the continued ignorance of medical knowledge, at least until the Renaissance and beyond, was the belief that disease was the punishment for a personal or moral transgression against a supernatural god. In other words, disease was considered as retribution. Also, the manner in which physicians practiced surgery did not advance knowledge of human anatomy and physiology. The medieval physician directed a barber/surgeon who proceeded with the cutting and then exposed the organ. The physician could then point and identify the organ according to the precepts of Galen. This traditional type of surgery, the metaphysical use of drugs, and other forms of ineffective treatments, including astrology, continued throughout the Middle Ages, all of which impeded the advancement of medicine. Galen practiced medicine and wrote during the height of the Roman Empire in the 2nd century of the Christian Era. Plagues and epidemics were a contributing factor to the decline of Rome, and Galen's methods did little to help stave off plagues and the eventual fall of the Empire in the middle of the 5th century C.E. There were many other factors that led to the demise of the Roman civilization—for example, the splitting of the Roman Empire into the East (Byzantine) and West (Roman) divisions, political corruption, oppression of minorities, poverty, and finally the assaults and attacks by the barbarian tribes from the north—but disease unquestionably weakened the overextended Empire.

A number of different diseases—smallpox, measles, dysentery, diphtheria, leprosy, bubonic plague, tuberculosis, chorea, ergotism, and influenza—caused widespread illnesses during both the Middle Ages and Renaissance, particularly among populations living in urban areas. When a sudden outbreak of a specific disease spreads among a large proportion of a population in a greater number than was expected, it is called an *epidemic*. When the epidemic is spread over a large geographic area, it is referred to as *pandemic*, meaning "all the people." When a disease is spread over a region, but with relatively few deaths, it is called *endemic*, meaning "local." Humans who contract diseases usually exhibit some symptoms, ranging from feeling unwell (malaise), high fevers (ague), skin eruptions, mobility problems, digestive problems, localized or general pains, comas, as well as many other less obvious symptoms such as high blood pressure, changes in body fluids, or malfunctioning of internal organs. From beginning of time the healer or physician tried to relate these specific symptoms to the cause of the illness and then provide a prognosis that included a treatment involving either drugs or some incantation or ceremony, or a combination of both.

There were epidemics and pandemics of a number of major diseases that significantly impacted on civilizations during the Middle Ages and Renaissance, mainly plague, leprosy, smallpox, St. Anthony's Fire, syphilis, and tuberculosis.

- **Plagues**—Historically, the outbreak of many diseases (epidemics of smallpox, for instance) were referred to as "plagues" in a generic sense.

However, they were not what is usually meant by the "Black Death," correctly known as the bubonic plague, in Asia and Europe from about the year 160 until the late 1600s. Bubonic plague is the name for a specific disease caused by the infectious agent bacillium *Pasteurella pestis,* also known as *Yersinia pestis,* after its discoverer, Alexandre Yersin (1863–1943), the Swiss bacteriologist. Using DNA, 20th-century scientists discovered that a simple, harmless human-intestinal parasite was transformed into the deadly plague bacterium over 2,000 years ago. This new organism was encoded to invade human organs, particularly the lymph glands, and was also adapted to live in the fleas that lived on rats. The term *bubonic* refers to the development of *buboes,* which are infected and swollen lymph glands. The term "black death" characterized the faces of the victims that turned dark after death.

A number of plagues (or epidemics) occurred before the Christian Era. In the year 430 B.C.E. one-third of the population of Athens was killed by some type of infectious disease, most likely the bubonic plague imported from Ethiopia, into Egypt, and then into Greece. Several hundred years later, in 160 C.E., the bubonic plague was known as *Barbarian boils* since northern invaders brought the disease into China. The large number of deaths resulting from this epidemic in China led to the fall of the Han Empire. One of the first recorded European epidemics of plague was the *Antonine plague,* which was brought west from India to Syria by invading Huns. Then, in 166 C.E., returning troops brought it to Rome. It is reported to have killed between four to seven million people, or one-third of the entire population of that region. This particular epidemic weakened the expansive Roman Empire. The next series of epidemics, possibly including measles, and smallpox as well as bubonic plague, occurred after 211 C.E. and devastated Rome and the populated area around the city.

After this last plague outbreak in the Greek/Rome Period of the 3rd century of the Christian Era, western Europe was essentially free of the bubonic plague for the next millennium. But as crusaders, traders, and others traveled between Eastern and Western countries, so did diseases. It is speculated that traders from the Middle East became infected with bubonic plague from rats in northern India and northern Africa and then carried the disease to China in about 1330. After nearly 1,000 years of a plague-free Europe, the scourge moved westward from Hong Kong and China. It appeared in southern Russia and the Crimea in 1343 or 1346. When a trading vessel from Genoa landed in Italy, everyone on board was infected or dead of the Black Death. Between 1346 and 1350 the disease moved north from Italy to western Europe. In 1348 bubonic plague was imported to London with deadly effects. Within two years about one-half of the European population died of the plague. Cities were evacuated, commerce halted, harbors closed, and quarantines were established. Although medieval Europe had few concepts of public health or personal hygiene, night crews plied the streets filling wagons with the remains of the hundreds that died each day. They filled the graveyards and burned piles of bodies in a futile attempt to quell the epidemic. Over the next 200 years Europe was even more devastated due to the long-term European pandemic. There were periods when the disease subsided, only to spring up again with a vengeance in another region. Each time the plague appeared to subside in Europe, it was reintroduced from the East, creating another epidemic. There were outbreaks of the plague in Spain in 1596 and again in 1648. Still, the great plague of London in 1665 was one of the last to devastate the continent. And while Christianity taught that it was a work of charity to care for the sick, the hospitals and physicians were overwhelmed and at a loss as to how to cure the ill or prevent the epidemics. As the concept of contagion became better understood, physicians insisted that the dead and their clothes be burned as a preventive measure to halt the spread of the plague. However, the plague epidemics in Europe did not deter the English, Spanish, and Portuguese from exploring and exploiting the New World in the 15th and 16th centuries. Consequently, during the later years of plague epidemics, Europeans exported the plague and other diseases (e.g., smallpox, measles, diphtheria, influenza, typhus, and possibly syphilis) with devastating results to the native populations of the Americas who had no immunity to these new scourges. There is some evidence that historically (14th or 15th century), a catapult containing the bodies of dead plague victims was launched against the enemy who, in turn, became infected and carried the disease to other countries. The aerosol version of the bubonic plague is extremely virulent, which makes it a potential bioweapon for modern warfare.

The rod-shaped bacteria that causes bubonic plague (*Yersinia pestis*) is initially spread to humans by the bite of fleas that infest rodents, usually rats or ground squirrels. Once infected, the person becomes contagious, and the bacteria can then be passed from one infected person to another by mucous aerosol droplets (sneezing, coughing, etc.). After becoming infected, a person may feel ill within a few hours, and a high fever, chills, and headaches develop in one to six days after exposure. Death can occur within two to four days if not treated within 24 hours. Today, there is a vaccine for the flea-carried version of bubonic plague that can prevent the disease if administered in the first day or two after exposure. But the vaccine is not effective for the aerosolized version, which can be successfully treated with antibiotics if administered early after exposure. The disease spreads rapidly throughout the body, and victims who live more than a few days may contract pneumonia or septicemia (blood poisoning) that is usually fatal. But, obviously a number of people have survived, possibly because they developed a resistance or immunity to the disease. Plagues still strike in remote parts of the world, but modern medicine and improved public health has been generally successful in preventing massive epidemics. During the Middle Ages it was not known who first concluded that the fleas from rats were the carriers of the bubonic plague bacterium, or that the deaths were even caused by a bacterium (not yet discovered). Also not known was the fact that aerosols spread the disease from person to person. But then, as now, it is impossible to eliminate all rodents from cities. Since the times when humans were cave dwellers, people and rats have always lived in proximity. It is well known that there are more rats now living in New York City than there are humans. Physicians in the Middle Ages and Renaissance, more by instinct than by science, instituted quarantine as a preventive measure against the spread of the plague, which led to awareness that public health played an important role in disease prevention. One mystery was why the plague never became endemic throughout all of Europe. Another was why it would die out and then return after several years or decades. The answer seems to be that after the plague ran its course as epidemics in Europe, it never established a permanent base in the western countries. Rather, it was reimported from Asia from time to time.

There were no effective treatments for the bubonic plague before and during the Middle Ages and Renaissance, other than caring for the sick and suffering. In addition to invoking the help of God, Christians of the time provided shelter for infected outcasts, anointed patients with oils, prayed and practiced the "laying on of hands." Superstition abounded, some psychic as well as physical. One group, called the Flagellants, wore long cloaks and carved scarlet crosses on their breasts. As they passed though towns and cities, they flogged themselves with three-tailed whips in the belief that the Lord would be compassionate and end suffering from the plague. Obviously, nothing was effective. Nevertheless, quarantining cities, harbors, and travelers, and burning cadavers did help to control the epidemics that would periodically die out only to return. These cycles continued until causation of plague was discovered along with the development of new drugs and therapies, and improved public hygiene.

* **Leprosy**—This disfiguring disease was known in ancient times, but it did not reach epidemic proportions until the Middle Ages. Leprosy disease spread rapidly from the 6th century C.E. until the 1300s. It was one of several deforming diseases that people considered a mark of the devil. Other diseases, some noninfectious, such as psoriasis and eczema, and some very infectious diseases, such as smallpox, plague, and syphilis, that produced skin sores or growths were often confused with leprosy, and victims were forced into isolation. It was assumed that lepers exhibited a moral defect and thus were singled out by God and forced to live

outside communities so as not to contaminate others, either physically or spiritually. Lepers were required to ring a wooden clapper or bell as they approached other people since they were considered unclean and apart from regular society. To some extent, this prejudice that has existed for hundreds of years still exists. However, around the end of the 15th century, except in Norway and in the Mediterranean region, leprosy ceased to be endemic in Europe. The reasons for its recession at that time are still debated and are still unknown.

The physical manifestations of leprosy develop slowly. The disease is caused by a tubercular rod-shaped bacterium called *Mycobacterium leprae.* Armauer Gerhard Henrik Hansen (1841–1912) discovered this bacterium in 1873; thus it is known today as Hansen's disease. Hansen's discovery is significant in the history of diseases since it was the first time that a specific microorganism was associated with a specific disease. Although it can be spread by aerosol droplets from the mouth and nose during close contact between individuals, it is not highly contagious. Most people have developed immunity to the disease, and thus relatively few people who are exposed to the bacterium actually develop leprosy. Leprosy patients exhibit whitish or copper-colored skin patches that may be raised but do not itch or create pain. Accompanying the thickening of the skin, face, ears, or hands is a loss of feeling. Before the advent of microbiology and the discovery of the leprosy bacillus, the disease was diagnosed purely by external physical symptoms. Leprosy can be cured by a regimen of multiple drug therapies consisting of three effective drugs that, when used in combination, prevent the bacteria from mutating and developing a resistance to the drugs. The last remaining leprosy hospital in the state of Louisiana in the United States was closed during the latter part of the 20th century. There is some indication that the armadillo, a small burrowing animal with a covering of jointed, bony plates found in the southwestern United States, may be a carrier for the leprosy bacillus. It is now known that leprosy is neither highly contagious nor resistant to drugs.

- **Smallpox**—It is not known exactly when smallpox first became a pestilent, but it can be assumed this disease, along with other pathogens and parasitic microorganisms, became infectious when humans altered their lifestyle from hunting and gathering to manipulating nature— that is, settled farming, domestication of animals, and ultimately living in congested communities. There is no question that smallpox was one of the most horrible diseases to afflict humans, and its eradication is one of the great successes of modern medicine.

    The highly contagious *variola* virus (*variola* means "pustule" or "pox" in Latin) is the cause of smallpox. Smallpox was also referred to as *petite veriole,* meaning "small pustules," which distinguished it from other diseases that formed larger poxes, or *grand veriole*. As with many other ancient diseases, smallpox was treated by applications of supernatural diagnostics and divination procedures. In other words, since some invisible foreign body was thought to be the cause of smallpox, this imposing spirit had to be removed either by an incantation or a physical process intended to achieve both spiritual and physical purification. Trances and dances attacked the spirit, while bleeding, flogging, vomiting, purging, bathing, or special foods worked on the body.

This virus is unique in that it is a large DNA virus that only affects humans. After one to six days, infected persons develop high fevers, chills, and headaches as a rash forms on their faces, arms, and legs. Severe scarring of the affected skin areas usually results if patients survive. Smallpox is highly contagious and spreads by person-to-person contact. Both quarantine and vaccination are effective as preventative measures against the disease. Smallpox was the first disease for which a vaccine was developed that provided immunity, which effectively prevents its spread; it is also the first human virus that is considered as having been eradicated from the Earth. The United States ceased compulsory vaccinations against smallpox in 1971, but today, with the possibility of the virus being used as a terrorist bioweapon, vaccinations are again being encouraged. The last reported natural case of smallpox occurred in Somalia in 1977. In the late 1970s the World Health Organization (WHO) of the United Nations (UN) claimed that a program combining vaccination and quarantine had eliminated smallpox.

Prior to September 11, 2001, when terrorists attacked the United States, a worldwide controversy existed concerning the disposition of stored supplies of the virus. The remaining samples are reportedly maintained in just a few laboratories in the United States and Russia. Should these samples be maintained for scientific study, and the possible development of future vaccines, or should they be destroyed, which would reduce the possibility of the virus becoming a bioweapon? Scientists and government officials must now make their decision within the threatening environment of bioterrorism.

- **St. Anthony's Fire,** also called *ergotism*, was first reported in the mid-800s as an epidemic known as the holy St. Anthony's fire. Several more serious outbreaks of ergotism occurred into the 12th century, at which time it was finally realized that epidemics occurred in the same years that rye grain did not ripen as it should. Rather, it was damp and moldy. Later, in the 19th century it was found that the fungus (ergot) *Claviceps purpurea* produces several toxic alkaloids that cause this disease. These toxic alkaloid poisons affect both humans and animals. Since the poisons attack the nervous system, victims of serious cases of ergotism exhibit odd movements and behavior. Lameness, necrosis (death of tissue), gangrene, and the painful loss of fingers, hands, or toes are symptoms of ergotism. Lysergic acid, the main ingredient of LSD, a dangerous hallucinogen, is obtained from the ergot mold.

- **Syphilis** may have existed in ancient times, but it was not recognized as a specific disease until the late 1400s. The word *syphilis* stems from the title of a classical Greek myth that was translated and rewritten as the poem *Syphilus sive morbus gallicus* (Syphilus, or the French Disease) in 1530 by the Italian physician and poet, Girolamo Fracastoro (1484–1553). Syphilus, the main character in the myth, was a shepherd who supposedly was the first sufferer of the disease. Therefore, Fracastoro used his name as the name for the disease that was referred to intermittently as the "large" or "big pox," or the "Neapolitan disease." It was also called the "French disease" since French soldiers reportedly acquired syphilis from the Spaniards while they were garrisoned in Italy. By the early 1500s, it was believed that syphilis was an American disease imported into Spain by sailors from the ships of Christopher Columbus, whence it spread to the rest of Europe. (Note: The question of the origin of syphilis as being European or American remains in dispute.) What is not in dispute is that garrisoned soldiers in foreign lands contracted the disease during the time of occupation and then spread it to others upon returning to their homelands. Fracastoro's most important contribution to medicine was his theory of contagion through (a) personal contact, (b) contact with clothes of an infected person, and (c) transmission of the disease by air. His was the first scientific explanation of the transmission of diseases, although it was not confirmed until the development of the science of bacteriology. Jean Fernel (1497–1558) successfully treated King Henry II's mistress, Diane de Poitiers, of syphilis, but he failed to cure his father of the disease while he was still alive. It is not known how he treated syphilis, but it is assumed he used some form of mercury, as did most physicians of his day.

  Syphilis was a disease that not only interested physicians, but also moralists and writers who originally confused syphilis with gonorrhea and claimed it was God's retribution for sinners. Gonorrhea was known as early as the 13th century, and as late as 1496 infections of both the male and female genitalia were described as "mixed" gonorrhea and syphilis. In 1530 Paracelsus claimed that gonorrhea was an early stage of syphilis. Gonorrhea is caused by a round-shaped, buckshot-type bacterium known as *gonococci bacillus,* while syphilis is caused by a spiral corkscrew-shaped spirochete bacterium known as *Treponema pallidum.* The confusion is understandable since the "germ theory" was unknown at this time. Both gonorrhea and syphilis (as well as several other diseases) are infectious venereal diseases spread by sexual contact. A person can be infected by one or more of these diseases at the same time, as a result of one sexual contact or several contacts over a period of time. Syphilis progresses through three stages in the human body: primary, secondary, or tertiary (late or latent stage). Symptoms of primary syphilis are evident a few days to a few weeks after contact with an infected person. An ulcerated sore develops on the male or female genital organ, mouth, or throat. This sore is called a Hunterian hard chancre and is usually about the size of a quarter with distinct edges. If left untreated, the disease can infect the bones, nervous system, and all soft organs of the body. When it progresses to the secondary stage, smaller raised sores are distributed over most of the body. The third, or latent, stage, may not occur until many years later and may affect different areas of the body. If the disease has invaded the central nervous system and brain, it can lead to a partial paralysis, known as paresis, characterized by a shuffling gait, as well as dementia.

The actual origin of the bacterium that causes syphilis is unknown, but there are several theories. One is that yaws—which is not a venereal disease but a disease caused by a similar spiral, corkscrew-shaped bacterium (*T. pallidum*) that arrived in Europe from Africa—may be the ancestor of syphilis. Because of the cooler European climate, yaws may have mutated into a more virulent form of syphilis. Also, there remain two conflicting and still controversial theories related to the origin and spread of syphilis from or to Europe and the New World. One side claims that in 1492 Columbus's crew carried syphilis to the Caribbean islands off the North American coast where he first landed. From there the disease spread rapidly throughout parts of the Americas since the indigenous populations of some geographic areas had no immunity. This claim is related to what did happen later with smallpox, influenza, and a few other diseases brought from the Old World to the New World. The opposite argument is made that syphilis, at least the variety that became an epidemic in Europe in the 16th to 19th centuries, originated in the Americas and was carried back to Europe by early sailors and explorers, and then reintroduced to the Americas. Written records support this point of view, as evidence of syphilis has been found in the bones of ancient Native Americans who lived long before the arrival of Columbus.

For many years syphilis was not just a physical manifestation of a disease, but was also regarded as a punishment for disobedience to the laws of God. Thus, there were several types of treatments that reflected the diminished status of women at this time in history, as well as the prejudices against the afflicted. One example was the whipping of women who infected their husbands. (It was more likely that the husband infected the wife.) Another example is that prostitution was either regulated by laws or banned outright. Also, during epidemics infected people, or even undesirables who *might* be infected, were quarantined. These last two measures, regulating prostitutes and quarantines, did not work then, and do not work today as effective measures in controlling epidemic or pandemic diseases, including venereal diseases and HIV/AIDS.

During the Renaissance there were some treatments available for the wealthy, such as the concoction known as Van Swieten's Liquor, a mixture of mercury and brandy. The poor who suffered from syphilis were treated in public hospitals and more or less left to pay for their immoral behavior. Galen had several theories and therapies related to all types of pox infections. His theory was that pox, as with all diseases of the human body, was an imbalance of humors. Of these humors, phlegm was most important as related to pox. Next was blood and sweat. Therefore, Galen advocated mercury to increase the flow of saliva to induce spitting, bleeding the patient by **cupping** to clear the blood, and heat to promote sweating. Most physicians, including Paracelsus, preferred mercury compounds in the treatment of syphilis as well as other diseases. One reason for this belief dates back over many centuries and can be traced to Galen, who thought the copious amounts of saliva produced by the ingestion of mercury would be beneficial. Instead, it is merely a sign of mercury poisoning. During the early 1500s a therapy contrived from a New World hardwood plant, commonly called "holy wood," was developed in Europe. Known as the "guaiacum therapy," after the guaiacum or *lignum vitae* (tree of life), it was promoted by Ulrich von Hatten. Guaiacum was hung in churches and homes. As a result, it became more popular and the price increased, and thus only the wealthy could afford it. However, as with most drugs and treatments of Middle Ages and Renaissance, it was ineffective. It was not until the early 20th century that the Nobel Prize winner Paul Ehrlich (1854–1915) developed a synthetic arsenic compound called #606, now known as arsphenamine, but originally patented as Salvarsan. It was the 606th chemotherapeutic compound he developed to kill the spirochete that causes sleeping sickness. His assistant discovered it to be effective in killing another type of spirochete, *T. pallidum,* the syphilis bacterium. Known as the "magic bullet," 606 reduced the number of syphilis infections in Europe by almost half, but it still proved to be a difficult disease to eradicate worldwide. Although penicillin was discovered in 1928, it was not used extensively until near the end of World War II in the mid-1940s. Since that time, penicillin and other antibiotics have proven to be the most effective in treatment of the early stages of syphilis, as well as many other contagious bacterial diseases. Another treatment first tried during the 15th century was still in use in the early 1940s in United States army hospitals. This treatment consisted of mercury or arsenic compounds (poisonous heavy metals) injected into the blood stream, after which patients were placed in a "hot box." The army's hot box completely enclosed the patient except for the head. The treatment increased the patient's body temperature to an almost dangerous level and resulted in profuse sweating. The theory was that the high body temperature, along with the heavy metal injection, would kill, or at least **attenuate,** the spirochete bacteria. However, this therapy was not very successful, especially for secondary and tertiary (latent) syphilis, and with the advent of antibiotic drugs, the treatment became obsolete.

• **Tuberculosis**—The DNA analysis of a 6,500-year-old Neolithic skeleton found in northern Europe indicated that the individual suffered from tuberculosis. Various forms of tuberculosis have been found in Egyptian mummies, including the variety referred to in history as *phthisis* (wasting away) or *consumption,* now known as pulmonary tuberculosis, or TB. During the Classical Greek and Roman Period, both Hippocrates and Galen described tuberculosis as consumption. In 1037 C.E. Avicenna (Ibn Sina), an Islamic physician, was the first to write about consumption as being contagious, and in 1526 Jean Fernel (Ioannes Fernelius), referred to TB as *phthisis contagiosa* (contagious consumption) in his work *Pathologia.*

There was, and still is, a direct correlation between the increase in population density and the increase in tuberculosis, as well as other infectious diseases. Tuberculosis continued to spread well into the 18th and 19th centuries with industrialization, slums, overcrowding, and generally unsanitary living conditions. When it reached the middle and upper classes during these centuries, TB became romanticized, as in Alexandre Dumas's novel and play, *The Lady of Camellias* (1852), where the heroine, a courtesan, repents her lifestyle as she slowly dies of consumption. A slow, lingering, and wasting disease, it was often associated with something beautiful or inspirational partly because a number of famous people were its victims, for instance, Fredric Chopin, John Keats, Anton Chekov, D. H. Lawrence, and Eugene O'Neill.

Tuberculosis is an ancient disease caused by the microorganism *Mycobacterium tuberculosis* that infects the lungs, bone, and other body parts. During the Middle Ages and Renaissance, the only cure was rest, although many ineffective remedies were tried. Later, wealthy patients resided at sanitariums often located at upscale resorts at higher altitudes where it was assumed the cleaner air would aid in treatment and cure.

By the early 20th century, it was not the wealthy who were sent, by and large, to TB sanitariums. Rather, it was the poorer classes who, primarily through ignorance and poor hygiene, spread the disease among the residents of crowded tenement neighborhoods. Consequently, the resort sanitariums lost their luster and became clinical warehouses for the impoverished. In 1882 Robert Koch (1843–1910) identified the rod-shaped bacterium responsible for the disease, and the development of antibiotic drugs in the 20th century provided a cure for tubercular patients. The result was not only a dramatic decrease in the number of

TB cases, but the eventual closure of the depressing sanitariums. Before the 19th century the number of cases ranged from 500 to 1,000 per 100,000 people and was found on all continents. The number of cases dropped dramatically in most countries by the mid-20th century. This was related to better nutrition and public health measures as well as the use of antibiotics. However, there is some evidence that the bacillus has mutated, becoming resistant to most antibiotics and thus more virulent. Tuberculosis is again spreading in parts of the world.

A number of other diseases existed before and during the Middle Ages and Renaissance, including insect-borne diseases (e.g., malaria, yellow fever, sleeping sickness, filariasis), infectious diseases (e.g., cholera, typhus, typhoid, mumps, influenza, measles), and a variety of parasitic diseases (worms). Many of these were confused with other maladies and ineffectively treated. It was not until the 18th, 19th, and 20th centuries that scientists and physicians had the equipment and knowledge to recognize that specific agents caused specific diseases, which ultimately led to more effective treatments and cures.

Major Medical and Surgical Discoveries

Dissection

Until the beginning of the early Middle Ages, the dissection of human cadavers was neither permitted nor performed in European countries. Dissections were considered sacrilegious and barbarous, and those who dared perform them were coarse and unclean. Thus, those physicians who sought more information on the anatomical structure of humans conducted dissection on animals, usually pigs or monkeys. This practice seems incomprehensible given our knowledge today. At the time, it appeared reasonable. Therefore, Galen and many others made errors by confusing animal anatomy with human anatomy. It was not until the Renaissance that physicians questioned these errors and were guided in their dissections by their curiosity rather than tradition. At this time in history dissections were usually conducted during the winter to delay decay of the cadavers. In addition, many physicians had difficulty in obtaining bodies for study since both Christian and Islamic religions forbid the desecration of the dead. However, this did not deter the determined professors of medicine who used dissections for teaching as well as for their own edification. Since most professors considered it beneath them to perform the dissection, they hired barbers to do the cutting while the professionals observed. The first incision was a large T-shaped cross from the neck down and side-to-side, so that flaps of flesh could be opened up to expose the organs. The professors often compared what they saw with Galen's descriptions. The courageous anatomists of the Renaissance who performed human dissection were responsible for the turning point in the practice of medicine and surgery. Their experiments added to the canon of medical knowledge and led to the scientific revolution of the 17th and 18th centuries. Some of the most famous Renaissance anatomists were the following:

- **Mondino de Luzzi** (1270–1326). Although he belongs to the pre-Renaissance era, de Luzzi is included because he is known as the "Restorer of Anatomy." In 1326 he published *Anothomia,* in which he described his findings based on the public dissections that he personally performed. He was one of the first physicians who did not rely on the surgical services of barbers, bath attendants, hangmen, animal gelders, and quacks.

- **Guy de Chauliac** (1300–1370) was the physician to the pope at Avignon. He was considered an excellent surgeon who improved the quality of surgical operations for removing kidney and bladder stones, as well as for eye cataracts. Unfortunately he also promoted the practices known as *coction,* which means applying heat as in boiling or baking, and allowing "laudable pus" to develop to heal wounds.

- **Niccolo Massa** (1405–1469) was one of the first physicians to question many of the statements of early anatomists. He performed his own dissections and included his observation in his writings.

- **Leonardo da Vinci** (1452–1519) is considered an example of what is known as a *Renaissance man.* Although he was not a physician, he explored all fields of knowledge, including engineering, architecture, science, painting, sculpting, and medicine. His research in the areas of medicine, however, was confined to depicting the human anatomy in splendid drawings. As examples, his dissections and anatomical studies were carried out not from a medical point of view, but rather as an aid in painting and sculpting the human figure. Da Vinci also attempted to learn about how different parts of the human body function. Nevertheless, if his drawings and paintings had been published during his lifetime, the course of medical history may have been altered. Even today, his drawings of blood vessels, muscles, and soft inner organs of the body surpass modern photographs. His studies of the skeleton assisted later anatomists in determining how muscles, bones, and joints work together. It is said that he dissected over two dozen bodies during the nighttime when the light in the mortuaries was extremely poor.

Not surprising, the concept of eyesight and eyes has intrigued humans for many centuries. Euclid, the Greek geometer, (ca. 330–260 B.C.E.) proposed that the eye, not the object, was the point of origin of what was seen. He tried to use geometry to explain how the image came back to the eye. Alcmaeon of Croton (fl. 450 B.C.E.) discovered the optic nerve, which connects the eyes to the brain. He was also the first to claim that all sense organs were connected to the brain. Galen questioned how an image of a single object could be viewed by more than one set of eyes at the same time, and how the image of a very large object could get through the small aperture of the eye. Da Vinci not only dissected eyes to examine their internal structure, but he also made a hollow glass model of an eye. Da Vinci peered through a hole in the rear of the model in order to see what the eye would see. Since he was unaware of the nervous system and how the optic nerve connected the eye and brain, he claimed that the eyes are "the window of the soul." Although da Vinci was not the first anatomist—the Greeks Herophilus and Erasistratus preceded him—he is often referred to as the "father of anatomy."

- **Andreas Vesalius** (1514–1564) dissected the cadavers of condemned criminals in order to perfect his surgical skill and expand his knowledge of anatomy. Since refrigeration was not available in those days, anatomists often worked continuously for several days until the corpse began to putrefy. Also, Leonardo da Vinci's anatomical drawings had not yet been published, and thus Vesalius was unable to benefit from them. Nevertheless, his work was so accurate that he was able to challenge the anatomical misconceptions of Galen. When he recorded his actual observations of these dissections, rather than those presupposed by Galen (who performed animal not human dissections), Vesalius heralded the beginning of the end of the almost 1,500-year-long "Tyranny of Galen." Galen's hold on medicine, even into the Renaissance, was tenacious, as evidenced by a story concerning Jakobus Sylvius, Vesalius's university teacher. Even after conducting his own dissections of human cadavers, Sylvius continued to claim that Galen was correct and the reasons that his own observations disagreed with Galen's descriptions was that the organs of the human body had changed since the time of Galen. Notwithstanding this new and important information on dissection, Vesalius still revered the old master, Galen. For instance, Galen claimed that small pores in the septum, which separates the heart's left and right ventricles, allowed the passage of blood from one side to the other. Vesalius attempted several unsuccessful experiments to locate these pores. Despite this, he accepted Galen's incorrect assumption.

Vesalius wrote the famous book *De Humani Corporis Fabrica Libri Septem* (Concerning the Structure of the Human Body). This seven-volume work, usually referred to as *De Fabrica,* was published in 1543. Most European physicians at this time, who continued to be unyielding followers of Galen, criticized the truthfulness of the statements in Vesalius's books. Upset by these attacks, Vesalius resigned his position at the University of Padua and burned all of his remaining unpublished material. However, after his death, his published works were widely distributed. As Vesalius's results gained further acceptance, Galen's influence waned. Vesalius's publications became fundamental to the continued development of the fields of anatomy and surgery.

- **Berengario of Carpi** (Latin name: Berengarius) (1470–1550), known for his skill as a surgeon, produced excellent anatomical drawings based on the dozens of operations and dissections he personally performed, as well as on his observations of nature. He was one of the first to publish drawings depicting the appendix, the sphenoid sinus, the circulation of the blood in the liver, and other unknown structures of the human body.

  Berengarius' experimental procedures advanced the understanding of the functioning of the body's organs. For instance, he researched the origin of urine in the bladder and how it left the body. He first utilized fetuses of animals, and later used human fetuses. After numerous dissections and experiments, he correctly determined that urine from the bladder originates in the kidneys, and the only exit is through the penis. This was a major discovery at the time.

Surgery

In the Middle Ages and Renaissance, as before, surgery was a bloody and grisly practice. As mentioned, during this period barbers or butchers performed most surgery, usually under the direction of a physician. During the Renaissance the practice of surgery changed somewhat, and physicians began to conduct their own surgeries. Anesthetics were seldom used during amputations or surgery. Surgical instruments were crude, often no more than ordinary saws, hatchets, and sharp knives. To stop the bleeding after an operation, a hot iron or boiling oil poured over the wound was used to sear it. If this painful procedure was unsuccessful in sealing off the blood vessels, patients often bled to death. One seemingly inhumane story of the late Middle Ages describes how a gangrenous limb was removed by merely tying a tight ligature around the joint above the gangrene infection and then waiting until the limb fell off. A reasonably successful method of tying off blood vessels was only developed in the mid-16th century. Two notable surgeons of the Renaissance are:

- **Ambroise Paré** (1510–1590), a physician and surgeon of the Renaissance, was the first to develop more humane surgical methods. He invented the use of ligatures to tie off blood vessels rather than using the painful method of cauterization. A ligature is a surgical thread, wire, or cord that ties and binds a blood vessel or duct, thus preventing the flow of blood or fluid. This practice was not new. However, it had been unused for hundreds of years since the time of Hippocrates and Celsus, who sewed blood vessels shut in an effort to stop the bleeding of wounds. The use of ligatures was revived and improved by Paré, who also described other advancements in surgery in his work called *Ten Books of Surgery*. Unfortunately, the use of ligatures was not popular with surgeons during the Renaissance who preferred hot irons. The practice of using ligatures, along with the use of compression tourniquets, was again revived in the 18th century.

- **Giovanni da Vigo** (1469–1525) was one of the first surgeons to deal with gunshot wounds. Since the composition of gunpowder, an old Chinese invention newly imported to the West, was not well understood, da Vigo believed that bullet wounds were poisonous and that in order to prevent death, boiling oil should be poured into the bullet wound. Ambroise Paré, who was considered a great battlefield surgeon, doubted that gunpowder was poisonous. Paré based his belief on his experiments with gunpowder that proved it did not cause infection. Paré proceeded to prepare a digestive cure that included pouring a mixture of egg yokes, turpentine, and oil of roses into the gunshot wound. It proved to be an effective treatment, since many soldiers recovered without infection or pain.

Reconstructive Surgery (plastic or cosmetic surgery)

In the past reconstructive surgery was usually performed to correct a deformity or repair an injury. More recently, it has become associated with cosmetic enhancement or alteration of some part of the body, which is purely elective. From the time of ancient medicine until the 20th century this type of surgery was not referred to as "plastic" surgery. Rather, the techniques for reconstruction were described as far back as the ancient Egyptian physicians. Today, the term plastic surgery refers to the molding or reshaping of tissue, as one would reshape plastic— hence the name plastic surgery. The earliest account of reconstructive surgery is found in 6th century B.C.E. Hindu writings. They describe how skin from a person's cheek was used to reconstruct a new nose. The practice was fairly commonplace, since the punishment for adultery was cutting off the nose of the guilty party. Unfortunately, this procedure also left a scar on the cheek. Later in the 1st century C.E., the Greek physician Celsus described surgery to alter or repair ears, lips, and noses, and there are indications that reconstructive surgery may have been practiced in China and Egypt as well.

During the Middle Ages and Renaissance, physicians and surgeons debated the practice of reconstructive surgery while theologians debated whether the practice of altering one's physical features by surgery went against the will of God. One concern was whether a fully detached nose could be reattached. Two notable reconstructive surgeons are:

- **Brancas** (date unknown). In the late 15th century, the Brancas, a father and son living in Italy, successfully reconstructed a man's nose, thus improving on the ancient Hindu technique. The Brancas cut a flap of skin from the man's arm to shape a nose and applied it to the facial stump. To be successful, the nose stump was scraped, exposing a fresh surface. If the nose had just been cut off, a surgical cut on the arm was attached to the face and the arm was bound in that position until they grew together, whereupon a section of the skin from the arm was cut off and reshaped as a nose. The surgeons understood that the skin would be nourished by the blood vessels of the nose, and ultimately nasal passages could be formed.

- **Gasparo Tagliacozzi** (1546–1599) was a well-known Italian surgeon who specialized in rhinoplasty, that is, reconstruction surgery of the nose, first introduced in ancient Greece and Egypt and later by the Brancas. Unlike many of his colleagues, Tagliacozzi described his technique in detail. He began by making two parallel cuts in the upper arm of the patient. Then he lifted the skin flap and placed a piece of linen cloth under the skin cuts. When the wound healed, he then cut across one end of the two parallel lines to create a flap. After he renewed the surface of the stump of the nose, like the Brancas, he sewed the flap to the nose and bound the arm and nose together until the flesh grew together. In addition to the loss of a nose as punishment for adulterous behavior, the spread of syphilis in an individual often caused an infection of the bone and the nasal cartilage, resulting in a "syphilitic nose." This condition could also be repaired by reconstructive surgery. Tagliacozzi was aware of the psychological as well as medical needs of reconstructive surgery patients, and he understood the importance of facial appearance. Reconstructive surgery for aesthetic purposes was not generally accepted during the Renaissance, and reconstructive surgeons were often demonized for their "meddling with the work of God." Although plastic or reconstructive surgery is a common practice today, there are some who continue to raise objections based on moral issues.

Pharmaceuticals

Humans have always used drugs, in one form or another, as part of rituals or as medications. During the Middle Ages and Renaissance the pseudoscience of alchemy was practiced by most physicians. The term alchemy is derived from either a combination of the Arab words *al* (for an object) and *kimia*, which is thought to mean *chem*, or *khem*, an ancient term for Egypt; or it may have originated in Greece as the word *chyma*, meaning "to melt and cast metal." Chemicals, herbs, drugs, and minerals have been used to treat illnesses for thousands of years in many countries. The ill-defined philosophers' stone, in addition to providing the secret for transmutations, that is, changing base metals into gold or silver, was also required to prepare the elusive *elixir vitae* (elixir of life) which was believed to be the long sought-after cure for all illnesses, as well as immortality. A variety of substances were dissolved in acids, distilled, coagulated, and solidified by a series of processes used by chemists over the centuries. Often, if the end product did not result in the production of gold, the optimistic alchemists claimed it was an elixir with therapeutic or medicinal value.

• **Paracelsus** (1493–1541), full name Philippus Aureolus Theophrastus Bombast von Hohenheim, was a physician of the Renaissance who is credited with establishing the important relationship between chemistry and medicine. He renamed himself "Paracelsus" after the ancient Aulus Cornelius Celsus (fl. 1st century B.C.E.), an encyclopedist whose book *De medicina* consisted of eight volumes covering all fields of medicine and health. Paracelsus added "Para-" to indicate he was beyond or surpassed the learning of the original Celsus. Paracelsus was not a surgeon. Rather, he was oriented to the chemical aspects of medicine as promoted by alchemy. There are two aspects or orientations to alchemy—one is chemistry, the other medicine—and both relate to the "philosophers' stone." (Note: Only the medical orientation of alchemy will be included in this chapter. See Chapter 6 for the chemical aspects of alchemy.)

In some respects, Paracelsus lived up to one of his middle names, Bombast (although at the time that was not its meaning), primarily because his methods and relationships with other physicians were somewhat bombastic. Today, he might be referred to as an irascible misanthrope or curmudgeon. He was considered much worse by most of his medical colleagues. As a young man he traveled extensively throughout Europe, England, Russia, Egypt, Arabia, and the Holy Land. He received his baccalaureate degree in medicine from the University of

Vienna in 1510 and his medical doctor's degree from the University of Ferrara in 1516. His life's ambition was to reform medicine as practiced in those days, a goal that did not endear him to other physicians. He rejected the accepted precept that the planets and stars (astrology) controlled all aspects of the human body, including diseases. As a professor of medicine, Paracelsus burned a number of the works of Galen and Avicenna and accused other professors of incompetence, as well as overcharging for medical services while not curing their patients. He also challenged tradition by teaching his students in the vernacular (German) rather than in Latin, which was the required language of instruction in medical schools. These and other unconventional traits made many enemies. Although his approach to medicine was unorthodox, it was the first real attempt to understand and combine the treatment of specific diseases with specific drugs. He also claimed he could learn as much about cures from "old wives' tales" and "rough-speaking people" of the land, such as gypsies, travelers, outlaws, and hardworking peasants, as he could from university professors. Paracelsus asserted that during his 10 years of travel, he discovered diseases that were heretofore unknown and that he learned how minerals and chemicals could be administered in the form of medication. Thus, with each disease, he experimented and developed specific drugs and chemicals with which to treat them. He used sulfur, lead, arsenic, iron, copper sulfate, potassium sulfate, as well as the then accepted and extensively used mercury, to treat diseases. In 1530 Paracelsus described a treatment for syphilis that was potentially more successful than what was prescribed at the time. This involved orally ingesting small amounts of mercury compounds. His theory was that "what makes a man ill may also be used to cure him." Hence, he administered doses containing small amounts of the substance believed to be the cause of the illness in the belief that it would also prove to be its cure. He is often credited with establishing the alternative medical practice known as *homeopathy*. Some evidence exists that Paracelsus, as well as other physicians who prescribed mercury, in fact, poisoned many of their patients. He was also famous for his extensive use of opium in the form of what was then called "laudanum." (If it didn't cure the patients, laudanum at least made them feel better.) Paracelsus was also the first to produce and experiment with ether by anesthetizing animals. (Ether is made by the distillation of ethyl alcohol with sulfuric acid.) Paracelsus was the first to diagnose goiters and cretinism, and he developed cures for a number of metabolic diseases, such as gout. He studied and treated the diseases of miners, mainly silicosis and poisoning by heavy metals such as arsenic, lead, mercury, and antimony. He also insisted that "miners' disease" was not punishment for their sins inflicted on them by the "mountain spirits." Rather, it was caused by the inhalation of metal dust and vapors that were present in the mines. He successfully cured a few notable people who were his sponsors, but after their deaths, he was driven from the university. By 1528 he had accumulated enemies among the local doctors, apothecaries, and city officials and was forced to flee from the university for his life. He spent the next years treating miners in the Alpine region of Tyrol where he consolidated his writings. His main publication was *Der Grossen Wundartzney* (The Wounded Man), published in 1536. This book led to his resurgence and renewed reputation as a great physician. Still, many of his writings, as well as his chemical system of medicine, were unappreciated until after his death. Paracelsus was proficient in alchemy, the chemistry of his day, and the field in which his greatest contributions to medicine were made. He was neither an anatomist nor surgeon, but he is credited with introducing iatrochemistry, which is the use of chemicals for medicinal purposes—that is, chemotherapy. History also recognizes him for his concept that after symptoms have been identified and related to a specific disease, specific medication(s) can then be prepared and administered to affect a cure, in other words, the pharmaceutical approach to medicine.

Circulation of the Blood

Ancients were aware that blood traveled through the bodies of animals, and when blood no longer flowed, death occurred. They were also aware that the beating heart and heaving lungs were connected with this circulation, since the cessation of the heartbeat and/or breathing meant that death was imminent. Aristotle (384–322 B.C.E.) believed the heart was the center or seat of the soul of humans, and he assumed that blood was manufactured in the heart, which he described as consisting of two separate chambers (ventricles). He was also aware of a connection between the heart and lungs, but was incorrect in describing how the lungs mix blood and air. He was the first to ponder the question of how the blood passed from one ventricle to the other since a wall of tissue separated them. He postulated the theory accepted for generations that blood seeped through invisible pores through the septum (wall) between the two ventricles. Galen incorrectly believed that blood was made in the liver from food that was passed from the stomach and from there was passed by a system of blood vessels to all parts of the body. One of Galen's 200 factual errors or misconceptions was that blood circulation was based on three systems: the liver, heart, and brain. Each of these imparted three natural spirits (natural, vital, and animal) into the body by three channels: the veins, arteries, and nerves. Unfortunately, both Aristotle's and Galen's misconceptions were accepted for centuries.

There were a number of physicians who tackled the problem of how the blood circulates throughout the human body; some theories were more accurate than others.

- **Ibn an-Nafis** (ca. 1210–1288), an Arab physician who lived in Cairo, Egypt, came close to an accurate explanation for the circulation of blood. He stated that the blood, after being replenished in the right ventricle, rises in a vein to the lungs so that its volume can be expanded and mixed with air to clear it of impurities. The cleansed blood is then returned from the lungs to the left ventricle of the heart to be sent throughout the body in the venous system. Ibn an-Nafis still believed Galen's misconception that blood passed through the body in an ebb-and-flow pattern produced by the heart.

- **Michael Servetus** (1511–1553) was a Spanish physician who was also known as Miguel Serveto. Ibn an-Nafis of Damascus was the first in the East to correctly describe the flow of blood in the human body, while in the West it was Servetus. In his book *Christianism: restitutio*, Servetus proposed that blood circulated from the right side of the heart to the lungs and then back to the left side. This was not as Galen decreed. He stated that it passed through tiny, invisible holes in the septum, the dividing wall between the two sides of the heart. Servetus was correct, but he mistakenly believed that the human soul was comprised of human blood. He also incorrectly insisted that there were two different types of blood—one type of blood was formed in the liver and was carried by the veins to nourish all the parts of the body, and the other type of blood obtained "vital spirits" as it was formed in the heart and then surged through the arteries. Unfortunately, his book, published in 1553, contained as much religious criticism as it did medical knowledge. Servetus challenged the concept of and the word *trinity* since he said it was not included in the original Bible. Calvinists, a group founded by the religious reformer and French theologian, John Calvin, burned Servetus at the stake for his sacrilege in the same year that his book was published.

- **Giambattista Canano** (1515–1578) contended that valves caused delays in the flow of blood from the heart. He claimed that such a valve system was necessary to prevent blood from collecting in the body's extremities before returning to the heart. He was correct in that heart valves control both the volume and rate of blood flow pumped by the muscular action of the heart.

- **Matteo Realdo Colombo** (ca. 1516–1559) was an Italian anatomist and student of the famous anatomist Andreas Vesalius. Among other medical discoveries, he is credited with demonstrating that blood from the lungs returns to the heart via the pulmonary vein. He published a book in 1559 on human anatomy titled *De Re Anatomica* (On the Anatomy), in which he described correctly how blood was carried from the left ventricle to the lungs by means of the pulmonary artery, and after mixing with air it is returned by the pulmonary vein to the left ventricle of the heart where it is then pumped throughout the body.

- **Andrea Cesalpino** (1519–1603) anticipated Harvey's correct description of blood circulation. He proposed that there was an outflow of circulation from the heart through arteries, throughout the body, and the blood was returned back to the heart through veins. He demonstrated this by repeating an experiment formerly conducted by Francesco de la Reina by tying a band tightly around the lower arm, thus depressing the veins and observing that the blood backed up above the constriction. De la Reina also correctly discerned that the veins are near the surface, while arteries are deeper in the flesh in the arms and legs.

- **William Harvey** (1578–1657). Although Harvey is considered a post-Renaissance physician, his work on the circulation of blood draws upon the idea, concepts, theories, and writings of many scientists and physicians before his time. His most important book, *De Motu Cordis et Sanguinis in Animalibus* (On the Motion of the Heart and Blood in Animals), was published in 1628. This book explains Harvey's conclusion that if the blood did not travel throughout the body and return to the heart and lungs to be recycled, huge amounts of blood would need to be produced continuously by the liver in order to maintain a constant flow—and sooner or later, where would all that continuously produced blood be stored? Using ligatures on different body parts, he demonstrated that blood is pumped from the heart into arteries, which begin to fill (expand) during the heart's systolic contraction. Conversely, venous blood filled the heart during the diastolic action. This led to Harvey's famous statement that "the heart is no more than a pump with valves" that sends blood through the arteries and returns it to the heart through the veins.

Fallopian Tubes

- **Gabriele Fallopius** (Falloppio) (1523–1562) held the chair of anatomy at the University of Padua that was formerly held by Vesalius. Fallopius's careful and accurate dissections of both male and female cadavers led him to question much of Galen's writings, as well as his predecessor's conclusions in Vesalius's book *De Fabrica*. Vesalius did not generally have access to female cadavers, and thus his observations and conclusions were not only incomplete but also inaccurate. Unlike his predecessor, Fallopius dissected both male and female cadavers and was the first to accurately describe many details of the female reproductive system. He correctly described the end of the ducts connecting the uterus to the region of each of the ovaries as resembling the bell-shaped end of a trumpet. As a result, this organ, now known as the fallopian tubes, was named after him. In addition, he not only correctly identified, but named the vagina, placenta, clitoris, ovaries, and hymen as parts of the female reproductive system.

- **Bartolomeo Eustachio** (1524–1574) discovered the eustachian tube and described and named the parts of the inner ear known as the cochlea, labyrinth, semicircular canals, stapes, and ossicle. Eustachio also made contributions to the understanding of the nervous and vascular systems and along with Fallopius identified the female reproductive organs.

Contraception

It is believed that for many thousands of years humans made no connection between coitus (the sex act) and pregnancy—babies just appeared every year or so, as did offspring of other animals. By the time humans recognized and understood this biological relationship, they also developed methods to prevent pregnancy and thus the birth of more children than could be adequately raised to adulthood within the community. Two of the oldest forms of contraception (prevention of conception) are *coitus interruptus* (withdrawing before ejaculation) and *coitus reservatus* (suppressing an ejaculation). The Old Testament of the Bible contains a passage relating to Onan, who performed coitus interruptus to avoid impregnating his brother's widow. Over the ages and into the Renaissance, women were privy to any number of methods that supposedly guaranteed the prevention of pregnancy. However, most of these methods were in the form of old wives' tales.

- **Jakobus Sylvius** (1478–1555), a famous and well-respected teacher and physician who lived in the 14th century, claimed that if the union was boring and without love, there would be no children, while a passionate coupling would be fruitful. Given the biological principles at work, it was not so remarkable that some of these methods of contraception were successful, since the female menstrual cycle limits the time period of fertility, that is, the ability to conceive.

During the Renaissance a number of elixirs, salves, lotions, and other potions were prescribed for both males and females to prevent fertility, including contraceptive sponges, made from linen, called *preservatifs*. Throughout the Renaissance the Roman Church opposed any form of contraception that interfered with procreation. Even so, coitus interruptus and *reservatus* were practiced (and still are), despite the belief that they were violations of nature.

- **Gabriele Fallopius** (1523–1562) claimed to have invented the sheath (condom), but this is questionable since a variety of types were in use for hundreds of years. The term *condom* was not yet in use during the Renaissance, and its derivation is speculative. There are records that indicate the ancient Egyptians first proposed the use of a sheath-type condom about 3,000 years ago, and Roman soldiers used sheaths made from sheep's intestines while they were garrisoned in foreign lands during the Empire's many wars. During the Middle Ages, as in ancient times, males wore a sheath that covered the penis not for the purpose of contraception but to protect them from insect-borne and other diseases, as well as establishing a certain social status. Fallopius described the sheath as being made of a small linen cloth designed to fit the penis and said that it should be used to prevent syphilis. He claimed it worked best as a prophylactic, if moistened, to prevent sexually transmitted disease. The Chinese developed sheaths made from oiled silk paper. Slaughterhouses used sausage skins to fashion sheaths, while the Japanese developed a hard sheath that not only prevented contraception but also aided the impotent. However, it was not until Charles Goodyear discovered how to vulcanize rubber that his company produced the first modern latex condoms in 1844.

  With the invention of the microscope (1600 or 1603), sperm, ova, and bacteria became visible, along with a clearer understanding of fertilization and the causes of diseases. This knowledge made the use of condoms a universal possibility. Most world health officials consider the lack of availability of condoms, or even if they are available, the resistance by some males to use condoms as a prophylactic against the transmission of diseases such as gonorrhea, syphilis, and HIV/AIDS, as a major contributing factor to the increase of these diseases, particularly among the poor and uneducated.

Conclusion

Medicine is an ancient and prehistoric art. Ancient medical practices were based on common observations of nature, trial and error, and spiritual intervention. The spread of Islam throughout western Asia, northern Africa, and into southern Spain during the 7th to 12th centuries was known as the "Islamic Golden Age." While most of Europe was still in the Dark Ages, Damascus and Spain were Muslim centers of knowledge acquired from translations of classical Greek literature, poetry, science, and medicine into Arabic. Great libraries were built to store these thousands of volumes, and while many valuable books were destroyed by invading Visigoths and Christian Crusaders from the north during the 11th and 12th centuries, other great books continued to be translated into Latin. These translations from Greek into Arabic, then from Arabic into Latin, and finally printed using local vernacular languages were the impetus of the European scientific renaissance of the 14th to 16th centuries.

Modern medicine is still an art based on the observations of patients by medical practitioners. But more importantly, the art of medicine is also based on the knowledge gained from scientific and biomedical research. Medicine today owes its existence to the thousands of men and women of the past who struggled to both understand diseases and to cure patients before causations and effective treatments were known. The invention of the microscope (along with other modern technical medical instruments) was one of the first and most important inventions that provided the spark needed for the profession to progress beyond medicine practiced in the Middle Ages and Renaissance to the Age of Enlightenment.

CHAPTER 5

MATHEMATICS

Background and History

Soon after the development of localized formal languages, prehistoric humans were no doubt able to express either verbally or symbolically concepts related to "how many or how few," "how big or how small," "how far or how near," or "how long" (time). There is, however, a difference between the languages of humans and the language of mathematics. Spoken and written languages are descriptive of the *sorts* of things, while mathematics is essentially a planned language expressing *size* and *amounts* of things that may be represented as symbols or letters and has, over the ages, become a universal language, unlike spoken and written vernacular languages. Intuitively, ancient humans of all civilizations established relationships of magnitude (quantities) between actual objects and symbols. Conceivably, by displaying the number of fingers a person could express small numbers, but larger quantities may have required special words or signs. Since periodic and regular movements of objects in the sky have been observed from the beginning of human existence, special words and markings on objects were the symbolic representations of the celestial cycles. For example, a tally stick made from a wolf's bone that is about 750,000 years old contains 55 deliberately carved sharp nicks. The first 25 markings are arranged in groups of fives, followed by one long notch that seems to indicate the end of that series, and then continued with additional 30 marks. These cross-marks at regular intervals could have been related to the cycles of the moon, but more important, it infers that they understood the concept of "groups" or "sets."

In the late 20th century, archaeologists discovered two chunks of ochre in the Blombos Cave in South Africa. These 77,000-year-old clay-like artifacts exhibited unusual geometric markings indicating that humans made them, but scientists are not in agreement as to their purpose. As agrarian civilizations developed, so did the need for simple methods of counting in order to determine the best times to plant and harvest. Arithmetic was based on experience, and all civilizations independently developed some form of mathematical expression.

There are several ages of mathematics: the Prehistoric Age (2.5 mya–10,000 B.C.E.); the Age of Early Civilization (10,000–800 B.C.E.); the Egyptian Age (2000 B.C.E.–100 C.E.); the Greek Age (800 B.C.E.–500 C.E.); the Medieval/Arabic Age (500–1300); the European Renaissance Age (1300–1600); the Age of Enlightenment (1600–1800); and the Modern Theoretical Age (1800–present). Mathematics of the Middle Ages and Renaissance was based on the works of ancient Egyptians, Greeks, Hindus, and Chinese, as preserved by Islamic scholars and later translated into Latin. During the Medieval Dark Ages, western and northern Europe were not known for their mathematical scholarship. However, the Muslims who invaded and occupied northern Africa and southern Spain from ~700 to ~1200 or ~1300 C.E. were renowned for their interpretations of Greek and Hindu mathematics while adding innovations of their own. It was during the period from the late 1100s to ~1400 C.E. when books from Muslim libraries were translated from Arabic into Latin and became available throughout Europe. These translations, and the reinvention and spread of the printing press in Europe during the mid-1400s, quickened the pace of the awakening (renaissance) of knowledge, in general, and mathematics and science, in particular. Europeans acquired a number of important ancient mathematical concepts during this period.

(1) About 5,000 years ago the Sumerians of Mesopotamia (present-day Iraq) used cuneiform characters for both writing and to record numbers. Their system was a sexagesimal system (base-60) in contrast to our decimal (base-10) system. (Modern computers use a digital, or base-2, system.) It is assumed that this base-60 system may have influenced the assigning of 360 degrees to the circumference of a circle. Also, for measurements of the Earth's longitude, starting at zero at the prime meridian in Greenwich, England and proceeding westward around the globe, one returns to the starting point after 360 degrees. This ancient base-60 system may also be the origin of our 60-minute hour and each minute consisting of 60 seconds.

(2) Arithmetic and algebra were developed along with the concept of zero by either the Greeks or the Hindus of India. As the Muslims

invaded northern India they adapted these mathematical concepts and then shared these ideas with the Western world as they conquered north Africa and Spain. Islamic algebra was basically Hindu algebra, even though the name *algebra* means "restoration" in Arabic, as in the transferring of negative terms from one side of an equation to the other.

(3) Geometry was imported from Greece and translated into Arabic, and later into Latin. The best record of Greek geometry is Euclid's *Elements,* written in about 300 B.C.E. *Elements* is a complete survey and compilation of older Greek mathematical theorems of which there is no original record. But Euclid's *Elements* has been copied and summarized many times and has been used for the past 2,000 years. It is the first example of the concept known as *rationalism.* In other words, Greek mathematicians asked not only "how" to solve mathematical problems but "why" problems are solved in certain ways. This method of proof might be considered the beginning of the use of logical processes for scientific investigations. In addition to Euclid, there were many famous Greek mathematicians. Thales of Miletus is known as the "father of Greek mathematics." He discovered five geometry theorems: (1) A circle is bisected by its diameter; (2) Angles in a triangle opposite two sides of equal length are equal; (3) Opposite angles formed by intersecting straight lines are equal; (4) The angle inscribed within a semicircle is a right angle; and (5) A triangle is established if its base and the two angles at the base are known. Other famous Greek mathematicians were Pythagoras of Samos, Hippocrates of Chios, Archimedes of Syracuse, Apollonius of Perga, and Hipparchus of Nicaea. The Greeks expanded geometry by their study of figures formed from cutting conic sections. (See Figure 5.1.) These studies led to the invention of algebra used to depict the graphic representation of the different sections.

(4) The concept of pi (π) was another bit of ancient knowledge that found its way to Europe via the westward spread of Islam. The search for the perfect mathematical expression for the concept of π has a long history spanning many civilizations. It is not known when ancient humans first attempted to determine the relationship between the circumference and diameter (or radius) of a circle, but they surely realized that such a relationship existed and that it was the same for all circles of any size. The interest in π most likely developed 4,000 years ago in Meso-potamia and Egypt when it was realized that the larger the circle the greater its circumference (perimeter). The Rhind papyrus, dating from 1650 B.C.E., mentions mathematical data from two centuries earlier, including

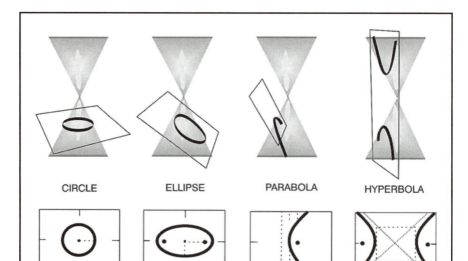

Figure 5.1 Conic Sections
By slicing solid cones at different angles, different plane geometric figures are formed. This form of geometry was a precursor to explaining figures by using algebra.

assignment of a value for π. Knowledge of the value of π was required in geometry to determine the area of a circle and the volume of a sphere. One of the most ingenious methods for approximating the value of π was more or less perfected in ~250 B.C.E. by Archimedes of Syracuse (ca. 287–212 B.C.E.) who proposed his theory of *perfect exhaustion*. He placed polygons with just a few sides inside a circle so that the points of the polygons touched the inner circumference. Another polygon was placed adjacent to the outside perimeter. Using plane geometry and fractions, he compared the areas of the inner and outer polygons and determined that the figure for the outer polygon was slightly larger than that for the inner polygon by a specific amount. To increase the accuracy of his calculation he proceeded to use polygons with up to 96 sides and arrived at ($3\ 10/71 < \pi < 3\ 1/7$), or 3.14163, which is very close to the accurate relationship between the diameter and circumference of any circle. Similar calculations were made in different countries. From the year 2000 B.C.E. to 1946 there were over 30 pre-computer calculations made of π. From 1946 to the year 2000 there were at least 40 computer-based calculations made of π. But since π is an irrational number, it can be calculated ad infinitum, as demonstrated by modern computers that have worked out π to several trillion decimal places.

(5) At least 5,000 years ago the Egyptians developed a hieroglyphic writing technique for both words and numbers. They developed an early measuring system based on the body sizes of the ruling pharaoh—for example, the *cubit* was the length of his forearm, and the width of the palm of the hand was also used as a unit of measurement. However, the Egyptians soon recognized that not everyone's body parts were of equal size, leading to standardizations for length, as well as weights. Early Egyptians used practical forms of geometry long before Euclid summarized this math form. Egyptian mathematics was the first to use a base-10 numbering system (decimal), which was adopted by Muslims and later all of the Western nations. They were also aware of the basic concept of algebra, in which sizes (quantities) can be expressed by letters. As with most ancient civilizations, Egyptian mathematics was of a practical nature, not theoretical or abstract.

(6) About 3,500 years ago the ancient Chinese developed and used the base-10 numbering system, but it is not known how or if their decimal system influenced the spread of the same system in Arabic mathematics. The Chinese used mathematics for very practical purposes, such as for astronomical and astrological calculations based on the lunar month, for measuring distances and quantities of goods, and in the construction of buildings.

(7) There are two uses or concepts for the idea of *zero*. One use is as an empty place indicating a value as the zero's place in the number 5,709. Without the place value of something in the zero's place, the number would be 579, since the 5, 7, and 9 are no longer in the correct spaces. The other use of zero is to represent it as a numeral, as in 0, 1, 2, 3, and so on. The use of zero is particularly important when considering negative numbers. These ideas expressing zero are not natural and someone (or several people) had to envision the concept. Several ancient civilizations indicated the place value for zero with a hook-type symbol. Other systems left a blank space or some other symbol. The first use of the figure 0 as a number is not known, but it may have originated with the Greek word *ouden,* meaning "nothing," or it may have been derived from *obols,* which were Roman coins of little value that were used in counting. There is also evidence that in about 500 C.E., when the abacus was used for calculations, there was no way to determine the placement of a blank space on ancient abacuses—for example, to represent the zero in the number 5,709. An Indian mathematician introduced the idea of using a symbol

that is now called zero (0). A few hundred years later the Arabs adapted this symbol for use as a positional notation. Regardless of who invented zero, the concept has been an indispensable mathematical notation throughout history, particularly with the decimal system.

(8) The ancient Indian (Indus) civilization was an assortment of people from various regions. Even though mathematics was an important part of their religion(s) and life, not much is known about specific Indian mathematicians. About 150 B.C.E. the Jains, a religious sect, developed a form of mathematics that included number theory, arithmetical computation, geometry, fractions, cubic and quadratic equations, ideas about infinity, and an early concept of logarithms.

During the Medieval Period in Europe the study of mathematics was dormant and little in the way of new ideas or theories was developed. During the Renaissance mathematics was mostly recycled from what the Arabic Muslims acquired from the Greeks and Hindus. It was not until the end of the 16th century and into the 17th, 18th, and 19th centuries that mathematics became more than just a tool for science and commerce. It became more theoretical. During the last few hundred years the study of mathematics was established as an academic discipline with specialties. Nevertheless, there were a number of mathematicians during the Middle Ages and Renaissance periods of history who made significant contributions.

Egyptian Mathematicians

Egypt has a long ancient history of mathematics. Some of the well-known mathematicians who lived before the Middle Ages were the following: Ahmes the Moonborn (ca. 1680–1620 B.C.E.), known as the author of the Rhind papyrus, which was the chief source of historical information about Egyptian mathematicians before the year 2000 B.C.E.; Euclid of Alexandria, who is known as the "father of geometry"; Heron of Alexandria (ca. 10–75 C.E.), who applied geometry to mechanics; Ptolemy (ca. 85–ca. 165), an astronomer who based his proposed theory for an Earth-centered universe on mathematics; Diophantus (ca. 200–284 C.E.), known as the "father of algebra"; Apollonius of Perga (ca. 260–ca. 190 B.C.E.), best known for his book *On Conic Sections;* and Hypatia of Alexandria (ca. 370–415), daughter of Theon of Alexandria and one of the very few woman mathematicians during ancient history.

As the Classical Period of Greece declined, many Greek scholars relocated to Egypt—particularly to the great learning center of Alexandria, noted for its magnificent library. Even after the fall of Rome in 476 C.E. Egyptian schools of mathematics still existed. However, by the Middle Ages and Renaissance, Egyptian mathematics was in a state of decline. Despite this, there are a number of well-known Egyptian mathematicians from this period.

- **Abu Kamil Shuja** (ca. 850–ca. 930 C.E.), usually known as "Abu Kamil," developed several important aspects of algebra. His *Book on Algebra* covered three areas: solutions for quadratic equations; applications of algebra to regular polygons; and practical problems, including Diophantine equations, a 3rd-century Greek analytical method of arriving at solutions for algebraic equations. This book was an important link between Arabic algebra and European mathematics. Abu Kamil went beyond the simple x^2 to arrive at higher powers of numbers, up to the power x^8. For example, to express x^5 he used "square square root" (x^2x^2x), for x^6 he used "cube cube" (x^3x^3), and for x^8 "square square square square" ($x^2x^2x^2x^2$). He also developed a rule book for land surveyors based on geometrical calculations for various shapes, both two- and three-dimensional. He provided instructions on how to calculate the areas, perimeters, and diagonals for figures such as squares, rectangles, and triangles, as well as how to determine the volumes and surface areas of a variety of solid figures, including cones and pyramids.
- **Ibn Yunus** (950–1009) was named after his great-grandfather, his grandfather, and his father—thus his name indicated a long lineage. Two Egyptian caliphs supported his scientific and mathematical work and book that included astronomical observations and accurate trigonometric tables that were useful for astronomers. (A caliph was both head of a religious sect and a secular leader of a Muslim state.) He used his knowledge to determine the time of lunar phases that were important for Muslim prayers, and that was used later for calendars. Although Ibn Yunus was interested in astrology, his mathematics was mostly applied to astronomy and date calculations based on his observations. The developments of trigonometric functions in arcs rather than angles, as well as spherical trigonometry, were his most significant contributions.

Indian Mathematicians

Indian scholars are not always recognized in the history of mathematics. Briefly, some of their contributions are the invention of a number system, the use of zero in a place-value decimal system (which is also attributed to Babylonians, Greeks, and Chinese); elaborate descriptions of geometry (mainly for constructing religious altars); original scales of measurement (length as "inch," "foot," and "stride"); arithmetical and

Figure 5.2 Evolution of Arabic Numerals (from Brahmin [Indian] numerals)
Arabic numerals slowly evolved from the 1st to the 11th centuries following the Muslim invasions of northern India.

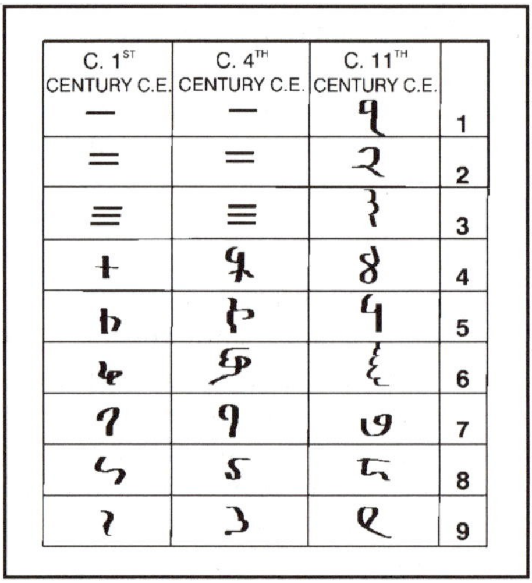

C. 1ST CENTURY C.E.	C. 4TH CENTURY C.E.	C. 11TH CENTURY C.E.	
—	—	৭	1
=	=	২	2
≡	≡	?	3
+	৭ৃ	৪	4
৷	৷৵	ৎ	5
৸	ৡ	ৼ	6
৭	৭	৬	7
৸	৫	৸ৢ	8
৭	৸	ৎ	9

algebraic procedures; procedures for fractions; plane and solid geometry; simple, cubic, and quadratic equations; and concepts of infinity. By about 300 B.C.E. Indian mathematicians used a base-10 numbering system, and by 700 C.E. they no longer used a place name for zero, inventing instead the zero or *sunya* (empty). It was about this time in history when Arabs who invaded northern India adopted the Hindu numbering system. (See Figure 5.2.) Later, the Hindu-Arab numbering system was transported to western Europe where it was translated into Latin. One difference between Hindus and other cultures was that many of the works of Indian mathematicians were either written in, or translated into, local languages, which made the texts available to the general population.

- **Aryabhata I** (476–550 C.E.) was born in Kusumapura (present-day Patna, India). He wrote 118 verses in a book called *Aryabhatiya,* which summarized Hindu mathematics up to his time in history. The book contained 66 rules, but no proofs, followed by 25 verses that dealt with the telling of time and the positions of the planets. The book's last section discussed the nature of spheres and ellipses. The math covered arithmetic, algebra, plane geometry, spherical trigonometry, fractions, quadratic equations, and a table of **sines.** Aryabhata invented his own numbering system based on letters of the Indian alphabet. It consisted of the 33 consonants representing the numerals 1, 2, 3, 4, 5, and so on to 25, followed by 30, 40, 50, and so on to 100. Higher numbers beyond 100 were represented by vowels, followed by consonants for smaller numbers. For instance, 434 is depicted by using the vowel representing 400, followed by the consonants for 3 and 4. Although cumbersome, his system did recognize the need for a place value (i.e., zero) and could be used to write very large numbers. He also investigated integer solutions to equations, for example, $by = ax + c$ and $by = ax - c$, where a, b, and c are all integers.

- **Varahamihira** (ca. 505–ca. 580). His most famous work is *Pancasiddhantika* (Five Treatises, or the Five Astronomical Canons), which detailed native Indian astronomy as well as Western astronomy. His interest in astronomy was mainly related to astrology, that is, casting horoscopes. Varahamihira was associated with the mathematics center at Ujjain where he developed theories for the epicycles of the motions of the sun, moon, and planets. He also gave examples for the use of the zero in the place-value system of decimals. His most important contributions in mathematics were in trigonometry. He developed several formulae using sine and cosine, and he also improved the sine tables prepared by Aryabhata.

- **Brahmagupta** (598–670) wrote a book titled *Brahmasphutasiddhanta* (Opening of the Universe), which included 25 chapters describing mathematical astronomy. The book covers the latitudes and longitudes of the planets; **diurnal** rotation; solar and lunar eclipses; the moon's phases and shadows; and conjunctions of the planets with each other and with the fixed stars. In other chapters, he addressed algebra, the **gnomon,** and spheres and provided some trigonometric tables. Brahmagupta used negative numbers and described how to sum a series. He defined the rules for the use of zero as follows: When adding or subtracting zero to a number, the number does not change, but when multiplying a number by zero, the result is zero. He tried to extend his rules to include dividing a number by zero but failed. Brahmagupta arrived at a method for solving quadratic equations by using continued fractions to find the integral solution of an indeterminate equation, for example, $ax + c = by$. He also solved quadratic indeterminate equations such as $ax^2 + c = y^2$ (or $ax^2 - c = y^2$). In his second book he reviewed much of the work covered in his first book on astronomy, while also developing a unique interpolation formula for the computation of the values of sines.

- **Mahavira** (or Mahaviracharya) (ca. 800–ca. 870) was born in Mysore, India, and as a young man became familiar with the mathematics of southern India. Mahavira wrote an instructional manual based on Brahmagupta's mathematics, designed to teach Brahmagupta's methods during the mid-800s. Unlike most other books by Indian mathematicians that discussed topics other than mathematics, such as astronomy, Mahavira's book was exclusively about mathematics. As a text, it included sections on operations for arithmetic and fractions, the rule of three, how to calculate areas, and use of the place-value number system. He provided methods for squaring numbers and determining the cube root of a number. He also gave a formula that approximated the perimeter and area of an ellipse.

- **Aryabhata II** (ca. 920–ca. 1000) wrote a book that included mathematics useful to astronomers. This type of mathematics was important to most Hindus of the Middle Ages since astronomy was a determinant within their religion. He covered the positions (longitudes) of the planets and their conjunctions with each other and the stars, eclipses of the sun and moon, and the phases and rising and setting of the moon and planets. He also included sections on geometry and algebra along with some practical applications. Aryabhata II provided rules to solve the indeterminate equations $by = ax + c$, when c is both positive and negative, as did many other Hindu mathematicians during this period. Much of the other content of his book was not new, but he did provide a useful sine table that was calculated up to five decimal places.

- **Sridhara** (ca. 870–ca. 930) wrote many books, but his two most important ones were written in verse form. The second one gave useful tables for monetary units and **metrological** terms. He provided instructions for performing arithmetical operations, squaring, cubing, and extracting the cube root—all with natural numbers. Following these instructions he provided rules for operating with fractions and provided problems in determining ratios, calculating interest rates, purchasing goods, how to determine wages, calculating arithmetical and geometric progressions, and a formula for determining the sum of a finite series of numbers. However, his short, clear rules for solving problems were not followed up by many examples, as neither proofs nor correct answers for the solutions of his problems were provided.

- **Sripati** (1019–1066) was interested in mathematics as a tool to understand astronomy better and how to apply mathematics to astrology. One of his major works consisted of 105 verses in which he calculated planetary positions (longitudes), eclipses, and planetary transits. A second volume dealt with arithmetic, algebra, and measurements. He provided rules for algebra; signs for addition, subtraction, multiplication, and division; square and square root; cube and cube root; and how to figure positive and negative amounts. He also developed an important method for solving a quadratic equation, that is, $\sqrt{(x + \sqrt{y})} = \sqrt{[(x + \sqrt{(x^2 - y)})]/2} + \sqrt{[x - \sqrt{(x^2 - y)}]/2}$. Sripati was better known as an astronomer than as a mathematician.

- **Bhaskara** (1114–1185), who is also known as Bhadkarcharya (Bhaskara the Teacher), was born in Vijayapura, India and became the head of the astronomical observatory in Ujjain, which was also a major center for mathematical studies. His book was written in the style of former Hindu mathematician/astronomers and covered standard problems relating to spheres, geography, planetary motions, calculating eclipses, and the nature of the visible heavens. One of his major interests was in the field of trigonometry where he made his most important contribution. For example, the discovery of: $\sin(a + b) = \sin a \cos b + \cos a \sin b$; followed by: $\sin(a - b) = \sin a \cos b - \cos a \sin b$. Bhaskara represents the end of the development of mathematical astronomy in India. It was not until the Renaissance and later that modern mathematics was introduced as a specific discipline.

Arab Mathematicians

Soon after the beginning of the Christian Era, the Arab numbering system was influenced by the numbering systems of several other countries. In time, their numbers evolved into the numerals we use today. (See Figure 5.2.)

- **Abu Ja'far Muhammad ibn Musa al-Khwarizmi** (ca. 780–850 C.E.) was thought to be from central Asia in the area south of the Aral Sea. He brought culture and the first significant library since the time of Alexandria to the court of Harum, who ruled the Islamic Empire from Baghdad. Scholars of the court, including al-Khwarizmi, transcribed Greek scientific manuscripts and studied algebra, geometry, and astronomy. Al-Khwarizmi's book *Hisab al-jabr w'al-muqabala* (Science of Reduction and Confrontation, or Science of Equations) was the first significant book whose subject was algebra. The first to use the word *algebra*, al-Khwarizmi is also credited with introducing the term *algorithms* (the Latin form of the author's name). Since his book was meant to provide practical applications of algebra, he provided six rules for solving quadratic or linear first- and second-degree equations. They are (in modern terminology):

1. Squares are equal to roots.
2. Squares are equal to numbers.
3. Roots are equal to numbers.
4. Squares and roots are equal to numbers ($x^2 + 10x = 39$).
5. Squares and numbers are equal to roots ($x^2 + 21 = 10x$).
6. Roots and numbers are equal to squares ($3x + 4 = x^2$).

He also used the concept of "completion" to remove negatives from equations and "balancing" to reduce positives of the same value on both sides of equations. He used geometrical figures to provide proofs for his methods of solving algebraic equations, which is possible evidence that he was familiar with Euclid's *Elements* (of geometry). There is some indication that his work can be traced to ancient Hebrew texts and possibly some Oriental works. Al-Khwarizmi's works are important since they were the conduit for Indian numerals, Arab and Oriental algebra, and to some extent Greek geometry to arrive in western Europe. Algebra and geometry taught in schools today have some histories of both sources. One of al-Khwarizmi's surviving works is a Latin translation, *Algorithmi de numbero Indorum*, which describes algorithms. In addition, it described the all-important Hindu number system, placement of decimals, and the concept of zero.

- **Al-Kindi** (ca. 801–873) was raised and educated in the city of Kufah, the 9th-century cultural and learning center. He was appointed to the House of Wisdom, the outstanding center of learning located in Baghdad (present-day Iraq), and was assigned the task, along with other scholars, of translating Greek scientific manuscripts into Arabic. He wrote several manuscripts related to Indian numbers, their harmony, multiplication of numbers, number quantities and measurements, and procedures for cancellations of numbers. He also wrote on space and time, which he considered infinite. Al-Kindi described the geometry of parallel lines and pairs of lines in a plane that are nonparallel. He was interested in optics, but as with many other scientist/mathematicians of this period he confused light with vision. He also commented on Archimedes' methods of measuring circles.

- **Al-Battānī** (ca. 850–929) was born into the Sabian religious sect of star worshippers. He and his father are best known in the field of astronomy as the makers of astronomical instruments. Al-Battānī was the first to use trigonometric methods rather than the Ptolemaic geometric procedures to measure and catalogue almost 500 stars. Al-Battānī was also the first to give the trigonometric formula of a right-angle triangle as such: $b \sin(A) = a \sin(90° — A)$.

- **Al-Uqlidisi** (ca. 920–980) traveled widely to learn as much mathematics from others and their writings as possible. Along the way he became a teacher and wrote several important books on mathematics. The one on arithmetic is related to the Hindu system of numbers. It is divided into four parts. The first section begins with an introduction to Hindu numbers, the decimal place value, addition, multiplication, integers, and fractions in both decimal and sexagesimal notations. The second section converts older mathematical systems into the Hindu system of numbers. He also describes the method of "casting out nines." In the next section, al-Uqlidisi provides answers to all the possible questions that his students might have. And in the last part, he suggests improvements in the use of the Indian "dust board" (the "dust board" was similar to modern-day chalk- or blackboards). For example, some progressive steps during calculations were lost when using the dust board, so al-Uqlidisi described how to solve this problem by using pen and paper for calculations.

- **Abu'l-Wafa** (940–998) was born during a new period in Muslim history when Islamic rule had already moved to western Iran and Iraq. The caliph of that time supported mathematics and astronomy, as well as efforts of Abu'l-Wafa and other scientists to build an observatory. One of Abu'l-Wafa's famous books describes the mathematics (arithmetic) required for business people. For instance, mathematics for the tradesmen at this time used both the Indian symbols and "finger-figuring" methods for everyday measurements and calculations. Abu'l-Wafa's book used only the latter method, using words instead of numerals and describing how simple calculations can be done mentally. This businessmen's book's title is *Kitab fi ma yahtaj ilayh al-kuttab wa'l-ummal min 'ilm al-hisab* (Book on What is Necessary from the Science of Arithmetic for Scribes and Businessmen).

  Abu'l-Wafa wrote another book on the practical uses of geometry, titled *A Book on Those Geometric Constructions That Are Necessary for a Craftsman*. It included 13 chapters explaining designing; drafting; construction of right angles; angle trisections; forming parabolas, polygons, and circles; and division of all types of figures, including polygons and spherical surfaces. Abu'l-Wafa is best known for his book *Theories of the Moon,* in which he was the first to use tangent functions as he compiled tables of sines and tangents at 15′ intervals, and sec and cosec related to his studies to determine the orbit of the moon.

- **Ibn Sina** (Latin name: Avicenna) (980–1037) received his education from his father, the leader of a small village. Even so, Ibn Sina claimed he was mostly self-educated. He became the physician for and cured a few important people. He was caught up in these unsettled times and wandered over much of the Arab world, and for a short period he became a court physician. Ibn Sina considered mathematics to be a wide, general field of study. He wrote two famous books. The first, *Kitab al-shifa* (The Book of Healing), was an encyclopedia of natural science, logic, and psychology. One section of the book covers mathematics that he divided into four parts: geometry, astronomy, arithmetic, and music. (He considered music a form of mathematics.) He made further divisions into **geodesy,** statistics, **kinematics, hydrostatics,** and optics. He also divided arithmetic into algebra, addition, and subtraction, and he included mechanics under mathematics. The second book was a history of medicine. Ibn Sina is also known as a philosopher, whose theory of knowledge is related to the concept that a person perceives the abstraction of an object rather than the object itself—something like new-age relativism. He considered mathematics and science as the causation of nature and the way to find the truth.

- **Abu Rayhan al-Biruni** (973–1048) was born on the shores of the Red Sea in an area now known as Karakalpakstan. Although he is most famous as an astronomer, he developed several important concepts in mathematics that he used to correct errors in the astronomical observations and calculations of former astronomers, including Ptolemy. He wrote several hundred books, including one about his contributions to the field of mathematics. It contains sections on practical arithmetic; the summation of series of numbers; geometry; descriptions of Archimedes' theorems; how to solve algebraic equations and trisect angles; descriptions of conic sections and stereo-projections; trigonometry; and a sine theorem for planes as well as spherical triangles. Al-Biruni also made contributions to the study of cartography, geodesy, and geography. He used triangulation to measure distances on the surface of the Earth and calculated the Earth's radius at 6339.6 kilometers, which is quite accurate.

- **Alhazen** (ca. 965 or 975–1038 or 1040 c.e.), known as "The Second Ptolemy," was born in Iraq and moved to Cairo, Egypt, as a young man under the sponsorship of the Caliph al-Hakim, who was interested in astronomy as well as other sciences. Alhazen provided the ruler of Egypt with knowledge of hydraulics and claimed he could control the floods of the Nile River. Unable to do so, he was almost executed for his failure. Alhazen made contributions to the fields of optics, developed a theory of vision and light and rules for lenses, and constructed a camera obscura. (See Chapter 7.) While he studied geometry and number theory, he did not contribute much to the field of algebra. He was one of the first to propose experimental scientific research. He wrote 96 books, half of which were in mathematics. His works influenced Renaissance men such as Roger Bacon, Johannes Kepler, René Descartes, and Christian Huygens, as well as others in the 17th and 18th centuries. Alhazen developed the first theory relating to solid angles leading to integrals, which was a very advanced form of mathematics using projections. Based on the work of Euclid and other Greeks, he developed a theory of conic sections and their use to form new types of figures. His work with conic curves led to the introduction of continuous movement into geometry, which developed into modern space-based geometry.

- **Al-Khalili** (1320–1380) was a mathematician whose most important work was applying geometry to the sphere in order to solve religious problems. One of the major religious problems was how to determine the exact location of Mecca (in Saudi Arabia) for the orientation of Islamic prayers from different points on the Earth's curved surface. Al-Khalili used the table from former astronomers to prepare more accurate versions for the terrestrial coordinates of Damascus. His tables related to time and the sun's position required for regulating Muslim prayers contained over 1300 entries. His main mathematical contribution was a method of solving spherical triangles using what is now known as the cosine procedure. The difficult task of solving the problem of latitude and longitude (direction) of Mecca requires the use of spherical trigonometry. Al-Khalili's tables, which were the most accurate for his time, helped solve that problem.

- **Ulugh Beg** (1394–1449) was born in Persia (present-day Iran) soon after his grandfather, sometimes known as Timur the Lame, invaded northern India and conquered Delhi in 1399. Ulugh Beg was brought up in his grandfather's court and traveled extensively with him. At age 16 Ulugh Beg became ruler of Samarkand, a city in present-day Uzbekistan in west-central Asia. Unlike his father and grandfather, he was more interested in making the region a cultural center than he was in politics and war. His primary interests were mathematics and astronomy, as well as history, poetry, and the arts. He developed a method of arriving at accurate approximate solutions for cubic equations, studied binomial theorems, and developed a formula for spherical trigonometry. Ulugh Beg produced accurate tables of sines and tangents at $1°$ intervals that were correct to eight decimal places. He also made the first extensive star catalog (*Catalogue of Stars*) since the one produced by Ptolemy. The data he collected from his observatory was used to estimate the length of the year as 365 days, 5 hours, 49 minutes, and 15 seconds, which is very accurate. However, Ulugh Beg was an ineffectual leader who was killed on orders issued by his own son.

- **Al-Umawi** (ca. 1400–1489) was most likely born in Andalusia, in southern Spain, after the "Golden Age of Islam" in western Europe. There were differences between the mathematics (as well as other areas) of Eastern and Western Muslims. At the time it was thought that Eastern Muslim scholars were superior to those in Spain, yet much of al-Umawi's work was superior to that of the East. His main text, *Marasim al-intisab fi'ilm al-hisab* (On Arithmetical Rules and Procedures), begins by describing basic rules for arithmetic and procedures for addition and multiplication. He follows with rules for calculating lengths and area; lengths of chords and areas of circles; areas of segments of circles, triangles, quadrilaterals; and the volumes of cones, prisms, and spheres. One unique and original aspect of his works is the rule indicating how to calculate a number n to be a square:

n must either end in 0, 1, 4, 5, 6, or 9;
if n ends in 6, the tens place is odd, otherwise the tens place is even;
if n ends in 5, then the tens place must be 2;
n must leave a remainder of 0, 1, 2, or 4 after division by 7;
n must leave a remainder of 0, 1, or 4 after division by 8;
n must leave a remainder of 0, 1, 4, or 7 after division by 9.

Figure 5.3 Ancient Chinese and
Arabic Numerals (both decimal
based)
Although both Chinese and Arabic
numbers evolved separately over the
centuries, both were base-10 decimal
systems.

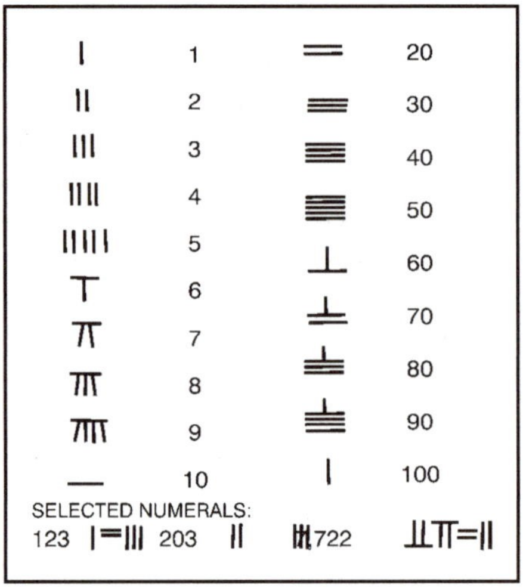

I	1	=	20
II	2	≡	30
III	3	≣	40
IIII	4	≣	50
IIIII	5	⊥	60
T	6	⊥	70
〒	7	≝	80
〓	8	≝	90
〣	9	I	100
—	10		

SELECTED NUMERALS:
123 I=III   203 II   丗722   ⊥〒=II

Chinese Mathematicians

In addition to adopting the Arabic numeral system, the Chinese also used ancient native Chinese characters to represent numbers. (See Figure 5.3.)

Both methods are base-10 systems but represent numbers in different ways. (Note: Some historians claim that the decimal place-value system originated in China and spread to India and the Near East along the caravan trade routes, while others claim "zero" originated in India.) For example, the Chinese system writes the number eleven (11) as "ten one," twelve (12) is "ten two," twenty (20) is "two ten," twenty-one (21) is "two ten one," and so on, up to 99. One hundred one (101) becomes "one hundred zero one," and for larger numbers an extra number must proceed the basic number; thus 212 would be "two one two." When a zero occurs in a large number, it must be included as the word, but only once. For example, 2003 would be "two thousand zero three." Higher numbers become more complicated. In addition, simple strokes are used for writing numbers in Chinese characters, but more complex characters are used for larger sums to prevent errors.

As with the Muslim and Hindu worlds, China also was aware of Greek mathematics, such as the Pythagorean theorem; using common fractions; calculating proportions, squares, and cube roots; determining the volumes of cubes, prisms, pyramids, tetrahedrons, cylinders, and cones; solving systems for three or more linear equations simultaneously; and rules for negative and signed numbers.

Ancient Chinese mathematicians understood and used the concept of negative numbers, while European mathematicians did not use this concept until the latter part of the 3rd century c.e. General use of negative numbers in western Europe did not occur until the 16th century c.e. The Chinese used counting boards (see Figure 7.1) to depict negative numbers as black rods and positive numbers as red rods, or if colors were not available, they used slanted rods to represent negative numbers and vertical rods for positive numbers. The Chinese were not confused by leftover numbers and made use of fractions, and later decimal fractions in their calculations.

As with most scientists of the world at the Middle Ages and Renaissance stage of history, many of the Chinese mathematicians were also astronomer/astrologers, physicians, and philosophers. Following are several important Chinese mathematicians of this period:

- **Wang Xiaotong** (fl. 625), an astronomer and mathematician, wrote *Xugu suanjing* (Ancient Mathematics), which contained 22 problems. He is credited with solving cubic equations by generalizing an algorithm for cube root.

- **Chou Kung** (Duke of Chou) (fl. 1100s) is connected to a dialogue style of writing dating to the 6th century b.c.e., that is now used as historical evidence of ancient Chinese mathematics. One entry is a demonstration of the proof (truth) of the Pythagorean theorem for triangles with sides of 3, 4, and 5 units. The proof consisted of diagrammatically stacking triangles one on top of the other. This writing also demonstrates how Chinese mathematics was not as advanced as was Greek mathematics. Even so, knowledge of multiplication and division of fractions, determining the common denominator of fractions, and calculating square root were evident during this period.

- **Li Chih** (Li Yeh) (1192–1279) wrote *Ceyuan haijing* (Sea Mirror of Circle Measurements), which included 179 problems in 12 chapters. All dealt with right triangles and circles inscribed within or outside their circumferences. He also described how to solve geometry problems using algebra.

- **Qin Jiushao** (ca. 1202–ca. 1261) wrote *Shiushu jiuzhang* (Mathematical Treatise in Nine Sections), which included over 80 problems in everyday mathematics along with some solutions for the more difficult problems. One was the treatment of indeterminate simultaneous linear congruencies for up to 10th-degree equations.

- **Chu Shih-Chieh** (or Zhu Shijie) (fl. 1280–1303) wrote two books of some importance. The first, *Suan xue qi meng* (Introduction to Mathematical Studies), gives the rules of signs for both algebraic addition and multiplication. It also explains how to treat negative signs, which was not understood in the West until many years later. His second book, *Siyuan yujian* (Precious Mirror of the Four Elements), described the treatment of simple and higher equations. He described how to sum several finite series, for example, the sum of $n(n+1)(n+2)/6$. He discusses binomial coefficients and was one of the first to use zero (0) as a digit (numeral) rather than as a place value. It is assumed that Chu Shih-Chieh is also the first to provide a diagrammatic depiction of Pascal's triangle and was also the first, at least in China, to develop a formula to achieve the sum of arithmetical progressions.

There were dozens of Chinese scholars who delved into mathematics during the Middle Ages and up to the end of the Renaissance. By the late 1500s Chinese mathematics was in decline and Western mathematics was ascending. This Western influence of mathematics in China continued until the closed-door policy of the country from the 17th to the 19th centuries, imposing self-isolation. It was not until the late 19th and early 20th centuries that Western mathematics was again translated into Chinese, resulting in a modern renaissance of Chinese mathematics.

Mesoamerican Mathematics

There are no records of Mayan mathematicians, but the ancestors of the current people of the Yucatán peninsula of Mexico developed their own complex base-20 (vigesimal) number system and two calendar systems. It is assumed that their number system was based on the fact that humans have a total of 20 digits, that is, 10 fingers and 10 toes. In about 700 C.E. the Mayans introduced zero, which they represented by a shell figure, but they did not use zero for place-holding. Their number system used only three types of symbols: (1) units represented by dots, which may have arisen from the use of pebbles; (2) horizontal lines used for computations, which may have represented small sticks; and (3) several special symbols. For instance, one dot is a single unit; two dots represents the number 2; four dots represents 4; while 5 is represented as a single horizontal line (possibly related to the five fingers). Thus, the line representing 5 is a major unit for larger calculations; 6 would be 1 dot above a single horizontal line; and 9 would be 4 dots above a line. After 20 the system becomes more complicated, particularly for other cultures that use a base-10 system. (See Figure 5.4.)

Mathematics was important for Mayan astronomy and calculations used for the development of their two calendars. The first type of calendar was the Tzolkin, a special religious ritual calendar consisting of

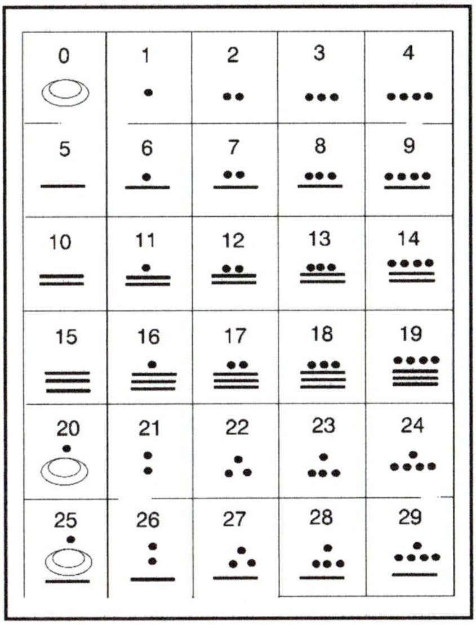

0	1	2	3	4
⬭	•	••	•••	••••
5	6	7	8	9
▬	•	••	•••	••••
10	11	12	13	14
▬▬	•	••	•••	••••
15	16	17	18	19
▬▬▬	•	••	•••	••••
20	21	22	23	24
• ⬭	• •	•• •	••• •	•••• •
25	26	27	28	29
• ⬭ ▬	• •	•• •	••• •	•••• •

Figure 5.4 Mayan Numerals
The Mayan base-10 numeral system was complicated. Even though the Maya used a shell to represent zero, zero did not serve as a placeholder.

only 260 days, divided into 13 periods (months named after 13 gods), each with just 20 days. The second calendar, called the Haab, consisted of 365 days and was designed for the conduct of civil, agricultural, and religious affairs. This everyday calendar was based on 18 months, each with 20 days, plus a short period of 5 days (known as the Wayeb), that was considered a time of bad luck. An example of how they used these calendars is the date found on one of their buildings, which is assumed to represent its year of construction. It reads: 9;8;9;13;0, which, using the Mayan numbering system, translates into $[0 + (13 \times 20) + (9 \times 18 \times 20) + (8 \times 18 \times 20^2) + (9 \times 18 \times 20^3)]$, which equals 1,357,100. This is the number of days between the date for the creation of the Mayan civilization (3113 B.C.E.) and 603 C.E., which is presumed to be the date for the construction of the building. There is no indication that they had a method of multiplication or division for their system. The Mayans developed astronomical instruments and made some accurate measurements of the movements of the moon and planets and star positions. By the end of Renaissance the Mayan civilization no longer existed—presumably for several reasons. One was the Spanish conquest. Other reasons for their demise may be human sacrifices, diseases, and changes in climate.

Before the Spanish conquest during the 1500s, the Aztec civilization included regions of Ecuador, Argentina, and Chile. Unlike the Mayans, the Aztecs had no written language, but similar to the Mayans they had a number system. There are no known Aztec mathematicians as such, but they did use mathematics in simple daily activities, in astronomy, and to develop their calendars. The Aztecs did have a very unique concrete base-10 numbering system, unlike our more abstract numbering system. Their system was a type of storage arrangement for numbers, represented by specific objects. It was called a *quipu* and consisted of knots tied in a vertically held string or cord. For example, for the number 597, seven touching knots were tied at the free end of the string to represent seven units. A space was left between these seven knots and the next group. Above the seven knots from the free end of the string, 9 consecutive knots were tied representing the tens. Further up the string, 5 closely tied knots represented the hundreds. For larger numbers, groups of knots and strings were used, and secondary strings were attached as offshoots to the main strings. For different types of objects, such as cattle, pottery, and baskets of grain, different colored strings were used. After the invasion of the Spanish conquistadors in the mid-1500s, the Aztec civilization, as with the Maya civilization, declined and ceased to exist, although descendants still live in both areas.

European Mathematicians

During the Medieval Period mathematics in western Europe was at a low point. Over the years of the Early Middle Ages, Muslims translated mathematics, as well as other sciences, from the Greeks, Persians, Indians, and Chinese into Arabic. In addition, they improved and added to original mathematical concepts, such as the use of the Indian concept of zero in a positional numeric system; invented Arabic notations for numerals; established algebra; and perfected the use of the Chinese abacus. By the Renaissance, European mathematics blossomed, as did learning in other scholarly areas, as access to Latin translations of the mathematics of the Greeks and other countries became available. Following are a few of the European mathematicians who not only compiled and explained other scholarly works, but contributed to the advancement of the field of mathematics:

- **Alcuin of York** (also Alhwin or Alchoin) (735–804) was born in York, England, into a well-to-do family. He became a monk and was appointed the headmaster of the school he attended. He wrote *Carolingian Codice,* also known as the *Golden Gospels,* in a very special script that was easy to read. This script is credited with the survival of much of the translations of ancient Greek mathematics. He is also credited with developing a pre-Renaissance revival of learning and interest in mathematics with the publication of texts on arithmetic, geometry, and astronomy. His books provided lessons in a question-and-answer format that became popular because it was easy to use and understand.

- **Gerbert of Aurillac** (955–1003) was born in southern France, where he became an excellent student. He was rewarded by studying the *quadrivium*—that is, a curriculum that included four scholarly areas: arithmetic, geometry, music, and astronomy—in the border area of France and Muslim Spain. This was the period in history when the Islamic city of Córdoba (in present-day Spain), with its large library, was the center of learning and thus an excellent place for Europeans of the Middle Ages to acquire an education in mathematics and science. Gerbert was one of the few mathematicians who could calculate figures in his head using Arabic numerals rather than the still used ancient Roman numerals. He constructed a large abacus on the floor of a church, using large disklike pucks instead of rocks or beads. He would sit up in the church loft and direct students to move the disks with sticks as he did the calculations—something like a mathematical game of shuffleboard. When Pope Gregory V died, Gerbert was appointed (not elected) pope. He assumed the name Pope Sylvester II. Since he was a foreigner, he was unpopular and was soon forced to seek exile.

- **Adelard of Bath** (1075–1160) was born in Bath, England. He studied in France and traveled extensively, as far east as Palestine. Fluent in Arabic, Adelard was one of the first to translate Euclid's *Elements* from Arabic to Latin. He accomplished this in two versions. The first was a strict translation of all 15 of Euclid's books, while the second version not only gave the prepositional statements but also how to construct proofs. Over the next hundreds of years many mathematicians used these translations of Euclid's work. Adelard also wrote influential books on arithmetic and geometry.

- **Gerhard of Cremona** (also Gerard or Gherard) (1114–1187) was born and educated in Italy, but moved to Spain to learn Arabic in order to read the *Almagest*, written by Ptolemy. He spent many years of his life translating this book into Latin. (A copy of his original translation is still in existence.) He was the first to translate the Arabic word and concept for *sine* into Latin as "sinus," which is the basis for the current English word *sine* used in mathematics.

- **Leonardo Pisano Fibonacci** (1170–1250) was born in what is now Pisa, Italy, and was educated in North Africa where his father held a diplomatic post. In his popular book *Liber abaci* (Book of the Abacus), published in 1202, he described the new system of Arabic numerals and how they could be used with the abacus to simplify calculations of complex problems. He also wrote books on practical geometry and worked on procedures for calculating roots of various types of equations. In addition, he developed a theory of proportions and designed techniques for finding the roots of equations. In the early 1220s he published *Practica geometriae* (Practical Geometry), which also became popular. In 1225 his most important book, *Liber quadratorum,* (The Book of Squares) explained second-order equations and his concept of the number theory. It included problems for merchants on how to calculate profits and convert between different currencies. Since this was before the days of the printing press, each of his books had to be hand copied. Most of his works were examinations of the books and concepts of former mathematicians, but Fibonacci also made several original contributions. For instance, he discovered (or invented) what is known as the "Fibonacci sequence" of integers, where each number is equal to the sum of the preceding two numbers: 1, 1, 2, 3, 5, 8, 13, 21, and so on.

- **Johannes de Sacrobosco** (John of Holywood) (ca. 1195–1256) was born in England and educated at Oxford. He became a professor of mathematics at the University of Paris, where his teaching spread the knowledge and use of Arabic arithmetic and algebra. His books were easy to read and thus widely used. *De Algorismo* contained 11 chapters, each covering a specific topic, including addition, subtraction, multiplication, division, and square and cube roots. He wrote several books on astronomy that applied mathematics and time for accurately calculating the length of days, weeks, months, and years. Johannes also developed an ecclesiastical calendar and insisted that the Julian calendar was off by 10 days and that this error could be corrected by counting each year as 364 days (instead of 365) for the next 288 years.

- **Roger Bacon** (ca. 1220–1292) was born in England and educated at both Oxford and Paris. His research interest was broad, covering mathematics, optics, alchemy, and astronomy, as well as philosophy. These and other topics, including gunpowder, were described in his major writings, *Opus majus* (Great Works), *Opus secundus* or *Opus minus* (Smaller Works), and *Opus tertius* (Third Works). In his *Opus majus* he made some unusual technological predictions for the 13th century. He predicted that machines would be developed to augment human power, and he foresaw automobiles, airplanes, and mechanically powered ships. He was also one of the first to insist that since the Earth was a sphere it is possible to circumnavigate it with sailing vessels. Bacon's major work in mathematics, *Communia mathematica* (General Principles of Mathematical Science), was never completed but nevertheless was published at a later date. It seems that Bacon was forthcoming with his ideas, of which some religious leaders did not approve. He ended up in prison for several years for teaching suspected "novelties," or new ideas many of which others did not understand.

- **Jordannus Nemorarius** (1225–1260) is credited with developing the formula for the law explaining the mechanical advantage of the inclined plane, one of Archimedes' simple machines. Nemorarius was also one of the first to use letters to represent numbers in simple algebraic calculations. He extended this concept to astronomy by designating the magnitude (brightness) of stars by letters instead of numbers.

- **Levi ben Gerson** (Gersonides) (1288–1344) was born and educated in France. His *Book of Numbers* dealt with arithmetical operations (addition, subtraction, multiplication, and division), permutations and combinations, as well as instructions on how to extract roots. His *On Sines, Chords, and Arcs* explained trigonometry, including the use of sines for the proof of plane triangles. Later, his *De harmonicis numeris* (Concerning the Harmony of Numbers) dealt mainly with descriptions and discussions of the first five of Euclid's books on geometry. He used his knowledge of mathematics to describe geometrical models for the motions of the moon and in his study of solar and lunar eclipses.

- **William of Ockham** (or Occam) (1284 or 1288–1348) received his early education from the Franciscans and later studied theology and philosophy at Oxford, where his unpopular opinions resulted in his leaving the university before completing his degree. He continued his studies in mathematics and logic on his own and today is best known for Ockham's Razor, also known as the "law of parsimony" (or economy). It can be stated in two ways: (a) *"Frustra fit per phura, quod fieri potest per pauciora"* ("It is vain to do with more what can be done with less"), or (b) *"Essentia non sunt multiplicanda praeter necessitatem"* ("Entities should not be multiplied beyond what is necessary"). This concept was devised by Ockham to eliminate many pseudo entities used to explain medieval theological positions. Scientists and mathematicians have interpreted this as meaning that the simplest theory (or hypothesis, or explanation) that fits the observations and facts of the problem should be the one selected. This is good advice for any everyday situation, where it is commonly referred to as "KISS" (Keep It Simple, Stupid).

- **Thomas Bradwardine** (1295–1349) did most of his work and writing on mathematics, logic, and philosophy during his tenure at Oxford in the early 1300s. He wrote books on many subjects over his productive years. In his *De proportionibus velocitatum in motibus* he addresses Aristotle's concepts of bodies in uniform motion and speed. Aristotle proposed that motion was only possible when a force acted on a body that exceeded the resistance to that motion, and when a force stops acting on a body, the resistance causes it to stop moving. (Aristotle, and many others, did not believe that a force could act over a distance and not be in contact with the body—for example, gravity.) Bradwardine used mathematics to show that if the resistance exceeds the force, the body cannot move, but, at the same time, the velocity could be zero. His argument was based on the arithmetical increase ($1 + 2 + 3$, etc.) in velocity as related to the geometric (2 + 4 + 8 + 16, etc.) increase in the ratio of the force to resistance, which was incorrect but solves the former problem. In his book *Speculative Geometry* he includes some basic non-Euclidian geometry. In *On the Continuum,* he accepted Aristotle's idea that "No matter is made up of atoms, since all matter is composed of an infinite number of substances of the same types." However, he still attempted to prove that this theory of atomism was incorrect. Bradwardine also studied "star polygons" and tried to use them to fill spaces by having polyhedra (many-sided figures) touch within the spaces.

- **Nicholas of Cusa** (1401–1464) used his interest in geometry and logic for the study of astronomy and theology. His main contributions in mathematics were his discussion of infinity, both large and small, and his claim that there is no constant motion, only relative motion, for the Earth as well as the rest of the universe. Nicholas of Cusa was one of the first to propose a pre-Einsteinian concept that all things are relative in relation to each other and that there is no center since all is in constant motion. He based this on his geometry of the circumferences of infinitely large and small circles. In his book *On Learned Ignorance* he pointed out that humans were incapable of conceiving of both the absolute and the infinite, which would be required to believe in a ordered cosmos since its complexity is beyond our comprehension. He was one of the first to propose that the Earth moves around the sun, that stars were also suns, and that space was without boundaries but not infinite, since that would equate it with God. Nicholas of Cusa also used Archimedes' age-old application of geometry to "square the circle" as an analogy for humans' search for the truth—one may come close to truth, but never achieve it.

- **Johann Müller** (Regiomontanus) (1436–1476), born in Germany, is considered a Renaissance man for his love of classical learning. He assumed the professorship in astronomy vacated by Purbach, his teacher at the University of Vienna. He later became the astronomer for King Matthias Corvinus of Hungary, where he set up an astronomical observatory. In addition to his scholarly work in astronomy and his lucrative lecturing circuit, he made contributions in the field of trigonometry. In *De Triangulis Omnimodis* he explained a systematic method for solving problems involving triangles and proposed the law of sines. He insisted that by learning the theorems about triangles, a person could understand the wonders of the stars, geometry, and astronomy. Regiomontanus established his own printing press, where he printed *Ephemerides*, a popular astrological almanac. He intended to print all the great treatises of Greek science but never completed the task. Later he gave up secular activities when he was summoned to Rome by Pope Sixtus IV as a consultant on the reformation of the Julian calendar. It is not clear if he was poisoned by his enemies or died of the plague, but he never arrived in Rome.

- **Luca Pacioli** (1445–1517) was born in Italy and educated in mathematics early in life. As a young man he became a tutor in the household of a wealthy Venetian family. After a short stay in Venice, he traveled and taught mathematics in several universities. One of his first famous books, *Summa de Arithmetic, Geometria, Proportioni et Proportionalita* (often referred to as just *Summa*) does not contain much original work. Rather, it is a summary of mathematics as known in his day. Well accepted by the public, it includes examples of arithmetic, algebra, geometry, and trigonometry. Pacioli also addressed the mathematics related to games of chance, including points, probabilities, and odds. Much of this work was inaccurate, but it preceded by many years Blaise Pascal's (1623–1662) more modern theory on probability related to gambling. While teaching in Milan, Pacioli met Leonardo da Vinci and they soon became friends. Pacioli was working on a new book, *Divina Proportione* (Divine Proportions), which included illustrations by Leonardo of regular Platonic solids and other geometric figures. (It is said this was the best-illustrated book in mathematics to be printed at the time.) Both men were interested in the "golden ratio" described in *Divine Proportions* as $a : b = b : (a + b)$, that means the ratio of a to b is equal to the ratio of b to the sum of a and b. This "golden ratio" was aesthetically satisfying both mathematically and artistically, and Pacioli later used the concept in his book on architecture.

- **Johannes Widman** (1462–1498) was born in what is now the Czech Republic and educated at the University of Leipzig, where he also taught mathematics. He is best known for his book on arithmetic that was printed in German in which he was the first to use (in print) the symbols $+$ ($_{\bar{p}}$ più or plus) and $-$ ($_{\bar{M}}$ meno or minus) for his examples.

- **Scipione del Ferro** (also Ferro or dal Ferro) (1465–1526), born in Italy, was not really well known, even though he is credited with inventing a method for solving cubic equations algebraically that was used by many mathematicians since his time. Today's formula for solving cubic equations requires the use of zero, which was not in general use in the early 1500s. A cubic equation is a polynomial equation with no exponent larger than three—specifically, $x^3 + 2x^2 - x - 2 = 0$ $(x-1)(x+1)(x+2)$, which can be further treated by algebraic functions. Ferro's formula involved two cases—$(x^3 + mx = n)$ and $(x^3 = mx + n)$, from which he extracted x^2 from the equations. He also extended Euclid's method of rationalizing fractions whose denominators were reduced from square roots to cube roots.

- **Niccolo Fontana Tartaglia** (Niccolo Fontana) (1499–1557) was born in Venice where, at age 12, his jaw and palate were cut by a saber during an attack by the French that killed his father. He was thought dead but recovered and learned to speak with an impediment and thus was given the nickname "Tartaglia," which means "stammerer." He made his livelihood as a simple teacher of mathematics in Venice.

There is a story relating to a mathematical contest between Tartaglia and the mathematician Gerolamo Cardano (1501–1576). It seems that on his deathbed, Scipione del Ferro revealed his secret method for solving cubic equations to a student whose last name was Fior. Fior considered himself a superior mathematician to his contemporaries and challenged Tartaglia to a contest to solve cubic equations. Each gave to the other 30 problems involving cubics. However, negative numbers were not used. There were several different types of equations, but Scipione del Ferro had only shown his student (Fior) how to solve one type, and Tartaglia had submitted several different types for the contest. Fior's 30 questions for Tartaglia involved the use of *cosa* (meaning "a thing," as used for an unknown in an equation) and cubic-type problems that at first stumped Tartaglia. He was inspired and ended up solving the 30 problems in about two hours, while Fior was unable to complete his set of questions. This seemed to settle the question of who was the superior mathematician, but the story continued with another mathematician, Gerolamo Cardano.

- **Gerolamo Cardano** (Latin and English name: Jerome Cardan) (1501–1576) was born in Milan, Italy. Cardano received his doctorate in medicine from Padua and became professor of mathematics in several universities, where he lectured in medicine, astrology, alchemy, and physics, as well as mathematics. When Cardano learned about the contest between Fior and Tartaglia, he worked on the solutions to these problems on his own. Later, he contacted Tartaglia and told him that he (Cardano) would publish Tartaglia's methods if he would reveal them to him. Tartaglia declined the offer, stating that he intended to publish the solutions on his own. Cardano again asked him to reveal the solution and promised to keep it a secret. Tartaglia again refused. Through politics and the possibility of a better job, Tartaglia left Venice and went to Milan where he met with Cardano. Cardano soon persuaded Tartaglia to reveal his method and again promised to keep it a secret. It was agreed, and Tartaglia wrote a poem revealing the secret, that was only to be known by Cardano. Cardano and his assistant, Ludovico Ferrari (1522–1565), used the method to find proofs for numerous cubic and quadratic equations. Once Cardano learned that it may have been del Ferro, not Tartaglia, who first solved cubic equations, he felt that his oath to secrecy could be broken. Therefore, Cardano published *Artis Magnae sive de Regulis Algebraicis Liber Unus* in 1545 (commonly known as *Ars Magna,* or The Great Skill), which revealed the secret solutions to cubic equations using radicals. Cardano also gave solutions to quadratic equations by radicals without permission to do so by Tartaglia. Even so, Cardano gave credit for the discovery to del Ferro, Tartaglia, and Ferrari in his famous book. As a result of this controversy, the policy was established that the first person to publish the results of an experiment, discovery, or invention, and not necessarily the first person that actually conducted the experiment or made the discovery or invention, is the one given credit. This rule is designed to get new scientific information into the open and available to all rather than kept secret. (See Chapter 7 for information on patents.)

- **Regnier Gemma Frisius** (Regnier Gemma) (1508–1555) was born in the Netherlands. Like many scholars, he adopted the Latin version of his name, thus Gemma Frisius. He studied for a medical degree at the University of Louvain in Belgium, but soon changed to mathematics and astronomy. As a theoretical mathematician, he applied his expertise to astronomy, geography, and cartography. In a revised book on astronomy Frisius indicated that North and South America were two distinctly separate continents. He also included information on geography, cartography, surveying, and navigation, as well as mathematics designed for use by the nonprofessional, common person. Frisius's book also described the mathematical and astronomical instruments made and sold by him—thus the book might be considered as much an advertisement for his instruments as it is a scholarly work. He also made and published information on how to use both terrestrial and celestial globes. He described how to calculate latitudes, longitudes, the meridians, poles, and eclipses, and how to identify the signs of the zodiac. He was aware of the necessity of establishing correct time for determining longitude at sea. Keeping accurate time on board a ship proved a problem for many years until the mid-1700s, when John Harrison (1693–1776) invented an accurate wooden "sea clock" (chronometer). Gemma Frisius's interest turned to medicine, but later he continued his work in mathematics and astronomy, and in selling his instruments.

- **Robert Recorde** (1510–1558) was born in Wales and educated at both Oxford (theology, law, and medicine) and Cambridge, where he received his medical degree. He was appointed surveyor of the mines in Ireland and was charged with the production of silver needed by King Edward VI of England after he established the coinage of English money in silver. The first coin to use Arabic numbers instead of Roman numerals for the dates was the English silver crown of five shillings. As a mathematician, Recorde is known for three elementary textbooks that he believed should be studied step by step in order to become an educated person. Among the first to be printed in the vernacular (English), his books introduced many new words that were substituted for the original Greek and Latin. His English mathematics books are the following: *The Grounde of Arts,* published in 1542, that covers arithmetic; *The Path-*

way to Knowledge, published in 1551, that explains geometry; and *The Castle of Knowledge,* published in 1556, that deals with astronomy. His fourth book, *The Whetstone of Witte* (1557), has an interesting title. Algebraists of the time used the Latin term *cosa* to mean a "thing" that could be substituted for a number. *Cos* translates in English to "whetstone," which is a stone used to sharpen knives and instruments—thus the use of his book could sharpen one's mind. In this book, Recorde is the first to use the equal sign (=). Thus, he might be considered as the inventor who explained that the two parallel line segments in the symbol indicated "no two things can be more equal." Recorde's symbol for equal (=) was not widely used until the 18th century; rather, the symbols (θ), or (*ae*), or the Latin word for equal, *aequalis,* were used.

- **Lodovico Ferrari** (1522–1565) was born in Italy and at the age of 14 lived in Gerolamo Cardano's house. Cardano recognized the potential of this exceptionally brilliant young man and taught him mathematics. By the age of 18 Ferrari was lecturing to the public on geometry. Later, he became involved in the dispute between Cardano and Tartaglia over the discovery of the solution to cubic equations. (See Cardano and Tartaglia.) It seems Ferrari also arrived at a method of solving cubic equations but could not publish his version because it involved using the secret solution developed by Tartaglia. Also, Cardano, Ferrari's mentor, had sworn an oath to Tartaglia not to reveal his solution. It seems Ferrari challenged Tartaglia to a debate that some historians claim Ferrari won, which led him to fame and fortune. He was the tax assessor for the governor of Milan and later became a professor of mathematics.

- **John Dee** (1527–1609), born in London, England, might be considered a true Renaissance man for his wide-ranging interests and influences during this period of history. He studied Greek, Latin, philosophy, geometry, arithmetic, and astronomy at Cambridge University. While still in college, he constructed his own quadrant and cross-staff that he used for astronomical observations. He continued his studies on the continent before returning to England. The publisher of a first edition of Euclid's *Elements* asked Dee, the best known mathematician in England, to write the preface, which explained the benefits of learning mathematics. Dee believed that mathematics was not only practical and useful for mechanical purposes, but also necessary for the study of all aspects of nature. Dee lived during the time when there was great conflict between the Protestants and Catholics in England. The Catholic Queen Mary imprisoned Dee's father, and upon release he was deprived of all of his assets, which affected Dee's ability to continue his studies. Dee was also arrested and charged with being a "calculator," meaning that he was suspected of using mathematics for magical purposes. He was later released, but as with many learned people during the Renaissance, he was judged guilty, since most of the population believed not only in magic, but also in astrology and alchemy. Dee stated that "...alchemy is the lord of all sciences and the end of all speculation." After Queen Mary's death, the Protestant Queen Elizabeth I became the monarch of England. Dee became the Queen's confidante, teaching her simple mathematics and casting horoscopes for her court. Dee also inspired the queen to explore new worlds and expand her empire, an effort that became the worldwide British Empire over the next several hundred years. Dee lived with his mother and received no funds from the Court to expand his library collection, which included books on a vast array of topics. He also accumulated an assortment of clocks, astronomical instruments, and a globe given to him by Mercator. His book *Propaedeumata Aphoristica*, which he presented to the queen, contains writings on mathematics, physics, astrology, magic, and alchemy. He proposed an 11-day revision of the calendar to the queen to bring it into synchronization with the astronomical year. It was rejected because it was too similar to the proposed calendar by the former Catholic regime. Even so, the rest of Europe at this time adopted the Gregorian calendar proposed by Pope Gregory XIII. One of his reasons for proposing a revised calendar was that he believed that mathematics and science would reveal the mechanisms for synchronizing the human body and spirit and explain the strangeness of human existence. His application of mathematics to physics in some ways predated Newton's laws of motion and gravity. Dee claimed that every object in the universe attracts every other object, which is the concept of gravity. But his statement that objects of different weights fall at the same rate is correct. Dee's wife and several of his children died of the plague, and Dee died a pauper several years later.

- **Thomas Digges** (1546–1595) was born in Kent, England. His father, Leonard Digges, was also a mathematician who specialized in surveying. After his father's early death at age 39, Thomas, at 13, became a student of John Dee. They were friends (Dee was a quasi father to Digges) throughout their lives. Digges's first publication was the completion of his father's book on surveying and included a section on using lenses to view distant objects. Digges is known for his application of mathematics, particularly trigonometric theorems, which he used to determine the parallax and positions of stars. Although Digges expressed his theories in mystical terms, he was one of the first to use mathematics when making his observations to expand the known universe to infinity. He wrote a book for soldiers that explained the ballistics of artillery and strength of fortifications mathematically. Both Dee and Digges influenced future mathematicians, scientists, astronomers, and philosophers, particularly the mystic philosopher Giordano Bruno, who was the first to distinguish between the solar system and universe.

- **François Viète** (1540–1603) was born in France and grew up in an age of political turmoil during the reigns of Charles IX and Henry IV of France. Although he was not a professional mathematician, he was a popular lecturer on mathematics and served Henry IV. He made several contributions to the field of algebra during his career. His major work on mathematics was never published, but several of his smaller manuscripts were preserved. They related several former theories about the planets to geometry. His *Canon Mathematicus* was an introduction to trigonometry and included trigonometric tables useful in the field of astronomy. One of his most important books, *In Artem Analyticam Isagoge*, introduced algebra in a way that differed from Arabic mathematics. He is credited as the first to establish rules for algebraic notations, such as using vowels for unknown quantities, and consonants for knowns (constant quantities). Although his system was not accepted in this form, it may have influenced René Descartes, who used letters at the end of the alphabet (x, y, z) for unknown quantities and letters from the beginning of the alphabet (a, b, c) for knowns (constant quantities). Basically, this is the system of algebraic notations used today. For example, in an equation similar to $ax = b - c$, it is assumed which letters represent the knowns (a, b, c) and which letter represents the unknown (x) to be solved in the equation—it's just an algebraic given.

- **Simon Stevin** (1548–1620) was born in the region of Flanders, that is now located in Belgium. A Dutch mathematician, he is called the "father of hydrostatics." (Hydrostatics is the study of liquids at rest with the forces exerted on them, e.g., gravity.) Hydrostatic equilibrium is the relation between the pressure and the geometric height and when stability is established. In his book *De Beghinselen der Weeghconst* (The Elements of the Art of Weighing), Stevin proposed his theory of the triangle of forces. He imagined a triangular surface with a base level with the surface of the Earth. He then imagined 14 identical balls connected at evenly spaced intervals on an endless chain that is draped over the triangle (see Figure 5.5).

  The idea was that the chain would remain motionless because the balls were in balance (equilibrium). Otherwise, if they moved around the triangle, perpetual motion would exist, which Stevin considered physically impossible. He incorrectly concluded from his work on stability that objects of different weights fall at the same rate. Stevin published a total of 11 books in which he contributed to the studies of trigonometry, algebra, geography, and navigation. In his book *De Thiende* he described in detail the nature and use of decimal fractions, including the use of decimals in coinage, measurements, weights, and time. In this small 36-page book he also investigated using decimal fractions to calculate the areas under curved surfaces. Thus, Stevin, at least

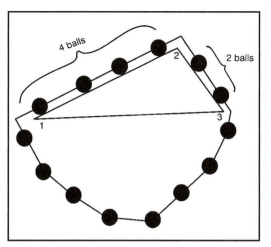

Figure 5.5 Stevin's Equilibrium Demonstration
Simon Stevin imagined a triangle of 14 balls on a giant endless chain that, once moving, would soon reach equilibrium and become motionless. He was considered an expert on hydrostatics and did not believe in perpetual motion.

by some historians, is considered the original inventor of calculus—years before Gottfried Leibniz's and Sir Isaac Newton's independent claims for the invention of differential and integral calculus. In a book on algebra, Stevin recognized the use of negative numbers but not imaginary numbers, and he was one of the first to use the symbol $\sqrt{\ }$ for square root. In *Appendice Algebraique* the method of using successive approximations for solving third-degree equations was printed for the first time. *Havenvinding* dealt with navigation, in which Stevin explained how to use a compass in order to find ports of call. Stevin published other books on trigonometry, geography, cosmography, perspective, business, and accounting.

- **Bartholomeo Pitiscus** (1561–1613), born in what is now Poland, studied Calvinistic theology and later became an instructor and court preacher to Frederick IV of Palatine on the Rhine. Pitiscus was a strong influence in Calvinist policies and hostilities toward the Catholic Church. His major mathematical work, printed in 1595, contained five volumes on plane and spherical trigonometry. His books were the first to use in print the term *trigonometry,* and his trigonometric tables were carried out to 15 places.

- **Adriaan van Roomen** (Adriannus Romanus) (1561–1615) was born in Belgium. After studying medicine and mathematics, he taught both fields as a professor at Louvain, Belgium. One of his famous mathematical feats was solving π to 16 decimal points using the old Archimedes technique with two 30-sided polygons. He also solved a 45-degree equation that was also worked out by his friend François Viète: $x^{45} - 45x^{43} + 945x^{41} - 12,300x^{39} + \ldots - 3,795x^3 + 45x = A$. Van Roomen used a special case that employed a complex sequencing of three combined square roots resulting in regular polygons. Viète gave van Roomen a special problem to solve. It involved three touching circles (known as the Apollonian Problem), which van Roomen solved using hyperbolas (a plane curve intersecting a circular cone with a plane parallel to the axis of the cone). Van Roomen also worked on methods of calculating chords and tangents for circles and trigonometric tables.

Conclusion

Mathematics developed over the ages as a human expression to explain the variety of quantitative aspects of nature. Prehistoric humans found ways that expressed the amounts, sizes, and distances of objects or events by simple symbols, gestures, and sounds. By the time of the Middle Ages all civilizations on Earth had devised systems to describe, in abstract terms, the concepts of how many or few, how large or small, or how far or close. Abstract numbering systems were expanded to delineate more intricate arithmetical manipulations such as addition, subtraction, multiplication, and division. It was not long after that more abstract reasoning arrived at geometric configurations, fractional amounts, decimal systems, algebra, calculus, and other higher mathematical notations. There really was no such science as mathematics in the early days. Rather, scientist/philosophers, physicians, alchemists, and other learned men and women not only used mathematics but also expanded this abstract field. The efforts of others related arithmetic to everyday life, thus educating tradesmen and the laboring classes in the use of calculations that would improve their lives. It was not until the Renaissance and beyond that mathematics bloomed as a theoretical science as well as a practical one.

CHAPTER 6

PHYSICS AND CHEMISTRY

Background and History

Physics is most likely the oldest of sciences. It had its beginnings as early humans interacted with their environment. Obviously curious about the matter found around them, and without thinking about it, they utilized energy and force/motion when they discovered fire and made tools to kill animals for food. Nor did humans consider the physics involved when they learned to pry loose a rock from the soil (lever) or build a slope (inclined plane) to reach their caves. Nor did they explain why a dislodged rock rolled down—not up—the hill (gravity), why light and images reflected off calm water (light/optics), or why water froze in cold weather (thermodynamics).

These discoveries about the nature of matter and how matter interacts are well understood now as physical phenomena, that is, the basic principles related to matter and energy. Today, laws of nature are considered the mother and father of all other sciences since, at some level, all matter and natural processes are based on universal physical principles (laws). In other words, basic physical laws are the same no matter where or when they are applied in the universe.

Chemistry, astronomy, geology, and biology all deal with very basic physical fundamentals related to matter and energy/forces and their interactions. Humans have arbitrarily assigned the boundaries and distinctions between the sciences, whereas nature makes no such distinctions between and among its sciences. Chemists are mainly concerned with properties of matter (atoms and molecules) and their interactions, and since these relate to energy (mainly heat, force, mass, and motion), physics is involved. Many biological processes involve intimate physical relationships of both matter and energy required to maintain life.

Early physics was qualitative (descriptive of what it is) in the sense that identity was more important than quantities (how much or how many). Even today the most popular physical sciences (chemistry, earth science, astronomy, etc.) are taught in public schools as *descriptive* rather than *theoretical* sciences. According to some scholars, the brains of early people evolved to the point where humans were capable of conveying information related to small numbers, such as two or three. It was not until many thousands of years later that systems of numerals and simple arithmetic led to the development of geometry by the Greeks, which could be used as descriptors of nature. Peoples of ancient Egypt, Mesopotamia, China, and Mesoamerica, as well as Greece, were ardent observers of the heavens and devised methods of measuring the motions, positions, and regular paths of the planets, comets, and eclipses. Beginning in about 600 B.C.E., and for the next 500 years, ancient Greeks described the physical nature of their environment in varying degrees of sophistication. A few examples: Thales of Miletus provided naturalistic descriptions of the physical universe and proposed that water was the fundamental ingredient of everything. Anaximenes of Miletus, another Greek philosopher, considered air as the original and major substance of the universe. Heraclitus of Ephesus taught that all physical things are in constant change due to tension between opposites. Anaxagoras of Clazomenae believed that all matter was infinitely divisible and yet maintained its own characteristics upon division, and that everything has some part of every other thing within it. Empedocles of Acragas originated the idea that all matter consists of just four elements—earth, fire, air, and water; this concept persisted for many centuries. He was also the first to demonstrate that air has weight. These and other early philosophers were originators of what, over the centuries, developed into the sciences of physics and chemistry. Euclid's compilation of all then known geometry, Archimedes' mathematical descriptions of simple machines, the use of "exhaustion" to determine pi (π), and the description of density and buoyancy are all examples of physics based on mathematics during the years before the Early Middle Ages.

Aristotle's physics, particularly his concepts of motion, were derived philosophically rather than by experimentation or even critical observations. Greek science, including physics, influenced others for almost 2,000 years. Physics, as we know it, was practically nonexistent in Europe during the Medieval Period (the Middle Ages, from about 500 to about 1300 or 1400 C.E.). Rather, the study of nature was based on philosophical and metaphysical speculations. People thought about and speculated on how things should be and how the world should work—experimentation, or investigating and measuring natural phenomena, was not considered a requirement of science, particularly the physical sciences. And to question was often seen as an affront to one kind of god or another and thus discouraged. By the Late Middle Ages a few exceptions to natural philosophy existed in the fields of astronomy, mathematics, motion, optics, and simple mechanics.

Then, during the Renaissance (~1300–1600), printing presses brought Latin translations of Arabic texts, based on classical Greek texts, to the Western world. These developments helped stimulate the concept of empirical experimentation and began to resemble what we know as physics and chemistry. Many years passed before the accepted philosophical and metaphysical explanations of nature were supplanted by the exploration of nature using several intellectual and procedural processes that became known as the "scientific method." And, it was not until the Age of Enlightenment following the Renaissance that, through experimentation, several universal physical laws that determined the interactions of atoms and forces were formulated. This period between the 17th and 19th centuries is known as the Age of Classical Physics. However, it was not until the 20th century that modern physics proposed that matter and energy are equivalent with quantum characteristics. There is no doubt that the 21st century will be a continuation and expansion of the age of theoretical and applied sciences.

Scholasticism

Even though scientific inquiry was restricted during the Middle Ages, there were a few original and rational thinkers. Most of these, at least in Europe, were monks who were responsible for the Age of Scholasticism. Scholasticism existed in the Medieval Period as a theological and philosophical school of thought founded on the works of Aristotle and other ancients under the authority of the Church. In other words, the best argument to explain something was to rely on accepted authority—and during the Medieval Period the highest authority was the Bible and the leaders of the Christian Church. In essence, scholasticism is based on revelation and faith, not objective reasoning. The monk scholars questioned concepts of natural philosophy and applied logic when examining theological problems, but almost always within religious boundaries of the authorities. St. Thomas Aquinas, Albertus Magnus, and Roger Bacon were three of the most famous scholastics of this period.

Although accomplishments in the field of physics in Europe during the Middle Ages were limited, interest in the fields of astronomy, medicine, and mechanics (the behavior of physical systems under forces, e.g., statics, equilibrium, and motion) grew during the Renaissance. Natural philosophy, which was broad enough to cover several areas of speculation, was the nearest concept to physics. Near the end of the Middle Ages many English mathematicians and scientists tried to apply geometry and algebra to Aristotle's "impetus theory" of motion. They explained that a falling body increased its speed because at any moment new speed was added to its motion resulting from the impetus (cause). Thus, the speed of the moving body could be measured even if the impetus could not. They extended this theory to moving bodies in the heavens.

One of the factors inhibiting the growth of objective science during the Medieval Period was the need to interpret and adjust scientific theories and philosophies to fit within the religious cant of the day. Mechanical theory that began with Aristotle and Archimedes remained dormant during the Middle Ages. Later, scholasticism was rejected during the Renaissance, and mathematical mechanics then became considered a science. This reawakening of early concepts of dynamics, statics, and motion (impetus) began at the beginning of the 15th century when scientists began reexamining mechanical concepts. Even so, a lack of understanding of impetus as related to force, momentum, and inertia remained until Sir Isaac Newton established the laws of motion in the 17th century.

Mechanics and Motion

Mechanics is the study of how physical systems interact with the forces of nature. Ancient mechanics might be considered simple mathematical physics based on the dynamics of simple and complex machines. The ancient Greeks considered mechanics as a balance between something static (unmoving) and dynamic (moving). This might be thought of as a type of equilibrium. Aristotle's concepts of motion might be stated as the following: (1) Heavy objects fall faster than light ones, and when falling their speed increases as related to their weights. (2) A falling object's speed is proportional to the density of the medium through which it is falling. Thus, a vacuum is impossible, since an object would fall at infinite speed. Aristotle was also responsible for another misconception of motion that existed for many centuries, namely, that action (force) at a distance was impossible. He taught that for an object to move, it had to be physically pushed by something in order to overcome inherent resistance of whatever it came in contact with (i.e., inertia). This later became know as the "impetus theory of dynamics." He did not believe that a spirit moved an object, but possibly an unseen force was involved. Thus, when the force was no longer in contact with the moving object, it just stopped moving. Aristotle reasoned that a falling body exhibited "natural" motion, and its position was merely displaced from its previous position as the body's "internal nature" resulted in the body seeking it natural place in the universe—that is, static or at rest. Obviously, experiments involving the throwing of objects horizontally were not considered necessary since everyone knew that an object would just naturally stop and find its own static state. Since no force could be exerted on an object at a distance, an accurate concept of gravity was delayed for many years.

Archimedes of Syracuse was probably the best physicist of the Greek Classical Period leading up to the Middle Ages. Simple machines, such as the lever, were used in prehistoric days, but it was not until Archimedes arrived at his mechanics for levers that it became part of academic physics. His law states, in essence, "The mechanical advantage of a lever is due to the ratio of the weight (load) to the action (effort) required to move the load, which is determined by measuring the distance the effort moves from the central point (fulcrum) divided by the distance the load moves from the central point." He actually used his knowledge of physical mechanics by raising a large ship by himself pushing down on a lever arrangement. Archimedes was also the first to understand the concept of buoyancy and fluid mechanics. He determined the physics of density of matter, which led to the now common formula for determining the density of an object: $d = m/v$ (d = density, m = mass, and v = volume).

It might be helpful to review some current concepts of motion. *Speed* relates to the distance a body moves over a given time, such as, miles per hour (mph). Speed is a *scaler* quality. *Velocity* involves the motion of a body in a specific direction. It is a *vector* quality, where both speed (distance and time) plus direction are stated, and velocity vectors (as with all vectors) can be treated mathematically, while scalers cannot. *Acceleration* is the rate of change in velocity of a body in motion, either slowing down or speeding up. It was a number of years before the importance of these dynamics of motion were understood.

Following are some of the philosopher/scientists who contributed to the growth of the physical sciences related to mechanics and motion during the Middle Ages and Renaissance:

- **William of Ockham** (or Occam) (1281–1347) was the first to question Aristotle's impetus theory of moving bodies. He argued that a moving body need not have anything in contact with it in order for it to keep moving. (He also claimed that planets did not require a group of angels pushing on them in order to keep them in their orbits.) He is best known for his cogent philosophical statement known as Ockham's Razor—"Entities must not be multiplied beyond what is necessary"— which is known in science as the "law of parsimony." (For more on William of Ockham, see Chapter 5.)

- **Jean Buridan** (ca. 1295–1358), a French logician and philosopher who studied under William of Ockham, made several important contributions. One was a more accurate interpretation of Aristotle's impetus theory. Buridan is credited with developing the theory of inertia, which states that once an object is set into motion, it will continue its motion at a uniform speed toward infinity without additional external interfering forces (unless acted on by another force). His insight predated Newton's first law of motion by several generations. Buridan used his impetus theory to explain that falling bodies accelerate as they fall, but he incorrectly stated that the speed that bodies obtain while falling is proportional to the objects' weight. Galileo Galilei (1564–1642) was the first to apply mathematics to the problem by experimenting with rolling stones down inclined planes and timing their rate of descent with his heartbeat and a pendulum device. Buridan also contemplated the decision between alternative courses of action. He developed a parable to describe this dilemma, known as "Buridan's Ass." In the story, an ass was placed equidistant from two equal sized piles of feed. As the animal had equal alternatives, it could not make up its mind and hence starved to death.

- **Leonardo da Vinci** (1452–1519). Although better known for his anatomical illustrations and inventions, da Vinci was interested in mechanics and engineering, particularly as related to the human body. He regarded the human body in terms of a single machine that could be explained by understanding the mechanics of how its parts worked in a holistic manner. Leonardo's limited understanding of mathematics, in some respects, weakened his theoretical basis for mechanics. Nevertheless, he did stress the importance of geometry as the correct foundation for painting, architecture, and the building crafts, as well as mechanics. Leonardo incorrectly used geometry to demonstrate that the speed of a falling body was related to the *space* it transverses, rather than the more correct factor of the *time* elapsed during the fall. Later, Galileo correctly discerned that a constant cause is responsible for a constant effect, and that a rate of change (time and distance) is not a constant value. Leonardo was an observant and insightful engineer who made the most of the forces and resources available to him. In the days before fire was used to produce steam as a source of energy, he devised a number of mechanical devices that converted human, animal, and waterpower to perform useful tasks. As a result of his work with light, Leonardo concluded that every action in nature occurs in the shortest possible manner. Later, Pierre de Fermat (1601–1665) stated the same principle as "Nature always acts by the shortest path." Leonardo had a fascination with water and how it behaves, particularly raging flood-water. His concepts of hydraulics were limited, but he foresaw how water could be used to benefit humans. He designed and drew plans to build a canal from a major river to provide water for farmers to use during the summer months. He also designed flour mills and water engines for driving several types of machines. Historians consider it amazing that Leonardo, who was an excellent engineer and designed so many different types of machines, never really understood the physical concepts involved.

- **Niccolo Tartaglia** (Niccolo Fontana) (1500–1557) was an Italian mathematician with many interests in the physical sciences. (See Chapter 5.) At one time he agreed with Aristotle that forced motion and natural motion were not compatible, but his work with artillery ballistics convinced him otherwise. It was observed that a cannonball increased its speed (accelerated) for a short time and distance after leaving the muzzle of the cannon and then slowed down as distance increased. From this he concluded that impetus was necessary for initial natural motion, which was opposite of Aristotle's theory. In the face of the fact that the trajectory of a cannonball was not a straight line, but rather curved toward the Earth's surface, Tartaglia still believed that the trajectory of the cannonball was a straight line. However, he did concede that gravity continuously exerted a slight pull on the projectile, which caused it to deviate from the straight line. His main problem was that he established a goal or end point for an accelerating body, which was also attributed to the impetus's terminal point.

- **Simon Stevin** (1548–1620), a Flemish mathematician, is also known for his contributions in the field of mechanics, as described in three of his books, *Principles of Statics, Application of Statics,* and *Principles of Hydrostatics.* Printed in Dutch, the three books used the same woodcut to print the title page, which included the inclined plane surrounded by a chain, that Stevin used to demonstrate the law of equilibrium (see Figure 5.5). In addition to inventing this proof, he also developed a proof for the equilibrium for levers with unequal lengths, the conditions for parallelograms of force, and other aspects of force that later led to the mathematical-mechanical representation of force by vectors. Stevin is also credited with arriving at the law of hydrostatic pressure as related to sailing ships. Sometimes referred to as Pascal's hydrostatic paradox, it is the proposition that any amount of water (large or small) can counterbalance any weight however great or small. Much of Stevin's work on hydraulics was an extension of Archimedes' work and a preview for some of Galileo's accomplishments.

• **Galileo Galilei** (1564–1642), best known for his work on astronomy, was also a mathematician and physicist. He provided the mathematics for moving bodies that described the physics of Copernicus' heliocentric universe. Galileo advanced the field of mechanics and motion beyond the knowledge of craftsmen to become specialized fields of physics. He was also one of the first scientists to base his conclusions on experimentation. During the Renaissance he wrote several books on the new science of "local motion," in which he explored the ancient concept of "impetus." Galileo, most likely, was aware of the theories of impetus as expressed by the Greek philosopher Aristotle and the Parisian philosopher Jean Buridan, who rejected Aristotle's theory that agitating air behind an object (for instance, a stone thrown into the air) caused it to move. Buridan contended that the moving force was within the stone and if a force pushed the stone, the stone would absorb the force and thus move faster until air resistance or something else slowed or stopped its movement. Another important concept which predated Newton, was that under ideal conditions, the stone once moving would continue to move in a straight line at a uniform speed for an infinite distance, until some other force interfered with its motion. Newton included this as his first law of motion (inertia). It was Galileo who first realized that the main function of a force was not to make something move but rather to produce a change of the object's motion once started. Neither Galileo, Buridan, nor Newton gave explanations of why these phenomena of motion (i.e., inertia, momentum, and acceleration) existed.

Galileo was the first to study motion by experimenting with "falling bodies." Since rocks or weights fall too fast from a tall building to be timed with clocks then available, Galileo slowed the action down by rolling balls down an inclined plane as he timed their acceleration with his heartbeat and crude pendulum clocks to determine their rate of descent. This great experiment in physics determined that the speed which bodies fall is independent of their weight. Contrary to popular belief, Galileo did not perform the experiment by dropping different weights from the Leaning Tower of Pisa. It is more likely that two years before Galileo's experiment, Simon Stevin dropped two pieces of lead of different weights from a height of 30 feet. It is said that when they landed, they sounded as a single impact. From his experimental work Galileo devised the classical physical law for falling bodies as such: $s = 1/2\ at^2$ (s = speed, a = acceleration, and t = time). This can also be expressed as the rate of acceleration of a falling mass in a vacuum as 32 feet per second squared. Although he conducted this experiment in the late 1500s or early 1600s, Galileo did not publish his conclusions until he wrote *Discourses on Two New Sciences* in 1638. Galileo came close to understanding and describing the concept of gravity, but that physical law had to wait until a later century. His experiments were an application of Stevin's theories, which led him to realize that movement, and in particular the concept of impetus, was descriptive (qualitative) of the motion of a body, and that motion was really a quantitative (measurable) vector force that could be ascertained. In his book *On Mechanics* Galileo analyzed the physics of five simple machines: lever, inclined plane, screw, windlass, and pulley. He was the first to explain that simple machines do not create work, but rather alter the way in which work is applied. Even today the physical concept of work is not generally understood. It is defined as ($W = f \times d$): when a force moves a body (mass) over a distance work is accomplished. Galileo considered both the inputs and outputs for simple machines in terms of force, power, distance, and speed.

Galileo's interest in oscillation motion involves a story (possibly a legend) of his fascination with the chandelier swaying in the breeze in the church he attended in the town of Pisa while a young man. He noticed that the arc or swing varied as the breeze became stronger or weaker, and that he could time the swings by using his pulse as a timing device. He discovered that the frequency of the pendulum (oscillations, or the number of swings in a given period of time) was dependent on the length of the rope or support holding the bob of the pendulum and not the length of the swing. He confirmed his theory that a constant frequency of a pendulum is inversely proportional to the length of its string by constructing and then measuring pendulums supported with different lengths of strings and different weights of bobs. He also realized that air resistance interfered with the constant rate of the swing and thus reduced the size of the arc until the bob came to rest. Galileo suggested that a pendulum could be used to measure the pulse rate of humans for medical purposes. Some years later Christian Huygens (1629–1695) invented the first grandfather clock, using weights and a pendulum to maintain the clock's escapement movement.

- **Johannes Kepler** (1571–1630). Kepler was introduced in Chapter 1 as an astronomer, but he also made contributions to the physics of motion, particularly mathematical harmonics related to the motions of the planets in their orbits around the sun. Kepler, along with Galileo, provided alternative physics for motion from that of Aristotle's "divine impetus." His first law states that the planets describe elliptic orbits around the sun, with the sun as one locus and the other locus being an imaginary point located opposite the sun in the ellipse. The second law asserts that

Figure 6.1 Kepler's Law of Areas This is a depiction of Kepler's second law, which states that a planet revolving around the sun in an elliptical path will cover equal areas of space over equal units of time. The spaces depicted by 1, 2, and 3 are equal in area.

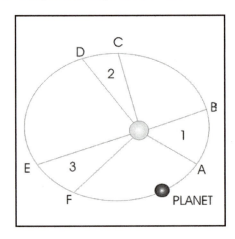

the line joining the sun to a planet sweeps equal areas of space in equal periods of time. (See Figure 6.1.)

Kepler later devised the third law, which states that the squares of the periods of any two planets are proportional to the cubes of their mean distance from the sun. The concepts of motion expressed by Galileo and Kepler were important stepping-stones for Sir Isaac Newton's development of his three laws of motion, published in his 1687 *Philosophiae Naturalis Principia Mathematica* (Mathematical Principles of Natural Philosophy).

The sciences related to mechanics and motion became important by the late 1200s, when gunpowder was introduced to Europe. Engineering and technical inventions became important for defensive weapons and fortifications, as well as offensive weapons and machines, during the Renaissance. As with modern times, past wars were both spurred by and resulted in the development of new theories, technologies, and engineered machines that utilized concepts of mechanics and motion.

Magnetism and Electricity

Magnetism can be created by charged particles, that is, electrons, moving through a conductor (copper wire), resulting in a magnetic field consisting of magnetic lines of force around the wire. Conversely, by cutting the magnetic lines of force in a magnetic field with a conductor (copper wire), a flow of charged electrons are induced in the conductor. (Note: The first example is the principle of the electric motor, while the second is the principle of the dynamo or electric generator.) This relationship between electricity and magnetism is called electromagnetism, which is a major universal physical law of nature. A similar relationship of the spinning Earth with its central core of iron produces a magnetic field that is responsible for the Earth's natural magnetism. Magnetism was known to the ancient Greeks, Romans, Chinese, and most likely others long before the Middle Ages. The mineral known as lodestone—magnetite ore, an oxide of iron—exhibits magnetic properties. Historically, it was well known that a piece of iron stroked with a lodestone could itself exhibit magnetic properties. It is not known who first determined that a lodestone exhibited attraction and repulsion forces and that a magnetized sliver of iron oriented itself with the Earth's North and South Poles, and thus could act as a compass. A small carved lodestone with a groove down its center was found at an Olmec site near Veracruz, Mexico. It is assumed that it was floated on a piece of cork on water, or it may have been suspended on mercury. This permitted free motion, as the stone's groove became oriented with the Earth's magnetic field. This device may have been used as a crude compass to orient buildings and religious temples toward the north. This period in Mexico was more than 1,000 years before the Chinese developed their compass.

Records indicate that the Chinese made an early version of a compass called a *sinan* in the 4th century B.C.E. It consisted of a spoon formed from a chunk of magnetite that was placed on a flat, circular plate of bronze, with markings around the edges. A modern working model of this early compass indicates that the narrower handle of the spoon orients itself toward the south. Other early compasses were constructed by piercing a piece of cork or soft wood with a magnetized needle that was floated on water in a small container. It was not until the 3rd century C.E. that dials and pointers were added to compasses. By the 8th century C.E. the Chinese used these slivers of magnetized iron floating on a piece of cork or soft wood in a small container of water for navigating the Silk Road and other trade routes. In 1070 C.E. Chinese sailors were the first to use magnetic compasses for seagoing navigation. Arabs and Westerners did not adopt simple compasses for navigation until the Late Middle Ages, sometime after 1100 C.E. More accurate and useful devices for navigation were developed in several countries. One innovation was to suspend the magnetized needle on a pivot instead of on water and enclose it in glass to form a more durable dry compass. By the 1500s another important innovation mounted the compass in **gimbal rings,** that act as a universal joint to assure the compass will stay upright and horizontal as the ship rolls in heavy seas. The use of a compass card improved the technique of **dead reckoning** navigation. By the end of the Renaissance a good compass became a standard and essential piece of equipment on all sailing vessels.

The following men experimented with the concept of magnetism and as a result made a number of contributions to the field of physics:

- **Shen Kua** (ca. 1031–ca. 1095 C.E.), a Chinese philosopher/scientist who traveled widely, maintained an interest in many areas of science. He is credited with discovering that the magnetic properties of magnetite (iron ore) could be used to make an accurate compass. This was several years before Europeans used compasses. In his book *Dream Pool Essays,* written in about 1086, he explained that by rubbing the point of an iron needle on a lodestone and suspending it from a single silk thread, it will point south. Conversely, the other end of the needle will point north, but in those days knowing which direction was south was more important. Lodestones are of different shapes but exhibit magnetic polarity just as do bar magnets.

  Chinese sailors used the "south-seeking" needle for many centuries to navigate at night and during cloudy weather long before the Arabs or Europeans used a compass. One reason the Chinese needle may have been successful is because of China's ability to produce high-grade iron that enabled needles to hold their magnetism over long periods of time. During this period of history Europeans produced only low-grade soft iron instruments incapable of retaining their magnetism. Shen Kua is also given credit for first recognizing the declination of a compass needle as it dips slightly from its level position. This was nearly 500 years before Gilbert's famous theory of declination due to the Earth's magnetism. Not much else is known about Shen Kua except he had interests in hydraulics, optics, as well as magnetism, and gave an accurate analysis of the nature of fossils.

  Shen Kua's book *Shi Lin Kuang Ji* (Guide through the Forest of Affairs), written sometime around the late 12th century, described the first dry magnetic compass, that is, one where the needle was not suspended on the surface of water in a container. A freely turning magnetized iron turtle or fish was suspended on a bamboo pivot above a plate that contained markings for directional points.

- **Petrus Peregrinus** (Picard) (ca. 1220–?), a French scientist with an interest in magnetism, described the attraction of opposite magnetic poles and the repulsion of like poles. He designed a compass needle suspended over a circular plate with lines representing directional points (N, S, E, W). A circular diagram with many multiple points connected by lines was called a *windrose.* European sailors used this type of 13th-century chart in conjunction with a compass for navigational purposes. He is also credited with determining that the two poles of a lode-

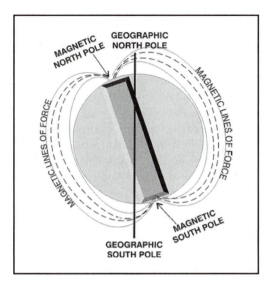

Figure 6.2 Earth's Magnetic Poles
William Gilbert envisioned the Earth's magnetism as resembling a large bar magnet with the magnetic fields projecting from each pole.

stone are the regions of the strongest magnetic force, and that like poles repel between two lodestones while two opposite poles attract each other. Peregrinus was a tutor of Roger Bacon.

- **Christopher Columbus** (1451–1506) and his contributions to exploring and discovery were discussed in Chapter 2. In addition to being a great Renaissance explorer he made two contributions to the study of magnetism. After making landfall on the Canary Islands off the west coast of Africa, he headed west and landed in the Bahamas in October 1492. Columbus used old Ptolemaic maps and figures resulting in his incorrect estimations of the distance from Europe to Asia. His crew lacked confidence in his navigation abilities, mainly for two reasons. In addition to underestimating the size of the Earth, Columbus's compass changed its north point position as he sailed west. Nevertheless, he is credited with realizing that the deviation of the magnetic compass needle from true north varies with the longitude of the ship. It was known for many years that a magnetic compass needle oriented itself with what became known as the magnetic North and South Poles—not the Earth's geographic poles. But Columbus was the first to realize that the Earth's magnetic North Pole was not at the *same location* as the geographic North Pole. Columbus also observed that the vertical declination (extent of dipping) of the magnetic needle changed with latitude. (See Figure 6.2.)

- **William Gilbert** (1544–1603) arrived at two important concepts regarding magnetism and electricity. First, he stated that the amber effect (static electricity) attracts small particles when certain materials are rubbed with different types of cloth, silk, or fur, and that this phenomenon is not the same phenomenon as natural magnetism exhibited by lodestones. Second, Gilbert realized the significance of a magnetized needle of a compass swinging horizontally as it seeks the Earth's poles, while at the same time the needle dips vertically at different angles at different locations on the Earth. He concluded that the needle's declination indicated that the Earth was similar to a giant spinning magnet. This phenomenon had long been observed by sailors as well as Columbus, but it was Gilbert who realized that the angle of declination of the needle varies in degrees with the latitude of its location on the Earth's surface, and at the (magnetic) North Pole, it would point straight down. Even so, he still believed that the Earth, though spinning on its axis, was the stationary center of the universe. He believed that magnetism influenced other bodies in the universe, and thus his idea is equated with the modern concept of gravity. In modern physics a single unit of electromotive force is called the Gilbert (Gb) after William Gilbert.

- **Wang Ch'I** (dates unknown) wrote a book titled *Universal Encyclopedia* in 1609, which described another way to magnetize and demagnetize iron. He discovered that in addition to magnetizing a piece of iron (needle) by rubbing it with a lodestone (a chunk of magnetite iron ore), an iron rod or bar held downward toward the magnetic North Pole and then struck a hard blow to its upper end with a hammer can also magnetize it. Another method involves heating the iron rod red hot and then while pointing it slanted toward magnetic north, squelching it in cold water. A more modern method is to place an iron rod or bar inside a coil of wire and then pass an electric current through the coil. The rod or bar is then magnetized by induction.

Today, the concept of electromagnetism includes electricity, magnetism, and radiation of all frequencies within the electromagnetic spectrum including light. (See Figure 6.3.) All frequencies of radiation that compose the electromagnetic spectrum, along with the interactions of beta decay and the neutrino, form what is known as the Electroweak Interactions, which is a major component of the Standard Model for particle physics, leading to a possible Grand Unification Theory (GUT), which is a single Theory Of Everything (TOE) that combines all the theories of the universe. In other words, electromagnetism is one of the basic attributes of the physical universe.

Optics and Light

Light is electromagnetic radiation of particular frequencies that forms a small but important visible section of the electromagnetic spec-

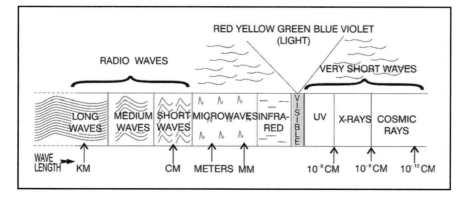

Figure 6.3 Electromagnetic Spectrum
The electromagnetic concept is a major principle of physics. Its spectrum represents radiation of the very shortest, high-frequency wavelengths, to visible light, and proceeding to the very longest, low-frequency types of radiation.

trum, while optics is the scientific study of light and vision. It is easy to assume that humans were always aware of light and the effects it had on their lives. Once they discovered that fire could be used as a tool, archaic humans no doubt used fire as a temporary source of light during nighttime. It is not known when humans first observed that a drop of water on a leaf seemed to enlarge the surface patterns of the leaf, or when the light from a solar eclipse projected through a small gap in the leaves of a tree produced an image of the partially blocked sun on the ground in reverse to the actual solar eclipse in the sky.

In the mid-19th century the British Museum acquired a polished oval-shaped rock crystal that was flat on one side and convex on the other (similar to a plano-convex lens). It was dated to about the 7th to the 9th centuries B.C.E. from the ancient world of the Babylonians. This artifact is assumed to be the oldest example of a magnifying or burning lens on record. It is not effective as a burning lens, so it is assumed craftsmen used it as a magnifying lens to assist them in producing small intricate carvings not visible to the unaided eye, or possibly as an ornament.

In the 4th century B.C.E. Plato claimed that the souls of humans were responsible for illuminating objects by projecting light from the eyes to the objects, which was then reflected back to the eyes as occurs with a mirror. The ancient Greeks believed that light was propagated linearly as a consistent stream of corpuscles (similar to Einstein's photons). They were confused as to the source generating these corpuscles—

whether they originated in the object or the eye. The accepted explanation was that the eye sent out rays that were then reflected back to the eye from the object. This concept persisted for hundreds of years. In about 300 B.C.E. Euclid of Alexandria published a work called *Optics* in which he described light as traveling in a straight line. This was most likely related to his geometry theorem that states that a straight line is the shortest distance between two points. He also described what later became the law of reflection of light from a shiny surface. Despite the fact that his theory of how the light from the image of an object reaches the human eye is incorrect, it was believed for many centuries. Euclid also claimed that rays of some kind were projected from the eye to the object and then were reflected back to the eye. Euclid also studied the connection between the actual sizes of objects to the apparent sizes that were received in the eye. He tried to use geometry to explain how such large objects could project an image into something as small as the eye. Aristotle was among the first to reject the theory that vision was light rays projected by the human eye to the object.

Around the end of the 1st century B.C.E. the Chinese were the first to use optical lenses to correct faulty eyesight. And at about the same time in history, Seneca, a Roman philosopher, observed that objects were magnified when viewed through a globe-shaped glass container of water. Hero (or Heron), of the 1st century C.E., was a great mechanical wizard who invented many devices. In his book *Catoptrica* he used geometry to demonstrate that the angle of reflection is equal to that of the angle of incidence, and that the actual path taken by a ray of light that is reflected by a plane mirror is shorter than any other possible reflected path. He used this knowledge to build and use a surveying instrument that is very similar to the modern **theodolite.**

Claudius Ptolemy, a Greek who lived in Alexandria, Egypt, in the 2nd century C.E., is best known for his theory of a geocentric universe. As an astronomer, he was also interested in the nature of light. He described several optical concepts and experiments in his final major book, *Optics.* He used geometry to demonstrate the nature of reflection, but he went further to demonstrate (prove) that the angle of reflection is equal to the angle of incidence. He also demonstrated that light that passed through water was refracted (apparently bent) from the light's original path. He made a set of tables to show that when light entered water at different angles of incidence, it was refracted at different degrees. (Light is refracted, or bent, when it passes through media of differing densities because light travels at different speeds in media of differing densities.) Ptolemy evidently did not understand why refraction occurred, but just recorded empirical observations for different situations. He continued to promote the belief that vision (light) was sent from the eyes to the object and then returned to the eyes.

These and other pre-Middle Age philosopher/scientists advanced the physics of light and optics by empirical observations, but not on the basis of experimental or theoretical physics. This natural-history approach continued until the Arab world translated Greek science into Arabic, which was ultimately retranslated into Latin by European scholars. Following are some developments in the Middle Ages and Renaissance that build upon the sciences of light and optics established by the ancient civilizations of Greece, the Far East and the Middle East.

- **Yu Chao Lung** (fl. 10th century C.E.) built a special pagoda, that is, a Chinese Buddhist religious tower of several stories, designed to observe images projected through a small hole in the building's wall. He demonstrated that light rays are not only reversed but diverge after passing through the hole.

- **Avicenna** (Ibn Sina) (980–1037) was a well-known Persian Islamic physician and scholar of many subjects, including interpretations of Aristotle and other Greek writers. Avicenna developed his own theory of the perception of light, which contested the ancient Aristotelian concept. He theorized that light is emitted from an illuminated source or reflected from an object and is therefore received by the human eye. He also speculated that the speed of light is finite.

- **Alhazen** (965–1040) wrote a book titled *Opticae thesaurus* (Treasury of Optics), which was published some 500 years after he wrote it. Like others before him, he studied the angles of incidence and reflection of light from plane mirrors. However, he went further. He devised spherical (convex) and parabolic (concave) mirrors. He compared these with the lens in the eye and how the eye must focus objects upside down. The geometry and explanations of how light is reflected and refracted from curved surfaces are much more complicated than light bouncing off plane mirrors. He was reasonably successful in arriving at geometric solutions for these problems. Alhazen's writings are among the first to mention the relationship of vision to the camera obscura. He also noted that the smaller the hole in the camera obscura, the sharper the projected image. Alhazen was one of the first to explore the anatomy of the human eye and to describe how the lens forms an inverted image of the object on the eye's retina.

- **Camera obscura** (Latin: *camera* = room or box, *obscura* = dark) has a long history and is one of the oldest inventions or discoveries relating to optics. The Chinese philosopher Mo Ti, of the 5th century B.C.E., was one of the first to mention a crude device that inverted the image of a light ray passing through a tiny hole in a darkened room. Such an image, although inverted and in color, was accurate in detail. At times the camera obscura was considered somewhat magical, and these rooms were usually off-limits to commoners. Aristotle was one of the first to understand some of the physics of light, such as that it travels in straight lines. He also explained that when these lines of light pass through a small hole, they do not scatter but rather are formed in an upside down image on the wall or panel opposite the hole in the dark room. As mentioned, Alhazen, a 10th-century Muslim scholar, wrote about a camera obscura that he invented. He said that when the image of the sun's partial eclipse occurs, its light can pass through a small hole in one wall of a room and be projected, upside down, on the opposite wall, and that the hole must be very small, or else the image is distorted. Roger Bacon claimed to be the inventor of the camera obscura because he described how he used one to observe a partial eclipse of the sun in the late 1200s. His claim as being the inventor was never accepted in the academic community. In the 15th century Leonardo da Vinci was one of the first to use the camera obscura to focus a sharp image on a piece of paper so that he could trace the image and even color the image as projected. In the 16th century Giovanni Battista della Porta published an account of how to build and use a large, dark room as a camera obscura for visitors making drawings of the outside landscape on the wall opposite a small hole. Some of his visitors, and many others over the next 100 years, saw this reversed image as magic and associated it with the occult and thus wanted nothing to do with it. Once lenses were developed and improved in the 16th and 17th centuries, the camera obscura with a lens substituted for the hole became a standard tool for artists. Over time, they were reduced from room-size boxes to handheld ones, and in time evolved into modern photographic and digital cameras.

- **Abu Rayhan al-Biruni** (973–1048) was a mathematician and an astronomer who corrected many mistakes by former astronomers, including Ptolemy. He also made contributions to geography and physics. He used triangulation to measure distances on the surface of the Earth and determined that the Earth's radius was 6,339.6 kilometers (Earth's actual radius is 6,378 km). In physics he advanced the studies of timekeeping, hydrostatics, motion (acceleration), and he determined the densities of several metals.

- **Shen Kua** was a Chinese philosopher of the late Middle Ages who wrote *Meng ch'i pi t'an* (Dream Pool Essays), in which he mentioned concave mirrors and how the reflected image was formed at what we now know as the "focal point." He also described how the reflected image from such a mirror was inverted and why the image from the camera obscura was also inverted.

- **Robert Grosseteste** (ca. 1168–1253), an Englishman, believed that observations should be used to validate theories. His concept of colored light was one of the first proposed. However, it was wrong. He stated that the intensity of colors is the result of their "strength," and that the colors vary from white (just above dark red) to black (just below deep blue). His idea of what caused the colors of the rainbow was also wrong. He believed that the colors are produced by reflection of the sunlight by layers of what he called a "water cloud" and not by the diffraction of light in individual droplets, as is the current theory. He also believed in the Greek theory that seeing an object involves the eye sending out rays to the object that are then returned to the eye. He was one of the first to correctly suggest that the Milky Way Galaxy is the diffused light of many, many close stars whose light is combined to appear "milky."

- **Roger Bacon** (1220–1292) was a philosopher/scientist who might be called an early Renaissance man for his contributions to science. He studied at both Oxford and at the university in Paris, where he was considered somewhat of an expert philosopher and alchemist. He believed in magical powers but at the same time professed that religious theology was the essence of knowledge. His interests were **eclectic** and included many areas of physics. His book *Opus Majus* (Great Works) was an encyclopedic presentation of the art and science curriculum of universities of that time. In his *De Multiplications Specierum* and *Perspectiva* he attempted to integrate and describe Greek and Islamic knowledge of optics, the eye, vision, and perspective, including convex lenses, reflection and refraction, and how to make magnifying glasses and, as mentioned, the camera obscura. Bacon studied small objects through his convex lenses and proposed that lenses could be used to improve eyesight. He also was one of the first to apply geometry to the study of optics. In his *Opus minus* (Smaller Work) he wrote about alchemy as well as optics. Bacon extended the work of his Oxford mentor, Robert Grosseteste, on concepts related to light, proposing that the speed of light is finite (thus constant) and suggesting that light is propagated as waves similar to sound waves. He also claimed that rainbows were the result of light reflecting (not refracting) off raindrops, although he could not demonstrate this phenomenon. Bacon is credited as one of the first to consider that speculations about phenomena may be different from how things actually exist in nature. This did not keep him from making many predictions related to how humans will control nature in the future. Most of his predictions were inaccurate, but some were rather prescient, for example, the telescope, the printing press, gunpowder, self-propelled boats, submarines, land vehicles, and flying machines. Although Bacon was neither an accomplished mathematician nor experimentalist, he claimed that both were necessary for the advancement of science and through science man could exert technical control over nature. In *Opus tertius* (Third Works) he presented new approaches to scientific education. Bacon is credited with being among the first to propose experimentation before making judgments about outcomes. Several authors claim that Roger Bacon relied mostly on Arabic science and that the experimental method was of Arabic origin already well known in Europe, even though not many philosopher/scientists of the Middle Ages and Renaissance bothered to set up experiments to validate their observations and theories.

- **Witelo of Silesia** (Latin name: Vitellius) (ca. 1231–ca. 1280) wrote *Perspectiva*, which is divided into 10 books and was based mostly on works of Alhazen. It was the most important work on Greek, Arabic, and medieval optics of its day. (Note: At this time in history the word *perspectiva* referred to the science of optics.) The book influenced other scholars in the fields of geometric optics and the science of vision. *Perspectiva* was used by Kepler to explore the function of the human eye and the process of vision. Witelo was the first to describe how to machine a parabolic mirror from iron and the first to demonstrate that the angle of refraction is not necessarily proportional to the angle of incidence.

- **Albertus Magnus** (ca. 1200–1280), a German philosopher and theologian, was also considered the patron saint of the natural sciences. He speculated that the speed of light was extremely fast but finite. He was the first to report on the darkening of silver nitrate crystals by sunlight, later used as a photographic process. He is more noted for his work in minerals and alchemy.

From the 1300s to the 1600s there were a number of people who speculated and explored the possibilities of using lenses to improve the camera obscura, making use of lenses in spectacles (eyeglasses) to be worn to improve eyesight, and using a convex lens to build microscopes and telescopes to magnify objects. These inventions and innovations will be discussed in the next chapter.

Sound

Almost all living organisms react to sounds, but not all can produce sounds. Some scientists insist that even some plants react to sounds. Sound is the vibrating motion of particles (atoms and molecules) in air or some other medium. (The word *vibration* is a derivation of *vibroare*, meaning "to shake" in Latin.) Thus sound is the disturbance of the particles of air that are transmitted outward from its source. If animate receptors are within the range of the sound waves, the vibrating particles may be received or detected by the eardrums or tympanic membranes and related auditory nerves. The production and reception of sound was a necessary environmental development for the survival of humans as well as for most animals. Both humans and animals used sounds for communication purposes. However, humans evolved the larynx, throat, mouth, and tongue required for speech. Ancient people were aware of the differences between pleasing tones and harsh ones, and enjoyed making and hearing pleasant sounds. Primitive people devised the means of producing pleasant sounds (music) by plucking string instruments, but it is doubtful they were aware of the physics of this phenomenon.

The first recorded information on the investigation of sound was by the Greek philosopher Pythagoras of Samos in the 6th century B.C.E. He experimented with stretched strings of various lengths to determine the sounds made when taut strings were plucked. The types of vibrations that produce music are called *simple harmonic motion,* which much later, in the 17th century, was described as "Hookes Law." Pythagoras discovered that when a stretched string was divided in half the result was a change in a musical octave, which was a pleasing sound of a different pitch (frequency). This led to a tuning system bearing his name, which states that by shortening the length of a string by whole-number ratios, a variation in pitch is produced.

Archytas of Tarentum (ca. 420–ca. 360 B.C.E.), a Pythagorean, was the first to state that when two objects are rapidly struck together, a high note is produced. Conversely, striking objects slowly produces a low pitch.

In the 4th century B.C.E. Aristotle was the first to suggest that sounds were waves propagated in air. His conclusion that air was a necessary medium for sound was based on philosophy, not scientific experimentation. He also incorrectly stated that high-frequency sounds travel faster in air than do sounds of low frequencies, but he correctly reasoned that sound would not travel through a vacuum.

Marcus Vitruvius Pollio, a 1st-century B.C.E. Roman builder, believed that when strings vibrated they not only caused the air around them to move but that the air also vibrated. He concluded that it was this vibrating air that we heard and not the vibrating string itself. He determined that sounds were transmitted differently through different types of materials and used this knowledge to design theaters that were acoustically advanced for their day.

There was not much scientific investigation conducted to understand the physics of sound from the Greek period of classical science until later in the Middle Ages and Renaissance. However, a few philosopher/scientists did make some contributions to the understanding of sound:

- **Anicius Manlius Severinus Boethius** (480–524 C.E.), born into a noble Roman family, eventually became a "Christian Thinker." He was the first to compare a sound wave in air to a wave created when, for instance, a stone is dropped into calm water. Although the two types of waves are not created exactly in the same manner, it was a useful analogy for its time in history. Although Boethius was not a scientist, his main contribution was his knowledge and translations of Greek science into Latin several hundred years before the Arab translations of Greek to Arabic were later retranslated into Latin. He was well known for his scientific translations of Aristotle, Seneca, Plotinus, Augustine, Euclid, and Archimedes. His translation of Pythagoras' work with sound and music provided medieval scientists with what little knowledge was known about the physics of sound.

- **Galileo Galilei** (1564–1642) studied the pendulum's motion in some detail, leading him to time the oscillations of the bob and to compare the length of the string supporting the bob with its length of swing. These studies led to his contemplation of how stretched strings of different lengths would vary in their rate of oscillation (vibrations). One experiment involved scraping a sharp piece of iron over a brass plate. He noted that the faster he dragged the iron over the plate, the higher the pitch of the scraping sound, and conversely the slower he moved the iron over the brass plate, the lower the pitch. He then scraped lines as shallow grooves onto the surface of the brass plate, and when he moved the iron across the grooves he noted that the closer the lines, the higher pitched were the sounds, while the wider spaced scratches produced lower sounds. (This is similar to the grooves in old wax/plastic phonograph records.) Galileo then made some measurements of the scratches and noted a ratio between the rates of vibration and fundamental notes, that is, octaves. Before this time all philosopher/scientists believed that it was only the length of strings that determined a sound's pitch. Galileo's experiments established that it was not just the length of strings, but also the *rate* (how many per second) of vibration, that caused different sound frequencies. He also constructed a device that produced frequencies that could be changed from consonants to dissonant sounds. He determined that consonant frequencies were rhythmic, while dissonant frequencies produced an unpleasant irregular "beat" on the eardrums. Galileo's work on oscillations, vibrations, frequencies, and pitch/tone provided the foundation for future studies and understanding of the physics of sound and music.

Until the 17th century it was believed that air was the medium that transmitted sound waves. In 1650 Otto von Guericke (1602–1686) constructed the first air pump strong enough to create a vacuum in a container. He conducted several experiments with his device and determined that sound does not travel through a vacuum (nor would a vacuum support animal life or combustion). Guericke was also the first to speculate that sound could also travel through water and solids as well as air. Nevertheless, his ideas were not proven experimentally for almost another 100 years.

Alchemy and Chemistry

As mentioned in Chapter 4, historically alchemy had two complementary goals—both leading to perfection. One was to produce the magical *elixir vitae* (elixir of life) that could cure all diseases and possibly achieve immortality. The second, and probably the more important goal, was the transmutation (conversion) of lesser base metals into silver and gold. Both of these objectives were thought to be possible if, first, the philosophers' stone (the essence of all substances) could be found to assist in these **metaphysical** activities. A third goal is sometimes referred to as a spiritual transformation of the practicing alchemist.

This section will consider the second goal—the transformation of one substance to another by changing the basic structure of different elements—a goal sought by alchemists from ancient times and into the Age of Enlightenment.

The *study* of chemistry deals with the composition and properties of matter, while the *practice* of chemistry concerns the interactions between and among different elements and compounds that exhibit endothermic reactions (take in heat) or exothermic reactions (give off heat energy) to form substances (compounds) different from the original constituent matter (elements). According to this definition, long before recorded history ancient humans practiced chemistry, even though they did not understand the physics of chemical reactions. Most likely, humans used fire as a tool to perform chemical reactions—the burning of wood to produce light or heat is a chemical as well as a physical reaction. Fire provided heat and light for their caves and transformed (cooked) food. Other examples of fire's early uses were the making and coloring of pottery and, along with water, modifying other substances. Alchemy has always had a spiritual and mystical foundation that related the transformation of solids into liquids and gases with the transformation of the human body into the soul. This connection between alchemy and the regeneration of the human soul was even stronger when it was observed that some solids sublimate, that is, change from a solid state directly into a vapor without going through the liquid phase. Also, humans have a conscious knowledge of impending death. Therefore, we have developed many rituals and spiritual initiations to address what for humans is an unacceptable but natural phenomenon by practicing mysticisms such as religion and alchemy. During and after the Renaissance, numerous initiate mystical and pseudo-religious orders, such as the Freemasons, Rosicrucians, and Knights Templar, were organized to address this spirituality of existence. This was done similar in manner to the ancient alchemists who claimed that one must first transform oneself through a spiritual awakening before transforming the spirit of matter, for example, common metals into more noble forms.

Following is a partial chronological list of events related to chemistry that occurred before the Middle Ages and Renaissance. These and similar historical events greatly influenced the future of that science.

Before the Christian Era

ca. 25,000: The first artifact of a baked ceramic figurine found in Europe, along with pigments used to color pottery.

ca. 5,000: Copper first smelted in Egypt.

ca. 4,000: The first colored ceramics made in Asia. Silver and gold smelted in Egypt.

ca. 3,900: Lime produced from limestone.

ca. 3,500: The chemical process of fermentation used to produce wine in west-central Asia (present-day Turkistan).

ca. 3,000: First records of the chemical processes used to produce beer and soap, and the alloying of copper with tin to produce bronze.

ca. 1,500: Iron first produced in the Middle East.

ca. 1,475–1,500: First liquor distilled in Asia.

ca. 1,400: First glass produced in Egypt and/or Mesopotamia.

ca. 900–1,000: Dyes produced by the Phoenicians who used alum as a mordant (fixative).

ca. 600: The year considered as the beginning of "scientific" alchemy.

ca. 500–800: Iron smelted in several countries; steel first made in India.

ca. 530: The ancient Greeks proposed the atomic theory.

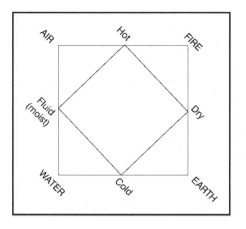

Figure 6.4 Four Elements of Alchemy
Several people over the centuries devised the intersection-box design representing the four elements and conditions basic to alchemy.

ca. 500: The ancient Greeks proposed the four elements of matter: earth (cold and dry), water (cold and wet), air (hot and wet), and fire (hot and dry). (See Figure 6.4.)

The word *alchemy* is derived from a combination of Arabic or Egyptian words—*al*, for an object or thing, plus either *kimia*, that is thought to mean *chem* or *khem*, an ancient word meaning Egypt, or the Greek word *kem* (*al-kemit*, or alchemy), which referred to Egypt's black earth and the black people of the Kemetic civilization who lived in the upper Nile River region of Africa. The word *alchemy* may also have been derived from the Greek word *chyma*, which means to pour or melt as in casting metals. Although alchemy was practiced in several countries as far back as 2,500 years ago, Egypt is considered the "mother earth" of alchemy and the concept of Hermetic mysticism is considered the "father" of the art and science of alchemy. There are archeological records indicating that a number of civilizations produced chemical reactions, including the Mesopotamians (Sumerians, Babylonians, and Assyrians), Greeks and Romans, Chinese, Hindus, West Africans, and Egyptians. Not surprisingly, each civilization had its own particular approach to this most ancient art-like science.

The Chinese used the preparation of gold more for the purpose of securing immortality and/or long life than for commerce. Magic and spirituality were a major part of alchemical practices in China. The story of a Chinese alchemist named Liu Hsiang, who was an imperial advisor to a Chinese emperor in the 1st century B.C.E., claimed he could improve on former magicians' efforts to prepare alchemic gold. The emperor supported Hsiang's efforts, spending considerable funds over a long period of time, but with no success. Liu Hsiang was impeached and sentenced to death, but later was ransomed by his brother. Other Chinese alchemists attempted to transmute metals into gold over several centuries, including Ko-Hung (254–334 C.E.), who claimed that Liu Hsiang failed because he did not fast (abstain from food) long enough and that only a few people should be present when a transmutation takes place. He also claimed that "disbelief will bring failure"—a self-fulfilling prophecy, but also a realistic one. Regardless, Ko-Hung became a famous alchemist who prepared elixirs from plants as well as from metals and minerals (mainly mercury and arsenic) with which he treated patients.

The goal of preparing alchemical gold from so-called base metals has a long history of attempts by both religious and honest men who had a spiritual aspect to their activities. Nevertheless, over many centuries numerous charlatan alchemists claimed to have actually produced alchemical gold. Obviously, none were ever validated.

Later, during the Middle Ages and into the Renaissance, Arabic and European alchemy developed into a nascent science where the efforts of alchemists actually contributed to the new physical science of chemistry of the 17th and 18th centuries. These alchemists discovered several new chemical elements, the nature of matter, and processes to control chemical reactions. Alchemists from 600 B.C.E. to the 1700s C.E. were a mixed lot of professionals, ranging from philosophers, physicians, astronomers, mathematicians, and theologians to scoundrels and quacks who believed in and practiced magic, astrology, and the occult arts, in addition to alchemy. Many claimed to have found the philosophers' stone and to have produced the elixir of life and/or converted lead into gold. None were able to substantiate their claims and some lost their lives because of their deception for not living up to their promises. In the years following the Renaissance—that is, the 1600s and beyond—most people did not correlate magic as espoused by alchemy and astrology with superstition. Rather, magic, alchemy, and astrology were logical systems that employed the imagination to gain some understanding of life and the universe. Even some well-known post-Renaissance scientists, such as Robert Boyle and Sir Isaac Newton, who were devoutly religious, believed in some aspects of alchemy and astrology. Actually it was Antoine-Laurent Lavoisier (1743–1794) who made critical measurements of the weights of ingredients before and after his chemical experiments, and who was instrumental in replacing alchemy with modern chemistry. Lavoisier, Dmitry Mendeleyev, John Dalton, Wolfgang Pauli, Ernest Rutherford, and many others ultimately were responsible for developing chemistry into a viable physical science.

Following are a few of more than 100 known ancient alchemical processes, some of which are currently important for modern chemistry:

1. *Ablution:* The purification of a material by washing with a liquid.
2. *Amalgamation:* The mixing of a metal without using heat to form an amalgam alloy, especially with mercury.
3. *Calcination:* The decomposition of matter by applying heat, or burning.
4. *Circulation:* The purification of a substance by circular distillation, using heat to separate and return condensed matter back to the distillation flask.
5. *Coagulation:* Causing a thin liquid to convert into a thick or solid mass by either cooling or heating.
6. *Combustion:* Raising the temperature of a combustible matter in air (oxygen) until it produces fire, heat, and light.
7. *Composition:* Joining together of two different material substances.
8. *Cupellation:* The joining or conjunction of two opposite substances; the union of something fixed with something volatile
9. *Decrepitating:* Splitting of substances.
10. *Desiccating:* Removal of moisture from a substance, or drying.
11. *Detonation:* Very rapid combustion; explosive burning.
12. *Digestion:* Slowly changing a substance by gently heating it.
13. *Dissociation:* Disintegration (breaking) of something into separate parts.
14. *Distillation:* Using heat, the separation of the components of volatile substances whose vapors are then condensed.
15. *Elaboration:* Separating the pure from the impure.
16. *Elixeration:* Changing substances into an elixir.
17. *Evaporation:* Removal of water by gentle heat.
18. *Extraction:* Purifying a substance by mashing it in alcohol and then separating the liquid.
19. *Fermentation:* Decaying of organic substance resulting in the release of gas (usually CO_2).
20. *Fusion:* Joining of powdery materials to form something new.
21. *Graduation:* Gradually purifying something through stages.
22. *Incineration:* Changing a substance by hot fire.
23. *Incorporation:* Mixing substances into a single mass.

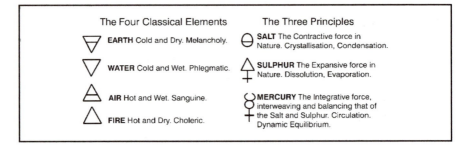

Figure 6.5 Alchemic Symbols
These represent a few of the classic alchemic symbols. A few are still used in modern chemistry.

24. *Lixiviation:* Oxidation of sulfides to form vitriols.
25. *Mortification:* The death of a substance, only to be reborn in a more pure state.
26. *Precipitation:* When a chemical reaction results in a substance descending to the bottom of the flask.
27. *Preparation:* Removal of unwanted matter to form a more pure substance.
28. *Rectification:* Purification by repeated distillations.
29. *Sublimation:* When a heated solid goes directly to a gaseous state without passing through a liquid state.
30. *Verification:* The making of glass by applying high temperatures to silicon minerals and possibly adding lime.

In addition to the numerous procedures devised by ancient alchemists long before modern chemistry used similar processes, they also formulated a series of alchemic symbols for some of the more common elements and alchemic principles. (See Figure 6.5.) Some of these symbols are still used. For instance, the triangle (Greek letter delta, Δ) sometimes represents heat in equations written for chemical reactions.

An important 11th-century alchemic document is called *Sage's Steps.* The original author(s) are not known since it passed through many hands. It is the first known parading of step-by-step instructions for teaching novice alchemists. The first step involved early training in Euclidian mathematics and Aristotelian natural science. The student was then instructed to use his hands to practice alchemic procedures, his eyes for observations of what occurred, and his mind to reflect nature's behavior. The second of the "sage's steps" was for the student to demon-strate how to use the laboratory and accurately follow instructions on how to purify gold and silver by cupellation (a high-temperature process used to refine silver or gold, in which the metal is absorbed into walls of the cupel, or cup containing the metal). The third stage involves the preparation of mercury oxide on a quantitative basis. This is one of the first examples describing how differing the weights of reacting sub-stances may alter the weights and characteristics of the final chemical compounds. It was not until the 18th century that measurable quantities of elements were compared with the quantities of the compound result-ing from a chemical reaction.

Following are some of the more important practitioners of alchemy who lived during the Middle Ages and Renaissance:

- **Geber** (Arabic name: Jabir ibn Hayyan) was the name for two different alchemists. The original Geber was an Arab named Jabir ibn Hayyan (ca. 721–ca. 815 C.E.) who was born in present-day Iraq. While hundreds of books on alchemy have been attributed to him, he may not have been the original author but rather a famous person under whose name oth-ers published. He upheld the Aristotelian concept that everything was composed of earth, water, fire, and air, which he believed combined in various portions to form sulfur and mercury, from which all other met-als are formed. Most alchemists accepted this belief well into the 17th century.

   The second Geber (also known as the "pseudo-Geber") was a 14th-century Spanish Arabian alchemist (original name unknown) who assumed the original Geber's name to make himself more famous. His best-known works were *Summi perfectionis magisterii* (The Sum of Perfec-tion), *De investigatione perfectionis* (The Investigation of Perfection), *De inventione veritatis* (The Invention of Truth), and *Liber fornacum* (The Book of Furnaces). He was the first to separate the sulfur-mercury con-cept of metals and to consider all metals as a separate entity. His books influenced European alchemy into the 17th and 18th centuries, partly because they were translated from Arabic to Latin, and later into the vernacular languages of Western Europe.

- **Al-Razi** (The Man of Ray) (ca. 826–ca. 925 C.E.) was born near Teheran in Iran (old Persia). While he is better known as a medical practitioner, author, and teacher, as most physicians of his time, he also studied alchemy. He wrote many books, few of which survive. One of the better known is *The Book of the Secret of Secrets,* later translated into German. He believed in the transmutation of base metals into silver and gold by using elixirs. (He never referred to elixirs as the "philosophers' stone.") In addition, he believed that worthless stones could be transformed into gems (emeralds, rubies, and sapphires) by using the correct elixirs. He was one of the first to describe the quantitative aspects for his alchemic reactions by stating the number of times a particular elixir, by weight, could be used to transmute base metals into a given weight of gold. He described several chemical processes used in alchemic transmutations, some which are well known and used today. They include distillation, calcination, solution, evaporation, crystallization, sublimation, filtration, and amalgamation.

- **Avicenna** (Arabic name: Ibn Sina) (ca. 980–1037) is better known as the Persian "Prince of Physicians." He was also a philosopher and scholar of mathematics and natural sciences based on Aristotelian Greek science. In his *Canon* on pharmacology he named over 760 drugs and chemicals, many used by alchemists and physicians (e.g., narcotics such as opium, cannabis, mandragora, and hemlock). A translation of one of his books, *De Mineralibus* (On Metals) acknowledges Geber's stated relationship between mercury and sulfur, but expounds more on the resulting compounds based on the corruptness as well as the purity of both. Avicenna was among the first of several medieval skeptics who questioned the transmutations of metals and minerals into gold. He believed that the results of such alchemic experiments produced mere imitations of gold. He stated that no matter how metals were treated they might gain induced qualities, but they remained basically unchanged. He also studied astronomy, mechanics, and the physics of motion, gravity, heat, and energy, and was one of the first to propose that the speed of light was finite.

- **Robert of Chester** (fl. 1150) may have received his name from the city of Chester, England where he went to school. After he and a friend translated the Koran into Latin that required two years of their time, Robert tackled the first translation of an Arabic alchemic book (*Book of the Composition of Alchemy*) into Latin. He completed the job in 1144. In the preface he tried to explain to the Europeans that alchemy was a new science with a new vocabulary. He translated many Arabic alchemic words into Latin that were later translated into English. We still use many of Robert's terms in modern chemistry. For example: *alcalai* = alkali, *azarnet* = arsenic, *carboy* = carboy, *elixir* = elixir, *naphtha* = naphtha, *natron* = sodium, *tutty* = zinc oxide, and *zaibar* = mercury. (See Figure 6.6 for an alchemic classification of substances.) Robert went on to translate other Arabic books into Latin, including mathematics and astronomy. Another great 12th-century translator was Adelard of Bath (1090–1150), who translated several Arabic mathematic texts, including one on Euclid. He also translated one on alchemy. Another was Gerald of Cremona (1114–1187), who translated 26 Arabic books into Latin, including Al-Razi's alchemy book on the properties of minerals. During the next several hundred years dozens of scholars translated many Arabic books, in all literary venues including the physical/chemical sciences, into Latin as well as into several local languages. These translations were the spark that ignited the scientific renaissance of western Europe.

- **Albertus Magnus** (ca. 1200–1280) was born in Germany of a noble family and later joined the Dominican order, where he was exposed to the translated works of the Greeks and Arabs and soon accepted scholasticism. His goal was to make all the great works in all areas of the natural sciences available to people who could read Latin. His major work was *De Mirabilibus Mundi* (On the Marvelous Things in the World), which was a hodgepodge of all kinds of information and much misinformation. Unlike many scholars of his time in history, he believed that there was much more to science than that which Aristotle had written about. In his *Little Book of Alchemy* he explained his belief that alchemic gold and iron lacked the same properties of natural gold and iron. After experimenting with samples of alchemical gold, he stated that after several ignitions it turned to powder. He also stated that alchemical iron is not true iron since it has no magnetic properties and that alchemy cannot alter a species (of metal) but only imitate the original.

- **Bartholomew the Englishman** (fl. 1250), a Franciscan, was one of several famous encyclopedists of the Late Middle Ages. His work *Liber de proprietatibus rerum* (Book on the Property of Things) included 19 volumes. He not only translated from Arabic, but also from original Greek and Hebrew sources. It was designed as a reference book for nonprofessionals. However, it contained only a short section on chemistry that included Aristotle's descriptions of the elements and translations of Avicenna's Arabic alchemy on how to transmute gold.

- **Arnald of Villanova** (or Arnold) (ca. 1235–1313) was a famous Catalan European physician who was born in Valencia in the region south of Catalonia (Spain). He relied on the efficacy of seals (markers) and amulets to heal patients and to defend them against witchcraft, natural disasters, brain disease, and financial ruin. He was one of the first to use alcohol to extract material from herbs to form tinctures he used to treat patients. Although Arnald was a physician, an astrologer, a reformer, and a diplomat, he also wrote several books on alchemy. He believed that he understood the deep knowledge of the Greek philosopher/scientists and accepted the general concept of metals as related to sulfur and mercury. However, he considered mercury to be more important than sulfur, which he considered harmful. He believed that gold and silver could be produced from pure liquid mercury simply by adding a very small sample of the gold or silver to large portions of the mercury. He gave specific instructions as to the ratio of portions of mercury to the small amount of precious metals required to prepare large amounts of alchemic gold and silver. Although there were the usual color changes, the chemistry related to his instructions made no sense. Nevertheless, he related the alchemic process to the conception, birth, crucifixion, and resurrection of Christ. He did perform the first known fractional distillation of blood to separate it into three fractions: water, yellow "fat" (most likely plasma), and red residue (most likely hemoglobin) that he related to fire. He used the "red fire" fraction to treat patients.

- **Raymond Lully** (ca. 1232–1325), a contemporary of the Catalan philosopher Arnald of Villanova, was also a Christian missionary to Spanish Muslims. (He was not too successful, since an Islamic mob stoned him to death.) Although he is credited with writing several books on alchemy, it is unclear as to how many he actually wrote. His credibility can be judged by the following story. He is said to have disbelieved in the premise that one metal can be transmuted into another, yet he claimed to be an alchemist. However, he asserted that he could, at will, change himself into a red chicken, and that he had changed 22 tons of base metals into alchemic gold inside the Tower of London. This gold was to be used by King Edward III of England to finance a crusade to the Middle East to reclaim the Holy Land. While there is obviously more to this story, it is a gross fabrication. His main contribution to alchemy is the system he designed for assigning letters of the alphabet to symbolize alchemical principles, materials, and operations. He then devised recipes and tables indicating how alchemical substances (by combining letters) could be manipulated to form new substances such as acids and reagents.

- **George Ripley** (ca. 1400–1490), an English alchemist who studied in Rome and later popularized the alchemy of Raymond Lully, was a well-known teacher of his own ideas on the science. His *Medulla Alchimiae* (Marrow of Alchemy) was published in 1476. He also wrote a sonnet as an allegory that explained the changing colors that occurred during alchemic processes, although the change in color seemed to be some type of secret only he claimed to understand. Ripley's reputation as an alchemist continued well into the 17th century.

- **Paracelsus** (1493–1541). In an earlier chapter Paracelsus, one of the most famous alchemists of the Renaissance, was mentioned for his contributions as a physician, particularly his founding of the field of *iatrochemistry* (chemotherapy). He was also a famous alchemist who provided a stimulus to the nascent field of chemistry, partly by relating particular chemical elements and compounds to the treatment of specific diseases. One example was including small doses of mercury for the treatment of syphilis several hundred years before the Salvarsan arsenic compound treatment was developed in the early 20th century. Paracelsus prepared a variety of elements and compounds as drugs (mercury compounds, sulfur, copper sulfate, iron, opium derivates, various salts, and minerals) in his laboratory, which he used for treating various diseases. He was the first to claim that a small amount of what made one ill could be used to make them well again. One example was during the plague when he applied a very small speck of a patient's feces to a piece of bread as a medicine. It is said that many patients were cured with this treatment. Thus, Paracelsus is credited with discovering the alternative form of medicine known as homeopathy, which uses microscopic amounts of drugs for treatments of specific diseases. (The modern alternative medicine known as homeopathy sometimes dilutes drugs over and over again to the extent that none, or very little, of the original drug remains in the medicine. Thus, there is little evidence, even today, that homeopathy is an effective form of medicine.) As a young man, Paracelsus worked in mines and thus was acquainted with different metals. He used this knowledge as an alchemist and physician. Later in life, he returned to the mines to treat workers who contracted lung diseases from breathing in the dusts of the mines.

By the end of the Renaissance skeptics questioned the transmutation of base metals to gold, as well as the existence of a philosophers' stone or the possibility of an immortality elixir. Nevertheless, alchemists classified and compared the physical properties of different substances and how they reacted with each other. When they measured and recorded

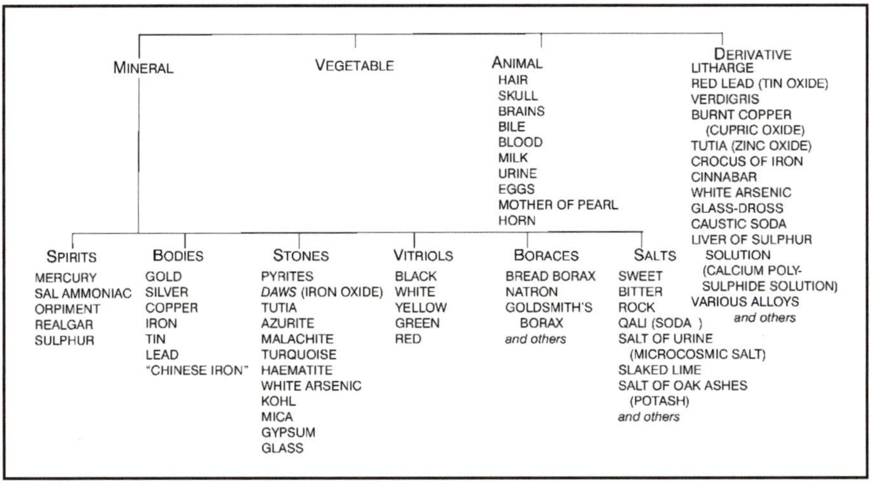

MINERAL		VEGETABLE		ANIMAL		DERIVATIVE
				HAIR		LITHARGE
				SKULL		RED LEAD (TIN OXIDE)
				BRAINS		VERDIGRIS
				BILE		BURNT COPPER
				BLOOD		(CUPRIC OXIDE)
				MILK		TUTIA (ZINC OXIDE)
				URINE		CROCUS OF IRON
				EGGS		CINNABAR
				MOTHER OF PEARL		WHITE ARSENIC
				HORN		GLASS-DROSS
						CAUSTIC SODA
						LIVER OF SULPHUR
SPIRITS	BODIES	STONES	VITRIOLS	BORACES	SALTS	SOLUTION
MERCURY	GOLD	PYRITES	BLACK	BREAD BORAX	SWEET	(CALCIUM POLY-
SAL AMMONIAC	SILVER	DAWS (IRON OXIDE)	WHITE	NATRON	BITTER	SULPHIDE SOLUTION)
ORPIMENT	COPPER	TUTIA	YELLOW	GOLDSMITH'S	ROCK	VARIOUS ALLOYS
REALGAR	IRON	AZURITE	GREEN	BORAX	QALI (SODA)	and others
SULPHUR	TIN	MALACHITE	RED	and others	SALT OF URINE	
	LEAD	TURQUOISE			(MICROCOSMIC SALT)	
	"CHINESE IRON"	HAEMATITE			SLAKED LIME	
		WHITE ARSENIC			SALT OF OAK ASHES	
		KOHL			(POTASH)	
		MICA			and others	
		GYPSUM				
		GLASS				

Figure 6.6 Alchemists' Classification of Substances
Medieval alchemists' method of classifying a wide variety of substances.
From E. J. Holmyard's *Alchemy*, New York: Dover, 1990.

the results of their experiments, they advanced the knowledge of the nature of substances. This was the beginning of modern chemistry. Figure 6.6 presents an example of the alchemists' organization chart of known chemical substances during the Renaissance.

Chapter 7

Inventions and Innovations

Introduction

Several factors separate humans from other species of organisms on Earth. Humans are not only conscious of their existence and the world around them, but are also aware of the uniqueness and consequences of their actions, which involve feelings of guilt, anxiety, and awareness that life can end. Humans are also free and can conceive of the creation of things new unto the world, including culture (ethics, morality, and values) as well as the design, invention, innovation, and construction of devices to make life more livable. Nature has no ethics or morality and only follows universal scientific physical principles and laws leading to constant change. The human attributes of culture, disquietude, and curiosity that lead to discoveries, inventions, and innovations distinguish us from all other animals. Although other animals may be curious and learn to use tools—for example, chimps use twigs to fish ants out of ground nests—it is not equal to the creative nature of humans. The learned habits and genetically innate behaviors of animals do not begin to compare with the human ability to conceive of and invent things designed to alter and improve their world. And humans have been inventing since their emergence as a distinct species in order to improve their lives in a harsh, uncaring, uncompromising environment. Early humans also made innovations and improvements to their original accomplishments. Most important, however, was the control and use of fire, for with this discovery, the rate of invention and innovation has grown rapidly over the short period of civilization's history.

Defining Invention and Innovation

One of the best ways to define an invention is to describe the process that identifies the uniqueness and usefulness of a device or that qualifies it for an exclusive monopoly for its inventor to commercially exploit the invention.

The history of patent monopolies for new inventions dates back to the 13th century in Venice, Italy, when such a monopoly was granted to the inventor of a new type of silk-making instrument. The first record of a patent being issued for an industrial invention was not until 1421 in Florence, Italy. This patent was a three-year monopoly to manufacture a barge hoist designed to lift and transport large stones. As printing became popular, Venice issued a monopoly to protect their printers in 1474. The first recorded patent in England was issued in 1449 to John of Ulynam for a 20-year monopoly to commercialize a new glassmaking process. Ulynam's patent included a clause that required the inventor to teach others the patented process of glassmaking. This was the first time the disclosure function of patents was used, and today it is required by all inventors when submitting their patent specifications. A detailed description of how the new device or process works, a detailed model or design, and detailed descriptions have been requirements of patent applications since the early 1800s.

During the Middle Ages and Renaissance the term *invention* had a slightly different meaning than today's technical descriptions used today by inventors who apply for patents. At one time anything that was discovered either by accident or deliberately, such as a new territory, or conceived by a person, such as a written poem, was considered an invention. Today, the U.S. Patent and Trademark Office considers only tangible property eligible for a patent. (For example, ideas and concepts, including scientific theories and poetry, are not tangible and thus are not qualified to receive a patent.) The concept of awarding a patent as a limited monopoly for almost any idea began at the end of the Renaissance under the rule of Queen Elizabeth I of England, when she sold unfair monopolies to support the crown. These monopolies covered all kinds of activities, including exploration rights, rights to produce or import and sell certain goods, and even rights to official positions. This abusive practice led the British Parliament to deem it illegal for royalty to sell such monopolies. Nevertheless, the Parliament retained the concept of "proto-patents" or "letter patents" enabling people to hold limited monopolies—that is, they were given exclusive rights to produce a commodity for sale up to 14 years. This was different from the patent of today, because it did not deal with a tangible technical invention. However, it did provide exclusive economic rights. For instance, during the Renaissance, if a person committed to building a mill in one location near flowing water, others would not have been permitted to build a competing mill in a neighboring location. In addition, weavers who produced a new design for their cloth were given exclusive rights to that design. These examples were outgrowths of the proto-patent concept, but unfortunately there was no mechanism at that time to determine what was an invention or who was the original inventor seeking protection rights. These questions were not settled for many generations. Thus, monarchies continued the practice of selling monopolies to the highest bidders.

Following the American Revolution, in 1790 North America adopted a patent system granting limited monopolies to inventors. The Patent Act reads, "The Congress shall have power to promote the progress of science and useful arts [i.e., technologies] by securing for limited times to authors and inventors the exclusive right to their respective writing and discoveries." Thomas Jefferson, Henry Knox, and Edmund Randolph were members of the first patent board set up by Congress. Jefferson, an inventor, was a strong supporter of the concept and understood the need for a monopoly granted for a limited time period to promote the infusion of technology in farming and manufacturing. Jefferson is known as the "father" of the U.S. patent system, as he was the first superintendent and the first patent examiner of the new U.S. Patent Office. Samuel Hopkins was awarded the first patent by the new United States Patent Office in 1790. Hopkins invented (discovered?) a new process for producing potash, an ingredient used in fertilizers. One year later, France awarded its first patent. Over the past 200 years the law has been clarified as to what inventions are eligible for patents.

Specifically, the invention must be *new, novel,* and *unique.* The invention must have *utility* and be *practical.* The invention must also be *nonobvious* (a change of color for an existing invention is *not* an obvious improvement that could receive a new patent). Only *tangible property* is eligible for patents. Examples of *intangible property* (intellectual property) are printed matter, music, artwork, and so on, which may be eligible for a copyright, but not a patent. In addition, scientific theories, laws, formulas, mental processes (ideas), and naturally occurring materials (natural resources, plants, and animals) are also not eligible for patents; however, new microorganisms and varieties of seeds may qualify for a patent. Computer software, depending on its nature, may currently qualify for a patent or copyright, but not both. The U.S. Congress is still debating how to treat the architecture for digital codes, programs, databases, and such.

The term *innovation* is sometimes confused with *invention* because it may be the first time some new or novel alteration was applied to an existing device or process. For instance, riding and balancing a bicycle in a high wire circus act might be considered an innovation of the original intent of a bicycle, but it is not eligible for a patent; on the other hand, a new and practical means of propelling a bicycle may receive a patent. Historically, humans have improved hundreds of thousands of inventions as well as scientific, technological, practical, and useful items, and this human behavior of innovation has accelerated over the ages. The U.S. Patent and Trademark Office has awarded over 6 million patents over the past 200 years. Even so, most so-called new ideas, inventions, and innovations have a basis in history—some very ancient history. (Only a few great minds have arrived at truly new ideas since the beginning of time.) For instance, Sir Isaac Newton's 17th-century superb laws of motion were a culmination of concepts related to motion as expressed by several classical Greek philosopher/scientists, including Aristotle. As Newton said, "If I have been able to see farther than others, it was because I stood on the shoulders of giants." The modern concept of the atom was first described by Leucippus of Miletus in the 5th century B.C.E.: "All matter is composed of very minute particles called *atomos*. They are so small that there cannot be anything smaller, and thus cannot be further divided." His student Democritus further developed this concept, as did many others over the ages. Additional innovations to atomism were made by John Dalton's 19th century atomic theory for the elements, and Niels Bohr's early 20th-century quantum theory of atomic structure. High-energy physics explored the innermost structure of the atom during the late 20th century, and modern nanotechnology of the 21st century continues to refine the original atomism theories. It might be mentioned that these examples, though important innovative concepts, do not meet the criteria for a patent. The wheel, although no doubt an invention (or discovery), has been around for thousands of years and thus is not eligible for a patent for use on vehicles, no matter how numerous the innovative uses for it as a practical device. But on the other hand, if an individual invented a *new, practical,* and *non-obvious* use for the wheel—for example, as a heavy flywheel to maintain spinning momentum that can be converted into propulsion as in an automobile—it might be granted a patent. But this idea has already been suggested as a means of propulsion, so it is not exactly original.

Strictly speaking, all inventions and innovations, no matter how important, are not really "original" because no inventor or innovator, or scientist or philosopher for that matter, works in a vacuum. They all have experiences and historical documentations relating to previous technologies, concepts, and ideas that become part of their creative thinking process. Due to Europe's isolation during medieval times, it was truly the "Dark Ages" as far as inventions and innovations are concerned. For the 1,000 years of the Middle Ages there are few documented records of inventions or innovations, with the exception of a few important innovations for the use of gunpowder and improvements of the printing press, which were introduced from the East. By the Late Middle Ages, when Arabic translations of Greek science and technologies were retranslated into Latin and imported into Europe, things began to change. These developments, along with the printing press and the introduction of more rational religious practices, are mainly responsible for the awakening during the Renaissance. Most of the following inventions and innovations of the Middle Ages and Renaissance have both antecedent and subsequent histories of their development.

Abacus

The abacus is an example of a useful device that was invented hundreds of years ago to meet the needs of traders of ancient days. Before the days when numbers were invented, merchants kept their transactions in their heads but needed some means to count what was bought and sold as well as to reckon prices and profits. This need resulted in lines drawn in the sand with pebbles used as placeholders for their calculations. Later, counting boards were developed as grooved stones or lines painted on boards to hold the pebbles or metal disks to represent transactions. The spaces between two lines represented units of tens, hundreds, thousands, and so on. The abacus was an innovation of the ancient counting boards. A typical abacus is the common Chinese version, which has 13 vertical wires, with 7 beads on each wire, contained in a rectangular wooden frame. The frame has a horizontal divider that places seven beads (heaven beads) separated by two beads (earth beads) above the divider and five beads below the divider. Historically, the Greeks, Romans, Japanese, Russians, Aztecs, and other societies devised their own versions of counting boards and abaci. Some used vertical wires or strings, while others used a horizontal arrangement. During the Middle Ages apices, coin-boards, and line-boards were invented as new forms of ancient abaci. (See Figure 7.1.)

The modern-type abacus was invented during the Middle Ages (500 to 1300 C.E.), and even today in some parts of the world, people claim that multiplication and division, as well as solving square roots and

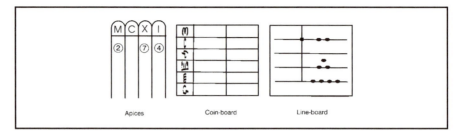

Figure 7.1 Three Types of Middle Ages Abaci
These are examples of many types of abaci that were developed more or less independently in several countries over the ages.

cubic roots of numbers, are easier to calculate on a modern abacus. By the time of the Renaissance the abacus was generally replaced by the Hindu-Arabic numerals written on paper, making it much easier to use this system for arithmetical calculations as well as higher mathematics. Although the abacus is not as popular as or used in the same manner as modern calculators or computers, these modern electronic devices might be considered an advanced innovation of the original counting boards and abaci.

Arches

Historians are unsure exactly when the use of dome-like vaults and arches occurred. Archaeologists have found evidence of these types of structures in the ancient Sumerian city of Ur in southern Iraq, in Mycenae in Greece, at Thebes in Egypt, at Mohenjo-Daro in present-day Pakistan, and in China. These arches and other dome-like structures date back thousands of years and have been found on several different continents. It is believed that the Romans borrowed their type of arch from the Etruscans in about 500 B.C.E. The Romans' innovative use of the arch was uniquely applied as a means of supporting the weight of aqueducts that carried water to their cities. They also invented the concept of an angular-shaped keystone at the top of the arch that helps distribute equally the force of the outward weight it is supporting. In other words, the keystone assists the arch to support itself. While this allowed variations in architectural styles, it also had limitations due to the horizontal stress on the two vertical columns of the arch. During the Middle Ages the flying buttress was invented to correct this problem. A buttress (external support) was constructed in line with each column to counter the outward horizontal force of the arch. (See Figure 7.2.) This innova-

Figure 7.2 Medieval Arch with Keystone and Flying Buttresses
The keystone placed at the top of arches distributes weight to each side of the arch, while the buttresses assist in counteracting the outward forces of the arch, thus permitting larger arches to form vault ceilings designed for the construction of larger ancient churches.

tion resulted in stronger and larger arches and thus the construction of higher enclosed spaces with greater areas designed to accommodate larger numbers of people in the massive medieval cathedrals.

Armillary Sphere

The armillary sphere has a long history of invention and innovation. This astronomical instrument can be a simple or very complex model representing the celestial sphere in respect to the horizon of the observer. Early armillary spheres were three-dimensional models of a skeleton-like globe composed on several rings (known as *armillae* in Latin) representing major celestial spheres, such as the celestial equator, the **ecliptic,** and the horizon. The first models were probably invented in either ancient China or Greece in about 100 B.C.E. or a few years later. Credit for inventing the three-ring armillary sphere goes to Zhang Hen (ca. 78–ca. 138 C.E.). One of the rings (armillae) represented the equator, one represented the paths of the sun, moon, and planets, and the third ring defined the poles of the Earth. At the pole of the sphere a tubelike device was attached to the center of the device. Astronomers used this sighting tube to view and map the stars and thus determine the directions, heights above the horizon, and estimated distances of individual stars. Over the years armillary spheres underwent numerous design innovations, mainly the addition of more rings and increases in the size of the rings.

The first major description of an armillary sphere appeared in Ptolemy's book *Almagest* in about the year 150 C.E. It was referred to as an ecliptic armillary sphere since the outer rings were designed to read ecliptic latitudes and longitudes. Larger, more complex armillary spheres were invented during the Renaissance. Some were several feet in diameter, constructed of wood or brass, and included a smaller, spherical terrestrial globe at its center. Numerous wire rings depicting the sphere of fixed stars and each of the known seven planets, as well as for the sun and moon, surrounded the central Earth/globe. Other innovations included additional rings that indicated polar and equatorial circles, polar caps, and meridians. The entire device would rotate upon turning a handle that controlled a vertical axle positioned through the center of the sphere. The Islamic version was designed to determine the celestial equator both in **ascension** and in **declination**. Similar armillary spheres are used today for demonstration and teaching purposes in school classrooms.

Astrolabe

Exactly when and who invented the first astrolabe is unknown, but it has undergone many innovations and improvements over the centuries. It is generally accepted that Greek astrologers, in either the 1st or 2nd centuries B.C.E., invented the astrolabe as an instrument to measure the altitude above the horizon of the stars and planets. It is believed that the Greek mathematician Apollonius may have conceived the basic idea of how to measure the positions of heavenly bodies, and the Greek astronomer Hipparchus may have used a crude astrolabe to measure the distance of the sun and moon from the Earth. Hipparchus also catalogued the positions of over a thousand stars. In the 4th century C.E. Hypatia of Alexandria (ca. 370–415 C.E.), believed to be the only female scientist in the ancient world, designed and built her own astrolabe and other astronomical instruments. The Arab Muslims adapted the Greek astrolabe for their own use and later, in the 10th century C.E., introduced it to western Europe where it underwent additional innovations and uses well into the 17th century.

The word astrolabe is a combination of two Greek words, *astro,* meaning "star," and *labio,* meaning "finder." The astrolabe is a flat, circular brass disk with degree markings on the outer rim, somewhat similar to a protractor. A sticklike moveable pointer called the *alidade* is pivoted to the center of the flat disk. When held vertically at eye level, with its baseline lined up with the horizon, the pointer is then moved to align with a celestial object whose position above the horizon could be measured in degrees. In a sense, the astrolabe is both a calendar and a calculator since it can measure the daily and annual movements of the sun, planets, and stars. During the Middle Ages and Renaissance additional disks and devices were added to the astrolabe. Many elaborate Middle Age and Renaissance models still exist. Surprisingly, these early astrolabes were not designed or used for navigation. A new type, the nautical astrolabe, was invented during the Middle Ages. This new and somewhat differently designed astrolabe along with the compass became important navigational tools. The astrolabe was also the forerunner of the navigational sextant, which uses a small telescope to provide more accurate measurements of the altitudes of the sun and stars above the horizon. Later, other astrolabe designs became useful for various purposes, since in addition to determining the time of day, it could assist in determining latitude, the time of sunrise and sunset, as well as the direction of Mecca. Astrolabes were also surveying instruments that measured the height of buildings and other objects. In addition, since the astrolabe could tell time by accurately measuring the sun's position above the horizon, it was not only an important instrument for navigation but also for telling the time of day for prayers in the Islamic world.

Blast Furnace

About 10,000 years ago ancient people probably learned by accident how to separate copper from its ore, ushering in the period known as the Copper Age. In time, the Copper Age became the Bronze Age after the ancients learned how to mix tin with copper to form the alloy called bronze. Bronze was harder than copper and could hold an edge, making it ideal for use as cutting tools and weapons. However, people at that time could not create fires hot enough to smelt iron from its ore. (Temperatures of at least 1482°C [2700°F] are required along with some carbon to separate iron from its ore impurities.) It is not known who first produced fires hot enough to smelt iron ore. However, archaeological evidence suggests that a crude form of iron was made in the Middle East as early as 3000 B.C.E., and the remains of a 2000 B.C.E. foundry was discovered in southern Africa. Iron was most likely a serendipitous discovery as the result of smelting other metals. By 1200 B.C.E. the Iron Age was in full swing, as humans learned to use charcoal to make hotter fires that produced a pulpy type of iron called a *bloom*. This impure iron was repeatedly heated and hammered to remove the **slag.** In time they learned how to mix carbon fuels and ores and use leather bellows to increase the airflow into the furnaces—thus increasing the temperature, resulting in a higher-grade iron.

In about the 1st century C.E. the invading Romans introduced England to a type of furnace that could produce wrought iron. Wrought iron is very brittle, since it still contains many impurities, but it was an improvement over former types of iron. The process of extracting iron called "roasting" was invented in the early Middle Ages. The ore was washed and heated in the presence of air to burn off impurities (roasting), resulting in compounds of iron oxide (Fe_2O_3 and Fe_3O_4) as oxygen in the air combined with the iron in the ore. This form of impure iron was broken into smaller pieces that were placed in a reduction-type blast furnace along with charcoal for "blooming." During this blooming process, the charcoal was burned as it combined with what little oxygen there was in the furnace to form carbon monoxide (CO). Carbon monoxide is an excellent reduction agent that removes the oxygen from the iron oxide. This resulted in a "bloom" of iron as a carbon-free spongy mass of red-hot but solid iron. The monks of the Rievaulx Abbey in Yorkshire, England, operated a foundry dating back to 1350 C.E. The monks were innovative people who mixed the iron ore with small chunks of charcoal (or soft coal). As the mixture began to burn, they forced a blast of air through it, thus the term "blast furnace." They were probably unaware of the actual chemical reaction involved—that is, the carbon that was added to the iron ore acted as a reducing agent, thus purifying the iron. The molten iron was then poured off from a tap in the bottom of the furnace, while the slag containing the impurities was scooped off the top. This resulted in what is known as "pig iron," a relatively pure form of iron that is often mixed with other metals and minerals to form different types of steel. Modern blast furnaces use essentially the same process, but often substitute pure oxygen for air.

Canal Lock

The law of gravity prevents water from flowing uphill. To their credit, this simple fact did not deter ancient humans from building canals, but it did impede boats from moving freely upstream. Sometime in the 1st century B.C.E. the first single-gate canal lock, referred to as a "fishlock," backed up water in a downstream-flowing canal. When a boat wished to proceed downstream, the gate was lifted and the boat was carried downstream with the current. However, this system did not work for boats going upstream. In 983 C.E. the creative mind of Chiao Wei-Yo conceived the idea of constructing two nearby fishlock gates that could trap a pond of water between them. This worked well for boats going in either direction, but only if the boats were low enough to pass under the beams supporting the vertically lifted gates. While a similar system was developed in 1373 in Vreeswijk, Holland, the problem of height limitations for boats still existed. The solution to this problem is credited to the genius of Leonardo da Vinci (1452–1519), who, during his service as a hydraulic engineer to the Duke of Milan, sketched and advised on the idea of using two locks (gates) that swung open horizontally instead of lifting vertically. Like many of da Vinci's ideas, it was some years before it was applied to real situations. The result was the invention of the "miter" lock-type canal gate. This consisted of two hinged gates that formed a "V" when opened upstream to release water, thus allowing taller boats to pass through the gates. Once the boat passed through the lower horizontal swinging gates, they were closed as the **sluices** at the upper gate allowed water to again fill the pond behind the lower gate as the weight (pressure) of the water made sure it closed firmly. This pressure on the lower gate was reduced once the pond was filled, and thus the gates could be swung open upstream to allow the boat to pass downstream. The miter horizontal-lock system worked equally well with boats going upstream or downstream. Similar canal systems were common in most countries and were one of the main modes of transporting goods and people during the Renaissance and well into the 19th century. The topography of England required numerous locks for its extensive canal system. The development of railroad and highway systems made the existing canal systems superfluous. Today, while some are used for transporting bulk materials, the canals are mainly tourist attractions. Even in the United States, sections of the old Erie Canal in western New York state and the Chesapeake and Ohio Canal through Georgetown in Washington, D.C., have been reclaimed as tourist attractions.

Cannons

Again, there are multiple claims of who, when, and where the cannon was invented. Since the Chinese are generally credited with inventing gunpowder and using it in fireworks and crude rockets, it is assumed that they should also be credited with the invention of the cannon. It is believed that in about 1288 C.E. the Chinese placed dart-like arrows (and later iron balls) into a pipe-like bamboo tube and used gunpowder to propel the projectile. This was the first use of a gun-type cannon. The Arabs, Germans, and English all made various claims for inventing the brass or iron tube cannon. One of them, the German monk Berthold Schwarz, claimed in 1313 to have not only invented gunpowder but also used it in a cannon-like weapon. In 1327 the English used primitive cannons to fire a kind of arrow similar to the Chinese design. By the mid-1300s the English designed both cast brass cannons ("casting" brass is pouring a molten mixture of copper and zinc into a cannon-shaped mold) and also cannons constructed from strips of iron welded together to form the barrel. They were not only heavy and required sleds to transport them into battle, but they also had a tendency to blow up. By the 15th century iron cannonballs replaced the carved stone balls that had been used for over a hundred years. Cannon innovations developed to the point that some weighed nearly 20 tons, with projectiles weighing over 800 pounds. The cannon soon gave the advantage in war to the attackers, as they could devastate the defenders' fortifications. While this led to improved fortifications to resist the cannon's power, it also required the development of more innovative siege weapons by the attackers. (See Chapter 8.) This back-and-forth conflict between the need for new offensive and defensive inventions and innovations of war technologies is still present in modern times.

Chess

The invention of the game of chess is credited to both India and China, possibly deriving from old Indian war games or old Asian board games. The game most likely spread westward by ancient trade routes, including the Silk Road from China. (See Chapter 2 for more on the Silk Road.) In the 6th century C.E. an Asian game called *chaturanga* developed into a two-sided game of skill using dice and moving figures of warriors (pawns), chariots (rooks), rajah (king), elephants, and horses. By the 7th century C.E. the Persians (present-day Iranians) adapted the game and called it *shatranj*. By the 9th century the game developed into much the same form of chess that is played today, with many strategies developed by players on all continents. Not surprisingly, modern computers have been programmed to play chess against humans. In 1996 the IBM computer Deep Blue lost its first contest to the chess champion Garry Kasparov, but in a rematch the computer won after being reprogrammed by several chess masters.

Clocks

Humans have always been concerned with time—that is, time in the sense of the one-directional "arrow of time" as time passes from the past to the present and the future. About 5,500 years ago ancient shadow clocks or clock sticks (upright sticks in the ground or a T-cross of one raised stick over another with markings on it) were an obvious method used in many parts of the world to track the path of the sun during clear days. (See Figure 1.7 in Chapter 1.)

Sundials (gnomons) of one type or another were developed by many civilizations, but they had the same problems of most early timekeeping devices, namely, tracking the passage of time at night. After their invention, candles were marked to indicate how long they burned between marks representing the passage of time. This may have been the first nighttime timekeeping device. In about the 14th century B.C.E., water clocks (clepsydras) were invented in Egypt. They were an improvement, but again they had problems. The rate of flow or drip of water was not consistent due to the change in water pressure resulting from the changing depth of the water in the holding tank as it emptied. Sand hourglass clocks were also early attempts to track the passage of time during the nights. These, too, were problematic. The texture of the sand and the size of the opening between the two bulbs of the instrument varied the rate of flow, depending on the design of the hourglass. Still, they were more accurate than water clocks since the change in the depth of the sand in the top bulb does not affect the rate of flow of the sand into the lower bulb (as the depth of water alters pressure on the drops of water). Even though sandglasses could measure units of time rather accurately, they had to be turned over every so often, which led to the need for a long-running mechanical device that would require the invention of other devices.

It was not until the Middle Ages, in about 725 C.E., when a Chinese engineer, Liang Ling-Tsan, invented the escapement movement as a device for regulating the movements in mechanical clocks. This breakthrough made the development of accurate timekeeping devices possible. In 1090 C.E. Su Sung was commissioned to build a large astronomical clock that used a revolving armillary (celestial) sphere to replicate the celestial movements of the planets, sun, moon, and some stars, as well as keeping time by indicating the hours. His clock was constructed in a 40-foot tall tower and was powered by a wheel consisting of buckets that were filled with constantly flowing water. His tower clock provided more accurate data needed to correct the problem with the Chinese calendar that was also a problem with the religious calendars in the Western world. Hanging weights drove the first mechanical clock in the West in the 13th century. Several countries developed clocks powered by weights hanging on ropes wrapped around a drum. As the drum unwound, a slow uniform motion was produced. A problem of all mechanical clocks at this time was an uneven unwinding motion. This was solved by an innovation of the original Chinese escapement movement now called the *verge escapement mechanism,* which allowed for movement of weights in short rather than continuous steps. Soon after the adoption of the escapement movement, the hands and the dial face were added. Early clocks merely had the hour-indicating hand. The minute hand and second hand were added in the 1500s.

Although a portable timepiece was desirable, it was impractical as long as a clock's power was derived from hanging weights. In the late 1400s the German craftsman Peter Henlein (1479–1542) invented the new type of spring that consisted of a thin band of steel (a form of steel later known as "spring steel") that could be coiled around a pivot. A key attached to the pivot allowed the steel band to be tightly wound up—thus the term "mainspring." After being fully wound up, the mainspring would slowly release its power to drive a small clock mechanism. The first versions of this clock-type watch were inaccurate due to the unequal exertion of the mainspring's power as it wound down. As the quality of spring steel improved, it became possible to manufacture improved versions of Henlein's clock. Two innovations that made small pocket watches and wristwatches possible were the oscillating balance wheel and the "stackfeed," which regulated the difference in the power of a fully wound and partially wound spring. These early watches and clocks were improved over the ages. Since the earliest pocket-watch models were made in Nürnberg and looked like flattened eggs, they were known as "Nürnberg Eggs." Spring steel for the mainspring, the stackfeed, verge-and-foliot escapement, and machine tools to make accurate small cogs and wheels all contributed to more advanced means of measuring time.

Following the Renaissance there were many innovations and improvements in timekeeping devices. In modern times many types of accurate timekeeping devices have been invented, including the chronometer, oscillating crystal timepieces, and the most accurate timekeeping device based on the vibrations of atoms of specific elements.

Concrete

The Egyptians were the first to discover a mixture of substances that could be used to bind stones for construction purposes. Gypsum, a sulfate mineral, was the basic ingredient that, along with lime, made this a special type of mortar. Some historians believe that the Egyptian gypsum mortar was more likely used as a lubricant to slide heavy stones rather than as a true mortar, since many of their structures used stones so accurately cut that they could be "dry-jointed" and did not need mortar.

The Romans are credited with formulating a mortar that used sand-like substances. They developed *pozzolana*, which was formed from volcanic ash and lime, and a concrete-type mortar that was made from regular sand mixed with lime. They also invented the process of refining lime by burning it in kilns to produce a substance called "slaked lime" that, when used with sand, improved the quality of mortar. The Romans soon realized that by adding an aggregate (gravel and small stones) directly to the slaked lime and sand mortar, they could produce a very hard substance. They called this substance *concrete*. Note: *Cement* and *concrete* are often confused with each other. Cement, also known as Portland cement, is the **hydraulic** binding or adhesive agent made from a mixture of pulverized clay and limestone baked in a kiln, while concrete is a mixture of sand, gravel and/or stones, Portland cement, and water that forms a finished, hard, dry stone-like material.

During the Renaissance, a French architect named Philibert de Porme rediscovered the old formula and patented it and is now considered the original inventor of concrete. The cement of today produces concrete far superior to that of ancient and Renaissance types. Cement was discovered by accident when Joseph Aspdin (1779–1855), a bricklayer from Leeds, England, made a fire too hot and burned his mixture of clay (silica and alumina) and pulverized limestone. When the results cooled, he mixed the crumbly stuff with water, which when dried became extremely hard. Aspdin named it *Portland cement* after a building stone it resembled found on the Isle of Portland located in the English Channel. He received a patent for Portland cement in 1824. Since then, Portland cement has been used worldwide in the construction of a great variety of concrete structures.

Crossbow

There is a main difference between a regular longbow and crossbow. When using a regular bow and arrow, the **fletched** end of the arrow must be inserted onto the bowstring, and the soldier must then be strong enough to pull the string back and release his fingers from the bowstring to send the arrow on its way. Crossbows are usually shorter and much more powerful than longbows, and require less strength since the bowstring must be cocked with a lever device that can be held in place until a trigger is pulled that releases the bolt (short arrow) on its way.

The date humans first used crude crossbows is not known. There is some evidence that ancient indigenous (native) peoples placed bows and arrows horizontally on the ground with pieces of wood used to hold the bows in taut, cocked positions. A long vine-type rope was attached to the piece of wood and then to a tree or bush, positioned so that a prey animal would trip it, knocking loose the piece of wood and resulting in the arrow being shot at the target animal. There is not much known about the crossbow from those ancient times until about 400 B.C.E. in China. By the year 341 B.C.E. Ma-Ling led an army of 50,000 Chinese crossbow soldiers into battle. The Chinese designed a repeating type of crossbow that consisted of a magazine located above the track that holds the arrow. Gravity fed arrows from the magazine to the track for shooting. They continued to use this repeating weapon into the 19th century. Early versions of crossbows were composed of a wood stock, a bronze lock mechanism, and lath crosspieces (bow) made of slats of wood, horn, bamboo, bone, sinew, baleen, or composites of these materials tightly bound, similar to the manner of the modern leaf springs of automobiles. (See Figure 8.16.) Ancient crossbows had a draw of about 165 pounds, more than twice as powerful as longbows. These crossbows could shoot bolts great distances (about 650 feet) at the exceptional speed of 260 feet/second, but they were slower to load and fire than were longbows.

Over the years the crossbow acquired various names. The term *crossbow* is derived from the Latin *arcuballista* (or arbalest). The Greek word for crossbow is *gastraphetes,* and the German term is *armbrust.* They are also known as the barreled crossbow and bullet crossbow. The "ballista" was a Roman siege engine similar to a giant crossbow. It might be said that the crossbow was reinvented during the Early Middle Ages, as it was not in general use from the time of the fall of Rome (~476 C.E.) until it was brought to England by the Normans in 1066 C.E. for hunting and later warfare. Improved crossbows using composite laths were used in 12th century C.E. during the Crusades. By 1139 crossbows were outlawed by the Roman Church as inappropriate and too deadly for Christians to use. Reason soon prevailed as their effectiveness in battle overcame theological queasiness, and the Christian Crusaders successfully used crossbows for years to defeat opposing armies equipped with longbows and armored horsemen. By 1314 the first steel lath was invented for crossbows, making them even more powerful (up to 700 pounds of pull). With this improvement, crossbows could shoot faster (about 140 feet/second), shoot farther (about 350–400 yards), and hit a target harder than most hand-to-hand combat instruments. A close shot could penetrate most medieval armor. By 1503 the English banned individuals from possessing crossbows—as they do today for firearms. By the late Renaissance firearms replaced crossbows as personal weapons of war. Even so, United States Special Forces in both World War II and the Vietnam excursion occasionally used crossbows.

Dyes

The dye industry is one of the oldest and continuously viable industries in history. It is not known who may have invented the use of dye to color hides and other materials, but numerous innovations for sources of dyestuffs and dyeing processes were developed over many thousands of years. The oldest written record of the use of natural dyes to color cloth was found in China in 2600 B.C.E. In the early 1920s red textiles were found in King Tutankhamen's (fl. 1358 B.C.E.) tomb in Egypt. This dye was most likely alizarin pigment extracted from the madder plant. Many other natural plant and animal matter were used as dyes and pigments over the centuries beginning at least by 3000 B.C.E. By the 8th century B.C.E. the Romans dyed wool, and India produced many beautiful colored textiles. The most expensive dye ever made was a deep purple shade extracted from a vein of a gastropod mollusk. It required about 8,500 of these snails to produce one gram of this dye, making it worth more than its weight in gold. It became known as Tyrian, or royal, purple since only royalty and the very rich could afford it. By the time of the fall of Rome in the 5th century C.E. many different types of natural dyestuffs became popular throughout Europe and Asia. New forms of deep blues and purples became less expensive, resulting in the emperor of the Byzantine Empire issuing an order on pain of death forbidding the use of certain shades of purple except for royalty. By the 8th century C.E. the Chinese had perfected a process of waxing designs on silk that would not take the dye, a process known as the batik wax-resist technique. And block-printing fabrics with dyes was a viable industry in India.

Following are a few examples of natural plants and animals used from ancient times, the Renaissance, and up until recent times when artificial dyes were invented:

Madder (*Rubia tinctoria*)—A Mediterranean evergreen plant whose fleshy roots produced an orange-red juice called *alizarin*, which is the parent form of many dyes ranging from brownish-yellow to orangish-red.

Indigo—a deep blue dye from a plant originating in India, but later grown in other countries. It is known as the *indigofera* plant or woad, a European shrub.

Murex—A marine gastropod (snail) found in the Mediterranean Sea that exudes a yellow secretion that becomes purple when exposed to sunlight. The purple dye was known as *Tyrian purple*, after the Phoenician town of Tyre in modern day Lebanon. It is also known as royal purple since only royalty and the wealthy could afford it.

Cochineal Insect—(Latin: *Coccineus* = "scarlet"). Also known as *kermes berries.* When the bodies of the female insect (*Coccus cacti*) are dried and crushed, they produce a brilliant scarlet-red dye. Tin (Sn) is sometimes mixed with the pulverized shells to brighten the color.

Cardinals' Purple—made from Kermes berries (Greek: *Kokkos* = scarlet) and was used to dye robes of Church officials. See above.

Safflower (*Carthamus tinctorius*)—a plant whose juices produce a yellow dye.

Cuttlefish (genus *Sepia*)—a squid-like mollusk that secretes a dark, inky fluid that produces a sepia pigment that can range from yellowish to tan and medium-brown tones.

Limonite—A yellow/brown type of pigment extracted from a type of iron ore.

Hematite—A red pigment from hematite iron ore found in clay.

Tannin—A brown/tan dye extracted from the bark of trees.

Galls—Dark brown to black pigments that are extracted from galls found on oak leaves. The galls result from insect damage. Iron salts were also used for black pigments.

One of the problems with early natural dyestuffs was the tendency of the dyes in the cloth to run and become washed out of the textiles. Soon after humans began dyeing cloth, chemical mordants were used to fix dyes to prevent running. One of the first substances used to fix dyes was a form of aluminum sulfate known as alum $[AlNH_4(SO_4)_2]$. Over the centuries many different types of chemical elements and compounds have been used as mordants. Dye from the indigo plant was an exception in that it did not require a mordant; textiles dyed with it were fast (did not run). During the Renaissance explorers discovered new plants and insects that could be used to make dyes. Some examples are types of bark (logwood and fustic), plants (lichens, berries, and seaweed), and insects (cochineal). Note: The cochineal bug lives on cactus plants found on the Canary Islands and in Mesoamerican countries where the natives would dry and grind up the female insects to produce a rich red dye. (The modern color carmine comes from the cochineal insect.) By the 19th century chemists learned how to make many types of artificial dyes and specialized mordants. Today, most dyes and pigments are derived from coal tar and petroleum products, and the price of dyed textiles has dropped considerably since the Middle Ages and Renaissance.

Eyeglasses (also known as spectacles)

Archeological evidence exists that suggests ancient people used small, thin rectangular pieces of thin wood or horn into which they cut two small slits to improve their vision as well as for protection from the sun's glare. The device, tied around the head with a cord to hold it in place, may have been used as an early form of sunglasses to reduce the glare from snow and ice. Also, if the slits were sufficiently narrow, they would improve the wearer's vision. Eskimos were known to construct such glasses before modern sunglasses became generally available. There is also evidence that during the early years of the first millennium the Chinese wore eyeglasses with colored lenses, possibly for aesthetic purposes since they were not designed for corrective purposes. The Romans used reading stones made from natural crystals that were laid over writing or objects to increase the size of the image. Glass reading stones produced in several countries were most likely convex lenses held in wood frames that magnified small objects and handwriting.

The actual inventor of corrective eyeglass lenses during the Middle Ages is unknown. The honor is shared by several people. Reportedly in 1267 Roger Bacon (1220–1292) conceived of a type of eyeglass that would correct farsightedness. He experimented with lenses and described how it was easier to read with these new convex lenses. There is some evidence that in 1285 an unknown craftsman in Florence, Italy, developed a process for making convex lenses—thicker in the center than the edges, that is, eyeglasses to correct farsightedness. (Note: Concave lenses—thicker on the edges than the center—to correct nearsightedness [myopia] were not invented until the late 15th century.) Eyeglasses were available by the late 13th century, as portrayed in an Italian painting from 1287 C.E. Salvino D'Armate, a 13th-century craftsman of Pisa, is sometimes considered the inventor of true eyeglasses in 1291, while Alessandro Spina of Florence is said to be the first to use convex lenses in spectacles to improve his longsightedness (now called "farsightedness") and reading. In 1300 a guild of Venetian glassmakers and crystal workers mentioned glasses for the eyes in their rules and regulations. They made it illegal for newly developed forms of glass to be used instead of the more valuable natural crystals as lenses for spectacles. In 1352 Tommaso da Modena painted a picture of a monk wearing spectacles perched on his nose. Once the process to produce transparent glass was perfected, natural crystals were no longer used to make eyeglasses. And once the principle of optics was understood, the number of innovations in the size, shape, and design of frames resulted in increased manufacturing of eyeglasses in the West. The credit for inventing eyeglasses for the nearsighted in 1450 goes to Nicholas Krebs. (See Figure 7.3 for typical eyeglasses of the 15th century C.E.)

Figure 7.3 Early 15th-Century Eyeglasses
The simple spectacles of the Renaissance were difficult to keep on a person's nose. Side arms that hooked over the ears were attached at a later date.

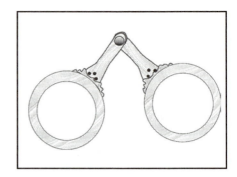

A major problem was that the glass lenses were heavy, and since they were usually perched or clamped on the nose, they kept falling off. It was also inconvenient to walk around while holding a reading glass on a stick. In 1730 Edward Scarlett, a London optician, arrived at a solution to this problem by inventing a system of two side pieces attached to the frames that would rest on the top of the ears next to the head and then wrapped around behind the ears to hold the spectacles in place.

The science of vision and lenses (optics) was not clearly understood until the 17th and 18th centuries. Since then, the drive to understand vision problems and devise more effective and safer materials is an ongoing enterprise. Non-shattering plastic lenses are often substituted for the glass lens, resulting in lighter spectacles. Lenses designed to polarize light can also be darkened for protection against ultraviolet light, as well as ground to correct far, close, and near faulty vision in the same (progressive) lens. And of course, glass, plastic and disposable contact lenses are available to those who want the same correction without the bother of wearing spectacles.

Flush Toilet

Disposal of human wastes was not much of problem for prehistoric humans. They could walk a few steps outside their caves or huts and deposit their wastes in a shallow ditch. If the area became too polluted, they just formed another ditch or moved on. There is some archeological evidence to suggest that at least as far back as 2500 B.C.E. wastewater drainage systems from individual houses leading into streams and rivers were used in the Indus Valley region of western India. But these facilities were very limited and remain so even today in modern India, where the population excretes over 900 million liters of urine and 135 kilograms of fecal matter each day. Over 900 million people in India openly defecate by roadsides, along railroads, or other public places. Only about 30 percent of the population living in the cities of India has access to modern sewage facilities, while only 3 percent of those living in rural areas have access to flush toilets. When going to deposit their wastes, people carry small brass pots (about 6 to 8 inches tall) filled with water for cleansing themselves. Indians always keep their cleansing hand below the table when eating. Much of the rest of the underdeveloped world is also plagued with the same lack of sanitary toilet and sewage disposal facilities as India is. Archaeological evidence and artifacts indicate the ancient civilizations of Egypt, Mesopotamia, Greece, India, Rome, and China constructed toilets similar to our modern-type sitting commodes and the necessary drainage systems to wash away the accumulated waste materials. These facilities were for royalty and the wealthy and did not exist for the general population. The Chinese also invented toilet paper in the 14th century c.e., primarily for use in the imperial Chinese household. As civilizations expanded and populations increased, concern about health and sanitation became more important, as did the means for disposing of both human waste and their garbage. Historically, the sanitary disposal of human waste was always a serious problem, and it still is a problem in much of the world.

During the Middle Ages and early Renaissance the populations of most cities simply threw their garbage into designated areas of the streets to later be picked up by "rakers," who usually dumped the trash into the sea or rivers. During the Dark Ages human body wastes were also dumped into the streets to be collected by the "Bucket Brigades." In the past, homes in eastern Asia were (and many still are) serviced by "night-soil" men who then sold human waste for fertilizer. Some castles, forts, and large homes were built on riverbanks, with overhanging rooms that had stools placed over holes in the floor so that human wastes would drop directly into the rivers below. The medieval cesspools, first located beneath the houses, were not much better, but outhouses and septic tanks located short distances from the homes were later improvements.

Sir John Harington (1561–1612), of Yorkshire, England, who was Queen Elizabeth's godson, is credited with inventing the first flush toilet, also known as the water closet (WC), in 1596. He attempted to market his invention (6 shillings and 8 pence), but the public thought his idea ridiculous so he constructed only two units. One he kept for his own use, and the other was given to the queen. Harington's concept of a flush toilet in every home was ignored for decades—a good idea before its time. Rather, during the late 1500 and into the 1800s a series of public toilets were instituted in several countries. In 1556 a public toilet for use by 100 families was established just outside Delhi, India. Not surprisingly, public managed toilets were very unsanitary, so the French contracted with private companies to manage their public toilets where a small charge for services was used to cover expenses. This concept of a commercial toilet service is now being used to provide sanitary facilities for many poor people of the world. Many innovations that improved on Harington's original design were made over the centuries. In 1775 Alexander Cumming invented a new type of sliding valve for the flush toilet and was the first person to receive a patent for a toilet. In 1778 Joseph Bramah received a patent for a toilet with a hinged sliding valve that could fill a tank with water, upon which the waste and water were dumped into the bowl, which had a trap at the bottom that prevented fumes from backing up into the toilet. In the 1800s a plumber by the name of Thomas Crapper made and sold toilets and received several patents for plumbing and toilet improvements. Contrary to myth, he did not invent the modern toilet. This myth may have been started during World War I when U.S. soldiers saw Crapper's name on toilets and thus introduced the slang term *crapper* for a toilet.

Flying Machines

It is assumed that humans have always been curious about flight and wondered what it would be like if they, like birds, could roam in the skies. For thousands of years humans attempted to effortlessly soar above the Earth—always unsuccessful until the early 20th century. There is even a story in Greek mythology that exemplifies this fantasy. No one was allowed to leave the island of Crete, so Daedalus made two pairs of wings, one for himself, the other for his son, Icarus. The large wings were made of feathers woven together and fastened with wax. Both escaped from the island, and as they soared, the father told Icarus to follow him but to be sure not to fly too close to the sun as the heat would melt the wax holding the feathers together, nor too close to the water as they would get wet. As they made their way to safety, Icarus enjoyed the freedom of flight and disobeyed his father about approaching the sun. As Daedalus looked back, he saw his son's body in the sea, recovered it, and buried him on an island forever named Icaria.

Leonardo da Vinci (1452–1519) is credited with proposing the first serious engineering approach to the mechanics of human flight. His sketches of designs for a heavier-than-air craft (as opposed to lighter-than-air craft such as balloons) indicated that he studied not only how birds moved but also how the physical properties of air were used to support them in flight. His first drawings were of devices that used flapping wings, but he could not fathom how to make them function as the wings of birds. Other drawings indicate a man lying prone and pumping his legs to rapidly beat the wings of the craft. Other models resembled, to some extent, modern helicopters, which were operated by a system of pulleys and paddles and even had retractable landing gear. His main problem, which he never solved, was how to get this heavy craft off the ground in the first place before it could be flown. He did sketch a type of motor that used the concept of energy stored in a pulled bowstring. Leonardo studied the dynamics and engineering of gliding objects, but there is no record that he ever actually built models of his flying machines. He also sketched his idea for a parachute, which may have been taken from old Chinese drawings.

Over the following years many others tried to build a flying machine. The Chinese are credited with several firsts. The first manned kite flight was in the 4th century B.C.E.; the first parachute and the first hot-air balloon both were in the 2nd century B.C.E.; and the first propeller and helicopter rotor (a toy top) was in the 4th century C.E. Ancient Indian texts of the 12th century C.E. describe an elaborate aerial machine that could fly great distances. It was described as a strong, well-shaped aerial "car" that was driven by a tubelike engine constructed of iron and filled with liquid mercury. A fire heated the mercury, and the latent power in the mercury was supposed to drive the airship.

Most of the flying machines constructed in China were either toys or some form of entertainment, such as flying kites and fireworks. Another example of a Chinese first occurred when Wan-Hoo is said to have built a rocket chair fitted with 2 kites and 47 gunpowder rockets attached to the back of the chair that were lit simultaneously with torches applied by 47 assistants. His goal was to fly to the moon. It seems that this first rocket-powered aircraft ca. 15th century C.E. was a disaster, since the machine exploded and killed Wan-Hoo. It was not until 1852 when George Cayley (1773–1857) built and flew the first heavier-than-air craft that was designed as a glider. In 1853 he also built an unmanned helicopter-type top based on a Chinese design that, when spun fast enough, reached a height of 90 feet. It was also during this period that lighter-than-air balloon flights became popular. Many others in several countries studied the aerodynamics and engineering of flight and built gliders and air machines that culminated in the success of the Wright brothers in 1903.

Flywheel

At some point in history ancient potters used a horizontal wheel mounted on a vertical axle or post to assist them in forming clay into pots. Also, at some point they learned that if an upper horizontal wheel was connected with a central vertical stick (axle) to a lower horizontal wheel, their feet could kick the lower wheel, which kept the upper wheel turning, thus freeing their hands to form the clay. It became obvious that the larger the wheels the less often they had to be spun to keep them turning. In 1122 c.e. a German monk, Theophilus, described this phenomenon of inertia and momentum that was first recognized by potters. This theory of moving bodies was not explained until Sir Isaac Newton's 1687 *Mathematical Principles of Natural Philosophy* proposed the three laws of motion. The second law stated that if something is placed in motion, it continues in motion at that speed and direction until some other force acts upon it. During the Middle Ages this concept was used with a large wheel that was slow in starting its rotation. However, once up to speed and turning at a steady rate, it would not only continue spinning for some time but also smooth out any jerky motions of machinery. In the 12th century British blacksmiths applied this flywheel concept to turn grindstones to sharpen their instruments. Later, large flywheels were used to smooth out the jerky motion of the reciprocating pistons in steam engines, as well as other machinery. Today, experimental flywheels are designed to supplement the power to propel automobiles and thus conserve fuel.

Glass

Glass was another serendipitous discovery of the ancient past and has now been around for about 5,300 years. Ordinary fires are not hot enough to melt sand (silica) unless the sand is mixed with a flux such as soda or ash that lowers the temperature required to turn silica into molten glass. Credit for its discovery is sometimes ascribed to Mesopotamian potters who experimented with sand and indigenous silicate materials used to create their colorfully designed glazes. About 5,000 years ago the Roman historian Pliny the Elder related the story of Phoenician sailors who, while camped on beaches, cooked on rocks and blocks of saltpeter (containing niter or sodium). After the fires cooled, they noticed that some of the silica in the sand that mixed with the soda had crystallized into glass globules. They recognized this as glass since they were familiar with obsidian glass from volcanoes and the glass-like deposits that sometimes form when lightning strikes a sandy area. Egyptian glass beads date back to about 2500 B.C.E., and glassware dating from 2000 B.C.E. was found in Turkey. Later, the Egyptians used a mixture of silica-sand, lime and soda, and malachite (copper ore) to produce colored glass. They poured molten glass into molds to form early glass objects such as seals, beads, inlays, and small figures. The revolutionary process of glassblowing with a pipe to form hollow vessels developed in Syria (Palestine) in about 100 B.C.E.

By the time of the fall of Rome in the mid-5th century C.E., most countries produced some type of glass. Innovative glassmaking as an art as well a science did not arrive until the Late Middle Ages and Renaissance. There is no record of when the chemical and physical nature of glass was first explored. But as far back as the 1st millennium glassmakers of several countries experimented with various minerals to form colored glass. Tin or lead oxide produces a milky-white glass. Yellow glass was made with antimony, lead, or iron; red glass by adding gold; orange glass was produced by adding oxides of copper; violet glass by adding manganese salts; greenish-blue by adding copper or iron oxides; dark blue by adding cobalt compounds; and black by adding a variety of metallic oxides. The production of colored glass led to stained glass during the Gothic Era of Europe as far back as the 5th century C.E. The invention of stained-glass windows reached its technical height during the Renaissance in Vienna, when glassmakers experimented with different combinations of pieces of colored and clear glass in ways to form pictures within one large window frame. As the picture was being formed, individual colored pieces of glass were held together by beeswax. When the picture pattern was complete, the sections were combined with lead beading that was soldered together as a single large window. During this period Venetian stained glass were famous as works of art much in demand by churches, wealthy homes, and as luxury glass products. Even today Venetian glass is highly valued.

Mirrors are inventions that are at least 8,000 years old, as evidenced by the obsidian mirrors found in Çatal Hüyük in Turkey. The Egyptians also polished stones and wood, as well as obsidian, to use as reflecting devices. However, it was not until the Renaissance that Venetians were able to form clear flat sheets of glass for use in producing mirrors. They placed a very thin sheet of tin on the back of a flat plate of glass, built a frame around the glass, and then poured mercury on top of the foil, which in time formed an shiny amalgam. The excess mercury was poured off, and the back of the mirror was coated with a type of varnish to protect the mercury coating and also to prevent oxidation. This is much the same as the process used for making mirrors today, except the reflective coating is usually added to the back of the mirror by vaporization of the reflective material.

The physical nature of glass was not understood until after the Renaissance, when the concept of atoms and molecules was accepted. Glass is considered a super-cooled fluid or solution that happens be a liquid below its freezing point and thus is solid. But it is only a solid in the short geological time frame because glass windowpanes (primarily those from ancient times) have been known to become thicker at the bottom than at the top due to the effects of gravity, as very sluggish glass molecules slip over one another—slowly but surely over eons of time moving downward. Some people claim that the idea of glass as a solution that "flows" over periods of time is a myth since it would take millions of years for a pane of glass to become thicker at the bottom than at the top. The answer may be that ancient glass panes were unevenly "flowed" into flat sheets and thus were manufactured unevenly.

Gunpowder

The term *gunpowder* is somewhat of a misnomer since guns were invented some years after the explosive substance now referred to as gunpowder. While historians agree that the Chinese invented black powder (gunpowder), they do not agree on the date of its discovery. Based on circumstantial evidence in ancient Chinese writings, its invention by alchemists occurred sometime between the 7th to the 9th centuries c.e. An alchemist's text dating from 850 c.e. mentions the mixture of saltpeter, charcoal, and sulfur. By the early 900s c.e. the Chinese invented "flaming arrows" that used gunpowder. The next written record of a formula for gunpowder appears in 1044 c.e. during the Chinese Sung Dynasty.

Arab alchemists, who called saltpeter "Chinese snow," learned about this new potential medicine and weapon in 1240. By the 13th century alchemists in the West also learned of the existence of saltpeter and gunpowder. Roger Bacon (1220–1292), who claimed to have invented gunpowder, wrote the first reference to gunpowder in Europe when he published his coded formula for its specific production in 1267. His formula used a low percentage of saltpeter, most likely due to its scarcity in England and Europe. The Chinese first used a formula of about 50 percent saltpeter because higher percentages in the mixtures were extremely explosive. Their later formula for bombs, grenades, and mines was six parts saltpeter (potassium nitrate-KNO_3), three parts sulfur (S), and one part powdered charcoal (carbon-C). European proportions for these three chemicals varied over the centuries depending on how it was to be used. Modern chemistry indicates that approximately 75 percent of the mixture should be the oxidizer (potassium nitrate), for a ratio of 75:15:10. During the Renaissance the ratio of gunpowder's ingredients for cannons was approximately 50 percent saltpeter; 65 percent to 67 percent for small musket-type guns; and 83 percent for pistols. The higher the percentage of potassium nitrate, the greater the production of rapidly expanding superheated gas, which, as the propellant, determines the velocity of the bullet or missile.

There has been some dispute as to how and when gunpowder was introduced to the West. Knowledge of gunpowder and its uses were most likely exchanged as a "transfer of technology" that resulted from the generally free commerce and trade between and among the Chinese, Arabs (Persians), and later the Europeans via the historically famous Silk Road of the 10th to the early 14th centuries. (See Chapter 2 for more on the Silk Road.) An early reference to gunpowder's discovery was found in a Florentine document dated February 11, 1326. This is about the same time as bronze cannons were used to fire iron balls.

Europe's desire for gunpowder was stymied by the lack of natural saltpeter. The European continent's geology is quite different from that of China's, which can produce saltpeter from the natural decay of organic matter in soil. Because the importation of saltpeter from Sicily and a few other countries in the early to mid-1300s was extremely expensive, the Europeans developed saltpeter plantations, designed to convert organic wastes by decomposition into crude potassium nitrate. Even with a limited supply of saltpeter, firearms were improved and produced during this period, but their use was somewhat restricted by the availability of gunpowder. On the other hand, the quality of most guns could not withstand large charges of gunpowder with a high concentration of saltpeter. While the Chinese never followed up on their discovery, once black powder was known and used in the West, several improvements in both gunpowder and firearms occurred during the Renaissance.

Serpentine powder—Black powder was the simple mixture of saltpeter, sulfur, and charcoal originally discovered by ancient alchemists. The Europeans attempted to standardize and improve the burn rate and thus the effectiveness and explosive power of simple black powder. This early effort, called *serpentine powder*, consisted of grinding each ingredient separately and then combining them. There were several problems with serpentine powder. First, when transported over some distance by wagons, the vibrations would separate each ingredient by their different densities and grain sizes, requiring the batch to be remixed before use. Sulfur, being the densest, would settle to the bottom; charcoal, being lighter, would seek the top of the pile. Remixing gunpowder at the front during battle proved to be a risky procedure. Second, the finer the grain size of the black powder, the more moisture it would absorb. Note: The lattice theory states that the smaller the size of particles for a given volume, the larger the total surface area—thus, more area on which moisture may collect. In addition, saltpeter [potassium nitrate (KNO_3)] in early gunpowder did not attract much moisture, but early serpentine powder always contained some calcium nitrate [$Ca(NO_3)_2$], which contaminated the saltpeter and readily absorbed moisture. This explains the saying, "keep your powder dry." Third, if serpentine powder was packed too tightly in the firearm, it could explode violently. Conversely, if packed too loosely, it could just fizzle like pyrotechnic fireworks. These problems resulted in unreliable gunpowder. Additionally, if the proportion of saltpeter was too great for the type of firearm used, the breech or gun barrel could explode with disastrous results. Much research and experimentation to improve gunpowder occurred during the Renaissance. One solution proposed reconstituting the ingredients similar to that which is done today with cigarette tobacco and with chicken parts to make "nuggets." The Renaissance solution resulted in corned powder.

Corned gunpowder—An example of a gunpowder innovation during the Renaissance combined all the ingredients in water or some other fluid to make a slurry (a liquefied paste), which assured a more complete mixture. Craftsmen experimented with several different types of liquids, including wine, vinegar, and urine, as well as water. The slurry was spread out to dry into large sheets that were like cakes or blocks, known as *knollen*. When dried, the cakes were then crushed into particles the size of wheat grains called *corned* or *crumbed* gunpowder. Corned gunpowder (wheat and other grains are known as "corn" in England) had several advantages over serpentine power. First, the mixture was more complete and would burn more evenly. Second, it was not subject to separation when transported, and thus it lasted longer. Third, the crushed pieces could be run through sieves with different hole-sized mesh, producing standard grain sizes. In addition, any leftover fine power could be reconstituted into a new corning batch. Finally, the large chunks or grains reduced the surface area and thus partially solved the problem of moisture absorption. At the same time a more controlled explosion was possible since the larger the grain, the less combustible surface area, and thus a slower burning rate, which was desirable for cannon powder.

Gunpowder is credited not only with changes in the nature of wars during the Middle Ages and Renaissance, but also with altering civilizations. As a chemical explosive, like all other scientific and technological inventions and innovations, gunpowder can be of both benefit and harm to humans. It seems ironic that in their search for the elixir of life, Chinese alchemists should instead invent a compound of death.

Microscope

Records indicate that as early as the 3rd century B.C.E. people noticed that objects appeared enlarged when viewed through glass globes filled with water. And, when a convex-shaped piece of clear natural crystal (most likely quartz) was placed over writing or objects, the image was magnified. Ancient Romans used "burning glasses," presumably made from convex-shaped clear crystals, to start fires by concentrating solar rays. Another example of early magnification involved water on a leaf's surface. When viewed up close through a drop of water, the fine network of veins in leaves was magnified. A crude type of microscope was constructed by placing a drop of water within a small opening in a flat piece of wood or bone. When viewed closely, the convex shape of a droplet of water magnified objects.

The Dutch spectacle-maker Zacharias Janssen and his father constructed the first microscope with ground lenses in either 1590 or 1595. The major component of his invention was the use of two lenses—a convex inserted was place in one end a tube while a concave lens was inserted in the other end (and used as the reverse of a telescope). Although the Janssens are credited with inventing the microscope, some years before their invention others had discovered that lenses could magnify objects.

The microscope was not as instant a success, as were Janssen's spectacles (eyeglasses), until two post-Renaissance scientists used similar instruments to reveal and illustrate some remarkable discoveries. Robert Hooke (1635–1703), a scientist, mathematician, and engineer, made several important discoveries with a compound microscope that he himself constructed. He described the structure of tiny divisions within bark (cork) and named them "cells" (a term still used) since they reminded him of the cells in which monks live within monasteries. Later, Antoni van Leeuwenhoek (1632–1723), who was not a scientist but was inspired by Hooke's microscope, constructed a single-lens microscope that enabled him to view bacteria and single-celled protozoa. He was the first person to describe single-celled organisms. Today, numerous instruments are used for extreme magnification. Some have improved complex lens arrangements that rely on light rays for observations, while others use shorter electromagnetic wavelengths of X rays and electrons to observe objects too small to be seen with the longer wavelengths of white light.

Paper

The Chinese are credited with inventing paper in the 2nd century B.C.E. However, during the Middle Ages and Renaissance, particularly after Johannes Gutenberg made several important improvements to the printing press, many papermaking inventions and innovations occurred. The earliest known writings were drawings in dirt or sand or on cave walls. Later, writings involved cuneiform impressions on clay tablets and waxed boards, or carvings in wood or bone. Ancient civilizations also used leather and papyrus as writing materials. The ancient Chinese invented, or rather developed, paper by concocting a basic pulp from the bark of the mulberry tree mixed with bamboo slivers and other plant fibers to make wood pulp. When rice stalks were added to the pulpy mixtures, the dried paper was soft enough to use as toilet paper, another Chinese invention. After the pulp was crushed, it was soaked, then filtered, pressed, and dried. Following the loss of a war with Turkistan in the 8th century C.E., several Chinese papermakers were captured and relocated to central Asia, where they made the first Arabian paper in Baghdad in 793 C.E. The Chinese papermaking process was dominated by Arabian influences and thus was not known in the West for at least another 500 years. During the Medieval Period, Europeans used expensive parchment, vellum, or silk for writing materials. They were slow to adopt the Chinese papermaking process, which was not brought to the West until about the 12th century, some years after the Muslims occupied Spain. Up until this time, paper was imported from Arabian countries.

There were at least two main innovations of European papermaking that made it more popular and less expensive than the parchment (sheep skins) and vellum (lamb skins) that were still used for official documents until the end of the Renaissance. The first innovation involved three steps: (1) Europeans used rags (linen) that were shredded, soaked, and then beaten into a pulp to separate the fibers. (2) The pulp was soaked in water inside settling tanks constructed with removable fine mesh screens at the bottom of the tanks. As the pulp settled, the water filtered through the mesh, leaving the fibers spread out evenly over the screens. (3) After the screens were removed, the water was pressed out of the matted fibers. The sheets were then smoothed out with a type of stone squeegee, then wetted with a sizing solution of gelatin and alum, and finally hung up to dry. This sizing step was an important innovation because it prepared the surface of the paper and prevented the ink used for the printing presses from running and blurring. With the reinvention of the printing press in Germany, the demand for affordable paper increased rapidly. The second innovation was the use of watermills. By the late 15th century watermills were used in the manufacturing of paper in England, and paper production increased to the point that it became difficult to find sufficient rags for papermaking. By the end of the 18th century a completely mechanized papermaking process led to another innovation that used less expensive wood pulp instead of rags. Wood is still the main worldwide source of paper pulp, but to save trees other sources of fiber, such as kenaf (Guinea hemp), are being introduced (primarily in Texas) to produce newsprint paper.

Plow

The heavy plow of the Medieval Period was preceded by hundreds of years of innovative use of devices to scratch and turn over the earth in order to plant seeds. Prehistoric people used digging sticks to open the soil for planting and to seek roots and tubers as food. In about 2000 B.C.E. farmers in the Near East used deer antlers and later devised bronze hoes to use as simple plows. By 250 B.C.E. the Romans had invented both a light and a heavy hoe (called *sarculum*, Latin for "a hoe for sowing"). Hoes were used in areas where a plow was inappropriate for sowing seeds and also to clean out weeds after planting—just as modern hoes are still used for these purposes.

During the 1st century C.E. Pliny the Elder of Rome described a wheeled plow that eventually became popular on the European continent. Agriculture practiced in Greece and in Italy (Rome) required only a light plow, while the heavy, wet soil of northern Europe and Asia needed a heavier plow with a larger blade (coulter). Heavy plows were in use during the 5th century C.E. in northern Italy about the same time as horseshoes, and by the 8th century both were introduced to northern Europe. One of the main differences between the hoe and earlier

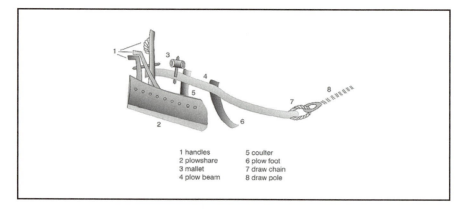

Figure 7.4 Heavy Plow of the Renaissance
The plow of the Renaissance was an improvement over ancient tools used to till the earth. The use of steel for the plowshare was an important innovation.

lighter plows and the heavy plow was that the heavy plow could not be pulled using human power. Sometime during the 8th or 9th century, heavy horse collars and the breast-strap horse harness were introduced to Europe. Both of these innovations enabled the heavier plows to be pulled by oxen or horses harnessed in tandem. (See Figure 7.4 for a representation of a heavy plow.)

It is somewhat amazing that modern plows are not radically different from those used in the Middle Ages and Renaissance, when the plow that cut the furrow and the moldboard that turned the dirt over from the furrow were made of wood. Two modern innovations have improved the plow: the use of steel for the plow, which made it more efficient in cutting deeper through heavy soils, and the use of gangs or sets of plows pulled by tractors.

Printing

The ancient practice of carving varying symbols onto the ends of sticks that were then used to make imprints in clay tablets (hieroglyphics) might be considered one of the first printing processes. Sometime during the Early Middle Ages, the Chinese and Indians invented a wood-block printing technique. This procedure involved the carving of negative **relief** images on the flat surface of a wood block (so that text would be reversed when printed). The block of type was then dipped in dye or ink and pressed onto cloth or paper. Around the 6th century C.E. civilizations in Japan and Korea, as well as China, first used clay characters and later wood blocks to print official documents, calendars, cards, and paper money. A history of mathematics, *Jiuzhang Swanshu* (Nine Chapters of Mathematical Art) by Lu Hui (fl. 260 C.E.), was the first book that was block printed in China. The *Diamond Sutra* (a book of Buddhist precepts) by Wang Chieh was a book printed in the form of a scroll using individual wood block letters in 868 C.E. By the 11th century, which was long before moveable type and printing was introduced to the West, China printed entire pages of books with type consisting of individual letters. This must have been a huge undertaking, since Asian alphabets consisted of thousands of characters. This is one reason that China preferred one-piece wood-block printing rather than moveable type—it was easier to produce as one piece rather than many thousands of individual wood characters. The Chinese did not use a press, but rather used a two-headed brush. One head spread ink over the wood block. After the paper was placed on the block, the dry brush on the other end was used to gently brush the paper to raise the characters. The next step involved carefully peeling off the paper from the block and letting it dry. The Chinese printed up to 2,000 sheets a day with this system. Even so, the Chinese are credited with developing a form of moveable type. In ~1041 C.E. Pi Sheng (ca. 990–1051) created a system whereby each character of Chinese writing was represented by a separate piece of material that could be assembled to print an entire page and then disassembled and rearrange for printing another page. He used baked clay to prepare the individual characters, which were glued to an iron plate. Since his clay type was fragile and unsuited for volume printing, it was soon replaced with individual wooden characters. Later, in the 14th century the Koreans used metal (bronze) moveable type. By the 12th century the Chinese used color printing of their money to prevent counterfeiting. By the 14th century the Chinese were printing books in as many as four colors.

As with the introduction of silk, paper, and many other products and ideas from the East, printing was also carried west by Asian, Arab, and European traders and travelers, such as Marco Polo. Moveable brass type and crude printing presses were in use in Islamic Spain at least 100 years before these innovations were translated into Latin and imported to Northern Europe. Europeans, as well as Americans, have long believed in the myth that Johannes Gutenberg (ca. 1400–1467) of Mainz, Germany, invented both moveable type and the printing press. Even in modern times, when asked, most journalists, politicians, and even scholars consider Gutenberg's moveable type and the printing press (1436 C.E.) as the most important invention of the past millennium because it spread literacy and general knowledge as well as economic power to both individuals and societies in the West. Granted, improved paper manufacturing and printing techniques have influenced more people in the past 500 years than any other invention, but in actuality, Gutenberg only made some innovations to these processes that were already well known in other parts of the world. Regardless, Gutenberg and others of the early 15th century did make some significant advancements in moveable type and the type of press capable of rapidly, and thus inexpensively, printing single or multiple sheets of documents and books. Gutenberg was rather secretive of his methods for producing and arranging metal type into frames that could be inked, then followed by a piece of damp paper placed over the block of type (called a typeset). He also adapted a wine or linen press that could evenly screw down a platen to press on the paper, transferring the ink to the surface of the paper. Gutenberg did not publish much about his various experiments and techniques because he was concerned that others working on the same problems might develop their own systems first. Most of the records regarding Gutenberg's press are from legal documents that indicate he sued others and they sued him. These and other expenses related to developing his printing system forced him to go into partnership in the early 1440s with a Mainz moneylender, Johann Fust (ca. 1400–1466).

Finding the correct metal or alloy for his first moveable type was a problem that required a great deal of trial-and-error research. Most metal alloys were too soft or too brittle, and the molds for even the same letters were often slightly different in size, and thus printing was uneven. First, Gutenberg formed about 270 characters in the ends of metal punches that were set up as type. He then invented a brass mold of each letter that could be filled with lead or **type metal,** thus duplicating exactly the same letter many hundreds of times. His next innovation with type was not implemented until after he printed the Gutenberg Bible. This new technique arranged the letters on a page within a printer's frame. They were then held firmly and evenly so that the type could be inked over and over again.

Making ink that produced a consistent image and did not smudge was another challenge for the printer using early printing presses. Ordinary writing ink and dyes did not work well since they did not adhere to the type or paper consistently. Gutenberg invented "printer ink," which consisted of common substances, such as resin, soap, and heated oil. Another formula used lampblack (soot from lamps) and a type of quick drying varnish. It is said he also experimented with other additives, including egg whites. Gutenberg's inks lasted over 500 years and still retain their black color. Modern printer's inks are basically derived from petroleum products and do not seem much improved from those of Gutenberg's time. He also made improvements in the paper to assure sheets of equal thickness and correct absorbency that would accept his ink without smudging.

Gutenberg's other invention replaced the screw wine and linen presses that were first used as printing presses by attaching a lever-type handle that could lower the platen onto the paper both faster and with a more even pressure. Once the process of printing a large book was begun, Gutenberg divided his workers into separate teams, each of which performed different tasks simultaneously. First, the molds were formed and individual letters of type were produced in Gutenberg's foundry. Second, as the type became available, another crew typeset a page or two. While this was being done, another crew printed the former typeset pages. At the same time, a fourth crew removed the moveable type from the frame of pages that had already been printed so the letters could again be set for new pages. This division of labor decreased the time required to print a book but required an original set of about 20,000 to 50,000 individual pieces of type (depending on the size of the pages and book).

Modern printing presses are remarkably similar in their basic operations as those of 500 years ago—albeit much improved with continuous roller presses, automation, photographic processes, electronics, and so forth. There is no question that printing had a tremendous influence and impact on the world over the past 500 years. Gutenberg's mass production of the Bible in 1456 is considered the high point of his career. Even so, he had difficulty in earning a living by printing books (as do some modern printers). However, he solved this problem by printing over 200,000 indulgences, which were one-way tickets out of purgatory that were offered for sale to lay persons by leaders of the Roman Catholic Church of his day. Gutenberg was plagued with legal and other problems, including the fact that by the end of the 15th century there were over 1,000 printing shops in Europe, which kept him from ever making a fair profit from his inventions and innovations.

Prosthetic Limbs

During prehistoric times and up until the Middle Ages and Renaissance the loss of an arm or leg usually meant death because the person either bled to death or, if they survived, they could not hunt or fight. Nevertheless, the first record of a prosthetic device appears in about the year 1000 B.C.E., describing a stump-like piece of wood or metal with a socket to fit the limb and held on with straps. These crude devices were used both for soldiers who lost a limb in battle and those with congenital birth defects. The numbers of people with missing limbs increased drastically with the invention of gunpowder and firearms. These weapons inflicted horrendous damages to the extremities of soldiers, necessitating the amputation of one or more limbs. The surgery was crude and often involved the painful procedure of using boiling oil or hot irons to cauterize the leg or arm stump to stop bleeding. The best a person could expect was a wooden peg leg fit to the stump that enabled one to hobble around. This changed during the late Renaissance.

A Frenchman, Ambroise Paré (1510–1590), became a leading surgeon of the Renaissance at a time when surgeons were still considered on a par with butchers. He knew neither Latin nor Greek and was unaware of the medical accomplishments of ancient physicians, which may have been one of the reasons for his success. He learned from experience, mostly as a military surgeon who developed a system of tying ligatures around blood vessels to staunch bleeding rather than utilizing the very painful practice of cauterization. Reportedly, he was the first surgeon to use this procedure. Instead of wood or metal stumps to replace missing arms and legs, Paré devised functional artificial limbs. In his 1575 book *Oeuvres,* he described and provided drawings in great detail of an artificial hand, arm, and leg designed for soldiers. He included specific instructions to craftsmen who could replicate his artificial prostheses. It also described how these were attached and held in place with a series of straps. He also invented prostheses for lower-limb amputees that were the first for above-the-knee amputees. These included bending hinges in place of knee joints that permitted the user to bend and sit down. His artificial hands had articulated fingers operated by springs that could be moved and used to hold objects. Paré's prostheses were great improvements in both mobility as well as aesthetics for amputees. Paré is also credited with several other firsts. In 1551 he described the "phantom limb sensation," which is the psychological awareness of the patient's missing appendage. He was also the first to discover that if the direction a bullet enters the body is noted, a straight-line path can be followed to help locate the site of the slug. This technique is still used in emergency rooms around the world. Paré was one of the first to develop prosthetics for cosmetic purposes, particularly for facial injuries or birth defects.

Other Renaissance surgeons made important advances in what we now call "plastic surgery." The difference is that early facial surgery or use of cosmetic prosthetics corrected congenital defects as well as injury, and was not just for concepts of beauty. Gaspar Tagliacozzi (1546–1599) was a Renaissance surgeon who practiced rhinoplasty. This operation involved restoring the nose with a strip of skin from the arm of the patient. The arm was strapped in contact with the nose so the patient could not move it until the flap of skin graft from the arm had taken hold. After several weeks, the flap was cut off the arm and formed into a nose. These types of operations were banned by the authorities and not revisited until the 19th century. There is a story of the famous Renaissance astronomer Tycho Brahe (1546–1601), who lost part of his nose in a sword duel and had a new nose made that consisted of putty and silver. It is not clear who did the surgery or if he made his own nose. Today numerous types of artificial body parts and organs are used as replacements in humans, and the art of plastic surgery for both reconstruction and aesthetics is quite common.

Quadrant

The ancient Indians, Chinese, and Arabs were all interested in how to determine the correct location in the sky of heavenly bodies, as well as their size and distance. They developed a number of sighting devices to measure angles that could then be used to calculate an object's height and size. One of the first was a simple instrument composed of a stick marked off in a 90° graduated arc that had a moveable arm (radius) that could be used to determine angles and thus altitudes. Ptolemy (ca. 90–170) invented or improved several astronomical instruments, one of which was a stick about 6 feet long with an upright block at one end, with a small hole in it, and another block that could slide back and forth along the stick. By sighting through the hole and moving the block, it was possible to measure the size of the sun and moon. One of his more advanced instruments was the *triquetrum* (Latin for "triangular") that consisted of an upright stick with graduated markings and pivoted arms at the top and bottom. The two arms were joined so their ends could slide. As a person sighted along the upper arm, the lower one changed its angle, thus determining an object's angle above the Earth's horizon and thus its altitude and distance by using geometry. (See Figure 7.5.)

During the Renaissance astronomical instruments became more sophisticated. In the mid-1500s Tycho Brahe, the famous Danish astronomer, was given unlimited funds by the reigning monarch of Denmark to build a large castle-type observatory in the city of Uraniborg on the Island of Hven, Denmark. Tycho insisted that the only way to obtain accurate astronomical measurements was to make the measuring devices large, and thus he constructed some of the largest instruments

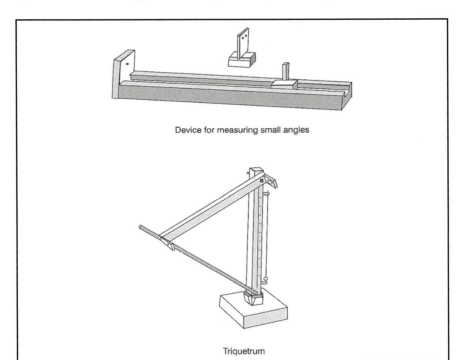

Device for measuring small angles

Triquetrum

Figure 7.5 Ptolemy's Device for Measuring Small Angles and Triquetrum
These are an artist's interpretation of two devices designed by Ptolemy and used by
him to measure angles.

of this period. He constructed an 11-foot mural quadrant large enough
for him to sit in and make his observation while his assistants recorded
his data. (See Figure 7.6.) Several of his other large instruments
included a large astrolabe and a celestial sphere that was 5 feet in diameter. Even though this was some years before the invention of the telescope, he made several important discoveries, including the accurate
position of 777 stars.

By the time telescopes were in use, the quadrant was more or less
replaced by the sextant in determining the angle between the horizon
and stars, sun, or moon. The sextant (from the Latin *sextus,* meaning
"one-sixth") was an innovative improvement over the quadrant. It was a
crucial navigational instrument in determining a ship's longitude and
latitude. The sextant consisted of a mirror mounted on a radial arm
that was moved until the **pole star** was reflected into a half-silvered mirror in line with its telescope to line up with the horizon. The angular

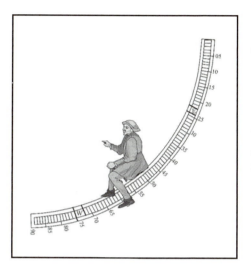

Figure 7.6 Tycho Brahe's 11-Foot Quadrant
Tycho Brahe claimed that large instruments were more accurate than were the older smaller ones. (This was before the telescope.) He sat inside the large quadrant he had built to observe the stars while his assistants recorded his measurements.

distance of the star above the horizon was then read from the graduated arc marked on the sextant. From this angle and using a chronometer (clock), the ship's longitude was determined. Today, most ships use the satellite Global Positioning System (GPS) for navigation purposes. (See the section on the astrolabe for more on early astronomical instruments.)

Siege Ladder

Despite efforts by ancient tribes to wall off their compounds for protection from various animals and their enemies, other hostile tribes would attack and find ways to breach the fortifications. In time, this led to what is known as siege warfare, where the defenders protected their ramparts while the attackers set up a siege as they attempted to force the defenders into submission by starvation or by breaching the defenses. As fortresses and defensive measures became more effective, so did the means of attack. The first documentation of a portable articulated siege ladder used as an offensive craft in siege warfare appeared in 537 C.E., shortly after the fall of Rome. (See Figure 7.7.)

Siege ladders were not used much during the early Medieval Period, but improved models were introduced during the Crusades of the 11th to 13th centuries. The siege ladder could be rolled up to a moat where it was used as a bridge for soldiers to cross as they were protected from a rain of arrows fired from the defenders on the ramparts above the ladder. Siege ladders were also designed to gain entrance to fortifications

Figure 7.7 6th-Century c.e. Roman Siege Ladder
The siege ladder was one of several innovative engines developed during the Middle Ages and Renaissance to breach fortifications.

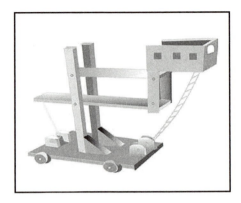

without breaching the walls themselves, since soldiers could climb the ladder to the protected platform and fire arrows down on the defenders inside the wall. One reason they were not used more extensively was the difficulty in moving the bulky structure into place.

During the Late Middle Ages and Renaissance more advanced siege engines, some of which could sling missiles over walls, were constructed on the site of battle. (The trebuchet, invented in ~1400 c.e., was designed to throw heavy stones against enemy fortifications.) The purpose of the siege was as much to starve defenders into submission by containing them inside the fortress as it was to breach their defenses. Long-term sieges, some lasting years, made conquest much easier and less bloody. Even today, cities and nations have built various types of fortifications in the belief that they can protect those within. The history of warfare has demonstrated that for every new type of defensive fortification, a new type of offensive weapon is invented to penetrate the new defensive measure. For example, a country can build deep tunnels and underground facilities considered to be impregnable, but an attacker can build deep-penetrating types of bombs ("bunker-busters") dropped from aircraft that explode inside these deep fortresses, or it can use "smart bombs" to breach the entrances to underground fortresses. Another example is the proposed "Star Wars" defensive missile system by the United States, designed to protect the country from an Intercontinental Ballistic Missiles (IBMs) nuclear attack. Undoubtedly, offensive methods of penetrating such a high-tech system will be developed. (See Chapter 8.)

Soap

Ancient humans scrubbed their bodies with the crushed leaves of soapwort plants, soapbark from the *saponaria* tree, as well as horsetail and yucca. While the sap and residue of these plants did provide a form of cleansing action, it was not the same as manufactured soap. It is not known exactly when soap was invented. Most likely, it was discovered by accident. One story relates how fat from animal sacrifice (probably goats) dripped into and leached lye out of the ashes of the fire and then continued to run down to the riverbank of Mt. Sopa in Italy. Here it mixed with clay, hardened, and was used by women to clean clothes in the river's water. There is some evidence that soap was known by at least 2800 B.C.E. in the "cradle of civilization" in Mesopotamia (present-day Iraq), where it was used as an antiseptic salve to heal wounds, as a hair dressing, and to clean clothes—but not for bathing. The Mesopotamians boiled animal fat and wood ashes in clay pots along with some water to form a type of crude soap. The Romans and Egyptians knew how to make soap but did not use soap for cleaning their bodies. Rather, they would bathe and then cover themselves with olive oil and pumice-like sand that was followed by scraping off the dirt and dead skin cells along with the oil. The Romans continued this practice while they occupied Europe.

Muslims introduced soap to the Europeans sometime around 1000 C.E. They used animal fat as well as olive oil mixed with lye derived from plant ashes, as well as natron, a compound of sodium. The Arabian soap made in Spain during the Muslim occupation was made from olive oil. Referred to as "Castile" soap, it was made from the olives grown in the Castilla region of Spain that produced solid, hard soap. By the 12th century the use of Castile soap, cut into small blocks, was widespread while much of the earlier soaps, rather than being hard, ended up more like modern liquid soap. Although soap was available to Europeans, it was a relatively expensive item. Thus, during medieval times few people bathed or washed their clothes with soap on any regular basis. From the time of the fall of Rome in the mid-5th century to the Age of Enlightenment, when the causes and effects of diseases eventually became known, personal hygiene was simply not practiced in Europe.

Spinning Wheel

The invention of the medieval spinning wheel has a long history. Primitive humans used animal skins for clothing, particularly in the colder climates. But lighter clothes were needed in hotter geographic regions. It is not known when the first woven textiles were produced. However, the art of weaving plant fibers to form baskets took hold in about 10,000 B.C.E. in Egypt, and possibly in Mesopotamia, China, and India as well. This basket-weaving process was most likely applied to weaving fabrics, as the basic technique was and is the same. Ancient basket weavers used fibers from flax, reeds, grasses, and the straw left over from growing grains. In more northern countries weavers used thin willow stems and roots to make baskets. Plant fibers such as flax and cotton, and animal fibers such as wool and silk, are basic materials that form yarn and thread for weaving cloth or fabric. But before weaving begins, the fibers must be formed into a continuous string. There is some evidence that the first crude instrument used for spinning fibers into yarn was a handheld device dating to about 8000 B.C.E. The more advanced hand-spindle, dating from about 5000 B.C.E., was an instrument that could be turned by hand in order to twist fibers together to form a continuous string of yarn or cord. The drop spindle was designed as a straight stick with a top-like wheel ("whorl") located near the bottom of the stick where the yarn was extracted. (See Figure 7.8.) (There is speculation that the rotation of the ancient hand-spindle gave someone the idea to invent the wheel in about 3500 B.C.E., while other historians speculate that the potters' wheel was the inspiration.) Hand spinning became an art, and even today, handspun fabrics are delicate and considered of high quality. Before mechanization to produce yarn and thread, textile workers in India could hand-spin thousands of yards of thin thread from one pound of long-fiber cotton.

Although the exact date is unknown, sometime in the late Middle Ages or early Renaissance in India, Egypt, or China, someone turned the horizontal top-like drop spindle on its side and invented the vertical spinning wheel. The wheel itself was larger than the handheld whorl and may have been turned by a hand crank. Later a cord was placed around the circumference of the large wheel that was attached to a small pulley or spindle that held the yarn. The device could be turned by crank, and eventually with a foot treadle-type pedal that allowed the operator to sit. Human power was replaced centuries later by water-power and steam power. Today, large automated spinning machines are powered by electricity. Soon after its introduction during the Renaissance, more productive types of spinning wheels encountered resistance from the hand spinners.

Some spinning wheels were over 6 feet in diameter and could out-produce hand spinners, although the wheels produced yarns that were

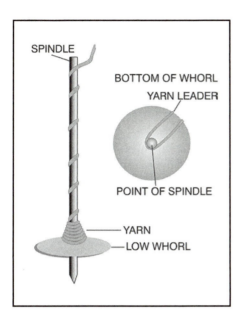

SPINDLE

BOTTOM OF WHORL

YARN LEADER

POINT OF SPINDLE

YARN

LOW WHORL

Figure 7.8 Low-Whorl Drop Spindle The drop spindle was an important invention about 7,000 years ago used to form yarn that was woven into fabrics. It also may have been the inspiration for the development of the wheel about 5,500 years ago, and it preceded the more advanced vertical spinning wheel of the Middle Ages and Renaissance.

less consistent and often uneven in texture. The incentive to solve these problems was keen because spinning wheels are much more economical than hand spinning. Today, in addition to the use of plant and animal fibers to form yarn and thread, humans have invented artificial threads such as nylon, which is spun as a viscous through a die. Many types of threads, cords, and ropes are now produced for a multitude of purposes, including braided steel cables.

Table Fork

Like all other animals, prehistoric humans transferred food to their mouths by using their feet, arms, hands, or fingers ("nature's forks"). This was the accepted method for millennia even after the introduction of spoons and table forks. It is believed that the fork was developed in the Middle East and then introduced to the West at the beginning of the 11th century. By the Middle Ages table spoons and forks made of precious metals, primarily gold, were valued as investments and passed on as heirlooms. There is a story that a princess from the Byzantine Empire (present-day Turkey) married the Doge of Venice and brought with her to Italy a collection of gold two-tined table forks that, up until this time, were unknown in Italy. Using a knife to cut up her food into small pieces, she then used her two-pronged table forks to eat the food. Using her forks instead of her fingers for eating enraged both the local population of Italy as well as the leaders of the Roman Catholic Church, who claimed, "God in His wisdom has provided man with natural forks—his fingers. Therefore, it is an insult to Him to substitute artificial metallic forks for them when eating" (Giblin). Soon after, the fork was mainly used by courtesans (high-ranking prostitutes), resulting in the Church's ban on the use of forks as blasphemous and immoral.

It was another 500 years before table forks were commonly used in northern Europe. Early flat, two-tined table forks were used to serve food onto plates or for individuals to stab a piece of food, let it drain, and then tear off smaller pieces with their fingers to eat. As the use of table forks spread, they were made from iron and steel as well as silver and sometimes gold. By the end of the Renaissance, they were curved and had as many as three, four, or even five tines. The handles of 16th-century table forks were made of carved wood, bone, antlers, and other materials. By the end of the Renaissance, at least for the tables of the wealthy in Spain and southern Europe, it was popular to display a great variety of decorative table forks as well as knives and spoons. It was not until the 1600s that table forks were used in England, and then only by gentlemen. The earliest known English table fork, now located in the Victoria and Albert Museum in London, is a flat, two-tined silver fork dated 1632–33. During the 17th and 18th centuries it became popular to collect pairs of knives and forks (and often spoons), some designed for travel, but almost always for use by the rich. Until relatively recently, the impoverished ate with their fingers, a practice that continues in many underdeveloped countries of the world. Oddly, fast food served in modern chain restaurants is now considered "finger food," and most people do not use table forks (even those made from disposable plastic) to eat their burgers and fries.

Telescope

During the Renaissance Period several philosopher/scientists conceived of an instrument similar to a telescope. Roger Bacon (ca. 1214–1292), a philosopher and alchemist, was interested in optics and lenses, among many other things. He invented a magnifying glass and proposed a design for an instrument with a combination of lenses that could be used for celestial observations. Leonardo da Vinci (1452–1519) considered the military applications for lenses that might be used as a spyglass during wars to spot troops at a distance.

Both the microscope and telescope were serendipitous inventions following the perfection of techniques for grinding lenses to make spectacles (eyeglasses). Both the microscope and telescope were invented near the end of the Renaissance. The credit for the development of the first telescope is usually attributed to either the Dutch spectacle maker Hans Lippershey (1570–1619) or Zacharias Janssen (1580–1638), who is also credited with inventing a microscope. Most history books, however, favor Lippershey as the main inventor, primarily because he applied for a patent for the telescope in 1608. As the story goes, Lippershey reversed the convex and concave lenses in a tube and just happened to look through the tube at a church steeple some distance away. To his amazement, the steeple appeared much closer when viewed through the tube than with the unaided eye. He kept his looking glass in his shop as a curiosity to attract customers, but neither he nor anyone else at the time considered using the device to view the heavens. The first practical use for the telescope was a military application, that is, during wars as a means of viewing the enemy's movement up close.

When Galileo Galilei (1564–1642) heard about the novel looking glass, he proceeded to construct one of his own with financial gain in mind. Galileo was an excellent lens maker who improved the curvature of his lenses to reduce optical aberrations. He built several telescopes, the most powerful of which was about 30-power, or equal to the power of a good pair of modern binoculars. Using one of his telescopes, Galileo observed two tiny objects that appeared to move around the planet Jupiter. Later he discovered two more moons of Jupiter's 16 satellites. After tracking and recording the changes in their positions, he realized that they were moons of the large planet and considered this evidence that the solar system was a heliocentric system as Nicolaus Copernicus (1473–1543) had theorized. After viewing the Milky Way with his telescope, Galileo concluded that this area in the sky contained more stars than any other section of the universe.

Telescopic technologies improved over the next hundreds of years. Today, there are not only telescopes that view objects in ordinary daylight, but there are others that can detect electromagnetic waves both longer and shorter than light waves (e.g., radio and X-ray telescopes). (See Figure 6.3.) And of course, the giant Hubble telescope continues to orbit the Earth as a satellite sending data back to Earth, which reveals the universe to be 13 to 14 billion years old.

Thermometer

The fact that the volume of air increases as it becomes warmer was known as far back as ancient times. However, this principle was not used in a practical sense until late in the Renaissance. There is some controversy as to who actually invented the thermometer as a heat-measuring device. It is likely that several people in different countries may have independently developed it. Even so, it is generally conceded that Galileo Galilei (1564–1642) was the original inventor of the air thermometer, called a *thermoscope,* in 1592, and there are records indicating Galileo used his new instrument in lectures in the early 1600s. His thermoscope consisted of a narrow, hollow glass tube about 10 inches long, with a glass bulb about the size of an egg attached to one end of the tube. The other end remained open. He would place his hands around the bulb to warm it. He placed the open end of the tube in a container with a few inches of water. When he removed his hands and the air in the glass bulb cooled, water rose up the tube into the bulb. He realized that the differences in the air temperature affected the volume of the air in the tube. Thus, if the air in the bulb cooled, water would rise into the bulb since the volume of the air in the tube was now less than when his hand warmed it. At first, Galileo was unaware that air pressure acting on the surface of the water in the lower container would also affect the level of water in the tube and thus the temperature reading.

Several others claimed to have invented thermometers. Robert Fludd (1574–1637), a Paracelsian physician who believed in the occult and devised medical devices, is said to have developed a thermoscope similar to Galileo's, even though he had never heard of Galileo's invention. Cornelius Drebbel (1572–1633), a farmer with no university education, became a skilled instrument and mapmaker. He received a patent in 1609 for a perpetual-motion machine that pumped the water, after it turned a waterwheel, up to a holding container where the same water could then be used over and over again to keep the waterwheel running continuously. His thermometer was designed as a J-shaped glass tube, open at the shorter end and closed at the longer end. It was filled with water, and later mercury, to measure air temperature. This design had the same flaw as Galileo's, in that air pressure affected the accuracy of the reading. Later, Evangelista Torricelli (1608–1647) used this principle when in 1643 he invented the first barometer to measure air pressure. Santorio Santorio (1561–1636), an Italian professor of medicine, claimed the invention in 1611 of a device similar to Galileo's. Instead of measuring air temperatures, Santorio placed the small bulb in a patient's mouth, thus inventing the clinical thermometer. Santorio also was the first to use Galileo's information on the pendulum to measure the heartbeats of patients. Many years later, in 1866, Thomas Allbut of England laid claim to inventing the clinical thermometer, but his design was similar to the modern, oral, mercury-based clinical thermometer and not the glass bulb used by Santorio.

Over the years scientists developed as many as 35 different scales on the glass tubes to estimate the temperature of air and liquids. Today, there are three basic systems: Fahrenheit, Celsius, and Kelvin scales. The first is the old English system, while the last two are basic to the universal metric system of measurements.

Waterwheel

Prehistoric humans were fascinated by the power of flowing water, yet terrified of raging floods. It is doubtful they tried to harness this source of power even after the Sumerians of Mesopotamia invented the wheel in about 3,500 B.C.E. Early humans built canals to control the flow of rivers and divert water to their fields. They used the wheel in building their carts and wagons, as pottery wheels, and to move engines of war, but not for waterpower—at least not until much later. The exact date and place of origin of the waterwheel are unknown. The first type was most likely the horizontal waterwheel. This type of waterwheel could be lowered into the water with a vertical shaft extending upward. As the flowing current passed the paddles of the horizontal wheel, it would turn the shaft that, in turn, was most likely used to operate a pottery wheel or even a bellows for a kiln, and later to turn millstones. This type of waterwheel was very inefficient and no doubt was not used very much since it relied on its placement within the current of a flowing stream of water. Centuries later, in the 1800s, the United States experimented with horizontal waterwheels that were driven by the in-and-out flow of tidal water, but again these were not a very dependable source of power.

Oxen-driven wheels used to lift water from a lower level to a higher level predated the first vertical waterwheels dating to the Classical Greek/Roman Period. Not long after, someone conceived the idea of placing a similar wheel with cups or cross-boards in a stream so that the flow of water under the wheel would turn a horizontal axle attached to the hub of the vertical wheel. This "undershot" type of waterwheel was also inefficient since it depended on the strength of the current. It was not until about 400 B.C.E. that the undershot vertical waterwheel made use of a horizontal hub/axle and drive shaft with gears. In about 200 B.C.E. an "overshot" waterwheel was invented and was much more efficient since it did not depend on the strength of the flow of water, but rather gravity acting on the weight of the water in the descending cups. It required many centuries before effective gears for vertical waterwheels were designed to convert horizontal motion into vertical motion that could be used to drive heavy millstones. There are recorded stories of the Early Middle Ages that tell of how people, particularly young women, were freed from grinding corn by the use of water-driven grain mills. There is a record of a 3rd century c.e. complex of 16 overshot waterwheels (8 on each side of 8 flour mills) built in southern France by the Romans. These wheels were spaced down a hillside so that water from a reservoir basin at the top of the hill flowed over the first (upper) wheel, and then the same water would flow over the next downhill wheel so that the same water would be used over and over as it passed from wheel to wheel until it reached an exit stream at the bottom. By the Late Middle Ages there were over 20,000 water-powered mills in existence. In addition to powering grain mills, vertical waterwheels were used to power sawmills, to turn grindstones, to sharpen tools and weapons, as pumps for both field irrigation and to drain mines, to forge bellows, as hammer mills used to forge iron, and to power massive looms to weave cloth.

There are four basic types of waterwheels: the horizontal waterwheel, the undershot wheel, the overshot wheel, and the breast wheel. (See Figure 7.9.)

The undershot waterwheel could only be placed in a fast and constantly flowing stream. Sometime in the early Middle Ages important innovations were made to the overshot type of waterwheel. When a stream or pond was located at a higher elevation than the waterwheel downstream, a new type of overshot vertical wheel became an important source of power. Usually a pond for storing water above the mill was built, and a sluice was constructed to bring the water from the pond down hill to the overshot wheel. The overshot waterwheel used the weight of falling water over its top to power the wheel instead of water merely flowing under the wheel. The continuously flowing source of water to the overshot wheel was the most effective type of waterwheel invented until modern times. During the Renaissance, overshot water-wheels were 25 or more feet in diameter because the larger the overshot wheel, the more effective it became, since the weight of the additional water would increase its power. They were also more dependable due to the steady flow of water from the reservoir. The question of which was more efficient, the undershot or the overshot waterwheel, was settled in the 1700s when a British civil engineer made some measurements and calculated that the overshot was twice as efficient as the undershot waterwheel.

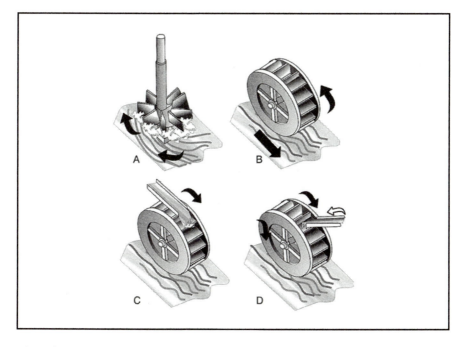

Figure 7.9 Four Types of Waterwheels
The four basic types of waterwheels are as follows: (A) the *horizontal wheel* set in a stream of water; (B) the *vertical undershot wheel,* where the water flows *under* the wheel; (C) the *vertical overshot wheel,* where the water flows *over* the wheel; and (D) the *vertical breast wheel,* where the water approaches the wheel from the front.

The horizontal wheel with its direct vertical drive shaft is the only type of waterwheel that does not need a set of gears to convert the direction of motion for useful purposes. Despite being ineffective when placed in a flowing stream, the horizontal wheel was the waterwheel of choice used by the Romans in the 4th century C.E. They built flour mills over streams and attached millstones to the vertical shafts attached at the hub of the wheels. The millstones were located on the upper floors of the mills that were built so the streams passed directly beneath them. Today, the horizontal waterwheel, known as a *turbine,* is the most efficient way to convert waterpower to electricity. The reason is that the force of the water driving the turbine blades is dependent on the depth of the water above the turbines. Therefore, the height of the dam holding the supply of water, and not the volume of water in the reservoir behind the dam, determines the force used to drive the turbines and thus the electric generators. The first hydro-electric plant to provide public electric power was built in England in 1881.

The technologies of the Greek and Roman waterwheels spread throughout Asia and southern and western Europe. During the Middle Ages and Renaissance the technologies and engineering of water-wheels, gear systems, and mills of various types were improved. Some historians claim that medieval Europeans were extremely power-conscious and became very inventive and innovative in constructing new powerful engines that used both water and wind. They saw nature as the provider of unlimited power, which could be harnessed to relieve human drudgery. This European fascination with new ways of generat-ing power resulted in the first civilization that was not destined to run on human power. Waterpower played an important role in both the development of agriculture and industry during the Middle Ages and Renaissance. Much later, the industrial revolution grew rapidly when the more efficient steam engine replaced water and wind power in the mid-1800s.

Wheelbarrow

A crude type of wheelbarrow that basically consisted of a box set on a wheel is believed to have been invented in China in either the 1st cen-tury B.C.E. or later in the 1st or 2nd century C.E. This crude device was not invented for agricultural use, but rather to carry weapons and sol-diers killed in warfare. The Chinese are said to have experimented with the placement of the wheel. The most common type had the wheel place at the center of the box to distribute the weight. This design was more like a litter than a wheelbarrow.

Even before the Middle Ages a two-person litter (two long poles with a board or cloth stretched between them) was the common type of device used to carry heavy objects. One disadvantage of the litter was that it required two or four people to carry the weight. During the 13th century, the shortage of labor in Europe most probably led to the importation of a type of wheelbarrow that is common today. The first wheelbarrows did place the wheel at one end, the load in the middle, and long curved handles on the other end. This design only required one worker, thus freeing up labor. (See Figure 7.10.) This basic design, still in use, provides a center of balance that transfers much of the weight of the load to the forward wheel.

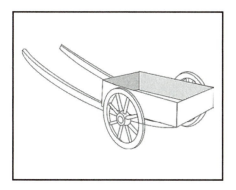

Figure 7.10 14th-Century European Wheelbarrow
The wheelbarrow of the Renaissance was one of the first to balance the weight on the front wheels. Later innovations used a single wheel in front of the load.

Windmill

The power of the wind was well known to ancient people. The use of sails to catch the wind and thus power boats was a major innovation, leading to centuries of seagoing exploration and discovery. Chinese and Greek literature recounts stories about different types of wind, to which various names were given. Before the general use of the compass, sailors from all countries partially determined their directions according to the type and directions of winds. Arabs followed the monsoon winds as they traveled to India, returning home when the monsoons reversed direction. Even Columbus's sailors depended more on the direction of the wind than they did on the captain's compass. Sailors exploring the New World were aware that at mid-latitudes winds blew to the west, and at northern latitudes they blew to the east. They used this phenomenon to plan their trips. Even in the Late Middle Ages men talked about the power of wind to move oceans, change climates, destroy homes, and cause earthquakes.

Most historians credit Hero of Alexandria (fl. 60 c.e.), a Greek physicist and inventor of the predecessor to the steam turbine (engine), with designing a machine to harness the power of the wind. His workshop notes describe a type of wind-powered "organ," which only existed in his laboratory. There is no record of using the wind as a source of power, other than for sailing ships, until the Persians invented a windmill that may have been designed from Hero's notebook. We do not know who invented it, but in about 640 c.e. the second caliph of Persia (present-day Iran) placed a new tax on windmills, so it is reasonable to assume they had been invented prior to this date. This new type of windmill was designed similarly to the old horizontal waterwheel. The tall vertical

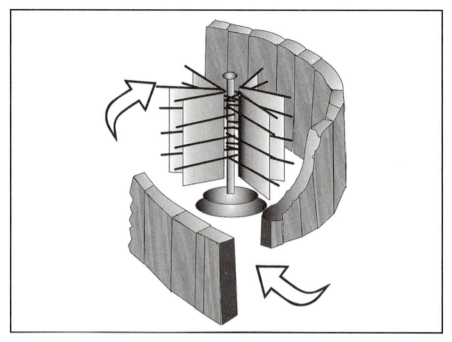

Figure 7.11 Persian Vertical Windmill
Ancient Persian windmills were designed with a series of sails attached to a vertical turning post/grindstone that turned when the wind blew through the open shutters of the flour mill.

shaft contained blades, or sails, attached to a vertical shaft housed in a vertical cylindrical brick building. The sides of the building had tunnel-like ducts (openings) that directed prevailing winds to the cloth sail-blades. These ducts were mounted on a vertical device that could be moved to catch the wind from different directions. (See Figure 7.11.)

By the 12th century C.E. the windmill had reached Europe. The vertical design proved rather ineffective, and a new type of horizontal windmill, similar to the more modern European windmill, proved more useful. The horizontal windmill placed four or six individual vertical sails around a horizontal hub/shaft that was connected by gears to a millstone inside the mill. The sail-like blades of the first horizontal windmills were difficult to move in order to catch the wind because the entire mill had to be turned. By the beginning of the 14th century a new innovative design solved this problem. The hub/shaft supporting the sail-blades were attached to the outside of a moveable cap on top of the larger base of the mill. But this cap still had to be moved by humans to catch the wind. In the mid-1700s this design was improved with the addition of a fishtail-like device that would automatically orient the blades so they faced the prevailing winds—something like the tail of a weathervane, which keeps the vane pointed into the wind.

During the Late Middle Ages and Renaissance, population pressure required the development of agriculture (as well as new industries) to provide adequate food. One use of the windmill was to pump water from behind dykes built on lowlands, thus reclaiming fertile land. There were great improvements in industries and agriculture from the early Medieval Period of Europe and the Renaissance and thus the need for sources of power beyond the availability of human power. Records indicate that by the 12th century there were over 6,000 windmills in Europe, and along with waterwheels the use of nature's power increased exponentially until the invention of the steam engine.

Today there is a move to reduce the output of carbon dioxide in the atmosphere created by the burning of fossil fuels by seeking alternative sources of energy. One source being explored, and in some areas exploited, is the use of modern windmills that house their own small generators that transmit their electrical energy into existing power grids. To date, they have been relatively unsuccessful due to the need to place them in uninhabited areas where there is a steady source of wind. Since they are noisy, people object to them being built near their homes, and environmentalists complain because the large blades kill birds. Although the electricity produced by wind is not, in general, competitive in price with electricity generated by other methods, the difference is lessening. Some countries in Europe and parts of the southwestern United States are investing in "wind farms" as the cost of wind electricity is becoming more competitive, and as an effort to reduce air pollution. Even so, it is estimated that even with government subsidies, wind-generated electricity will never exceed a few percent of the total requirements for a developed nation.

CHAPTER 8

WEAPONS AND WAR

There is evidence that from the beginning of prehistory humans developed tools that were used for everyday chores as well as for weapons of war between and among different clans, tribes, and cultures. New studies of the neurological combinational composition of our brains have indicated that there is an evolutionary genetic component, as well as acquired social factors in the makeup of humans, that drives them to cooperate and build positive relationships with individuals and groups, while they often compete and aggressively attack each other and other groups of people who are unlike their own clans or tribes. These conflicts were usually conducted for one of several reasons, including the capture of women, the enslavement of men for labor or cannibalism, access to territory or resources, or because another tribe had different beliefs and practices. Evidence also exists that simple differences between diverse groups, such as language, customs, religion, physical features, and even dress/ornaments, were perceived as a threat that often required a hostile response. (Even today, wearing different colors can lead to aggressive actions between rival gangs.) There are several ways to define weapons:

(1) The general category of *handheld* weapons can be subdivided into three groups: (a) simple shock weapons (rocks, clubs, sword, knife, etc.); (b) missiles that use muscle power (spears, javelin, sling, bow and arrow, etc.); and (c) mechanical power to launch missiles (trebuchet, catapult, crossbow, etc.).

(2) *Conventional* weapons may be thought of as those using chemical power (explosive chemicals) instead of human muscle power to propel fire balls, shells, and bullet missiles (rockets and guns—including cannons, muskets, rifles, pistols, etc.). Weapons that

use explosives are truly "chemical weapons" since their power results from the violent interactions of electrons of various chemical reactions that produce hot, rapidly expanding gases (explosions). These reactions do not directly involve the nuclei of the atoms of reacting chemicals.

(3) There are two classes of *nonconventional* weapons. The first involves two types of nuclear reactions. One uses the energy derived by the fission (splitting) of the nuclei of atoms of uranium or plutonium (e.g., atomic bombs), while the other involves the fusion (combining) of hydrogen nuclei to form helium nuclei plus energy (e.g., thermonuclear or hydrogen bombs). The other group of nonconventional weapons is also divided into two types. The first is sometimes referred to as "chemical weapons," although they do not depend on explosive chemical reactions, as does gunpowder. They are poisonous chemicals (sarin nerve gas, mustard and phosgene gases, etc.). The second type is biological weapons that are not chemicals, in the strict sense, but are rather living organisms used as biological agents to cause serious illness, disease, or death (bacteria, viruses, and toxins, such as anthrax, smallpox, plague, etc.).

Our knowledge of the actual origin and dates for the invention of most prehistoric weapons of war is skimpy, but we can surmise with some degree of accuracy that pre-*Homo sapiens* first used fists, then rocks and sticks, both handheld and hurled. The mace-like club and simple spear were most likely the first crude manufactured weapons. The stone ax is an example of human ingenuity dating to the pre-Bronze Age. It consisted of a rock that was shaped to smash bone or cut flesh attached to the end of a stout stick to form a deadly weapon as well as an effective tool. It was no simple task to figure out how to effectively attach the stone to the end of a club. It is assumed that this problem was solved when the stone was shaped with a groove that fit into the split end of the club (shaft). Strips of wet animal hide were wrapped around the stone and end of the shaft. When the leather strips dried, they shrank, thus firmly securing the stone ax head to the split end of the club. This concept of the stone ax was, and still is (e.g., the modern tomahawk and steel ax/hatchet), a technical marvel that employs several principles of physics, including angular momentum (the angular velocity of a mass around an axis of rotation). This is also one of the main physical principles responsible for maintaining artificial Earth satellites in their orbits—for example, the moon and artificial space satellites. The earliest recorded use of a bronze ax dates from 1300 B.C.E. in Egypt. By the

Figure 8.1 Mace
The maces of the Middle Ages and Renaissance were innovations based on the same principles as the ancient stone ax. They were deadly weapons when used in hand-to-hand combat.

3rd century B.C.E. in Mesopotamia and other Middle Eastern countries, clubs with mace-like spiked ends were formed from copper. By the 14th century the true mace was invented. It was short club with a metal mace attached to the end, which was enhanced with spikes or other protrusions that could inflict serious damage to an enemy. (See Figure 8.1.)

The progress of human civilization is directly related to the progress of the technologies used to develop weapons of war, as evidenced by the numerous and varied tools that have been adapted for that specific purpose over thousands of years. Most of the weapons, strategies, and tactics of war were invented or discovered long before the Middle Ages and the Renaissance. But during the period from the 5th to the 16th centuries ancient weapons and the methods of warfare were refined to become much more effective and deadly—only to be superseded over many generations by more advanced technologies and deadlier modern weapons.

There are several reasons why the technologies of war were slow to develop until the Middle Ages and the Renaissance. Early civilizations required human energy to hunt and gather food for survival. Only later was settled farming and animal domestication practiced. As a consequence, before 10,000 years ago there was little time or energy left for war. In addition, geographic topological restraints, such as mountains, oceans, rivers, and glaciers; poor methods of communication and transportation over distances; and unfavorable climates prevented the technologies of war from expanding beyond the small, closely defined groups that used simple tools and weapons. Early Chinese and Indian civilizations developed military technologies that were specifically adapted to their local regions. These weapons and technologies eventually were transported westward where they were adopted and also improved by the Greeks, Romans, and later by the northern European warlords during medieval times. This process of absorption began in the 7th and 8th centuries when Middle Eastern Islamic Arabs invaded northern Africa and southwestern Europe, bringing their weapons and techniques of war westward. The reinvention of weapons and the advancement of military techniques accelerated during the Early Middle Ages and the Renaissance when improved personal armor, substantial fortifications, transportation in the form of horse-drawn war machines, battleships, improved gunpowder, and guns with greater firepower dominated the world.

Body Armor

There are four basic types of body armor: *leather/cloth, chain mail, scale armor,* and *full body armor.* Armor dating to the 5th century B.C.E. was the first to be discovered in China. The Chinese developed a superior type of *leather armor* that was coated with lacquer, making it impenetrable. Lacquered leather plate armor that covered the upper torso and thighs was discovered in a Chinese tomb in about 433 B.C.E. The Chinese also developed *cloth armor* from layers of paper derived from the mulberry tree. Chinese warriors in the 11th century B.C.E. wore a type of personal armor made from the layers of animals hides, including the skins of rhinoceros. From the 5th century B.C.E. to about the 5th century C.E. Greek foot soldiers wore cloth body armor, called *cuirasses,* that was constructed from several layers of linen, while their leaders wore bronze chest armor plates. Warriors in India wore quilted cloth body armor into the early 19th century. The Aztecs of Mexico used thick cotton (about 1 inch, or 2.5 centimeters) quilted suits, soaked in salt brine to make the material harder, to fight the Spanish invaders. It is assumed that they learned this technique in the 10th century C.E. from the Mayans of southern Mexico, who had already developed powerful bows and arrows.

By the 13th century armor was made from layers of cloth or leather, often overlaid on each other and/or quilted for strength. By the late Middle Ages this form of arming was known by many names, including *arming coats, aketon, gambeson, hacketon, wambais, wambs,* any many others. This padded cloth armor was either worn beneath metal armor or as the only means of protection by foot soldiers who could not afford a suit of metal armor.

Figure 8.2 Protective Armor
Different countries invented a great variety of styles and types of body armor over the centuries. These examples are from the Late Middle Ages and Renaissance and are typical of the armor used in Western Europe.

A type of armor referred to as *scale armor,* composed of fishlike scales of metal plates, provided more protection. Scale armor was developed at about the same time in China, Egypt, and later in Greece. It was first made of wood, bone, and horn. Later, overlapping brass and iron plates were attached to leather jackets and pants. The early use of brass and, eventually, iron for body armor was limited because of the excessive weight and lack of mobility such armor provided. This problem was partially solved by inventing *metallic chain mail,* a flexible metal-cloth armor composed of overlapping wire rings sewn to the outside of fabric or leather tunics. During the 14th century great strides were made in creating *full body armor* by *armorers,* who were the metal workers and craftsmen who designed and constructed the body armor. (See Figure 8.2.)

The undergarments, entire suit of armor, helmet, spurs, sword, and other defensive gear worn into battle was referred to as the warrior's "harness." In the 15th century a number of innovations and improvements were made in the forming of metals, as well as the articulating of the hinged metal plates, that provided increased mobility and protection. Mail undergarments were sometimes worn beneath the joints of armor plates for increased protection, and later entire arms and hands were encased in fine steel mail rings that provided greater movements. As a suit of full body armor improved, it also became lighter, weighing about 55 pounds, which is about 10 pounds less than the backpacks and gear of today's United States soldiers.

The making of armor became a major industry from Italy to Germany, contributing to the spread of commerce throughout Europe. The types of personal armor improved as people and armies shared ideas as they traveled both westward and eastward over the silk and spice routes. The numerous military campaigns that were fought during the Middle Ages and Renaissance also resulted in the sharing and adapting of new types of military equipment. Personal armor improved over the ages, and by the time of the early Renaissance the quality of steel and the production of fine mesh offered considerable protection from head to hand to foot, as well as providing some protection for the knights' horses. Mail-type "hoods" were placed over the heads of warhorses, and breastplates protected them somewhat from frontal attacks, but they were still more exposed than were their riders. The importance of personal armor diminished after the reintroduction of gunpowder and improved handheld guns during the Middle Ages and Renaissance. Guns and gunpowder made war a more distant and less personal pursuit, as well as a more deadly way to settle affairs of state. Following is a description of some of the components of personal armor that were designed and used over several centuries, including the Early and Late Middle Ages and the Renaissance.

Byrne

A mail shirt originally constructed of open plates, the byrne evolved into a scale-type mail shirt with a hem that at first reached to the waist and later extended below the knees. (See Figure 8.3.)

Chain Mail

Chain mail was invented in the 5th century about the time of the fall of the Roman Empire and was later improved by craftsmen in northern Europe. At first, chain mail armor was bulky and heavy, but not as cumbersome as some of the full body plate. It consisted of a series of overlapping iron or steel wire rings riveted together and then sewn onto

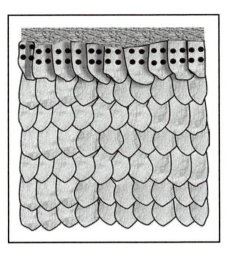

Figure 8.3 Byrne Scale Armor
Byrne-type scale armor was constructed of brass or iron overlapping
plates. Its design was similar to a
long T-shirt and evolved into protection from the neck to the knees.

leather or layered undergarments. Each ring was attached to four other
rings, two in the row above it and two in the row below it. This produced
an overlapping of rows that provided mobility as well as reasonable protection for the soldier. The *hauberk* was a mail shirt that extended low
enough to protect the legs. By 1300 C.E., a shortened version, the *hauber-
gen,* was laced up on each side with leather thongs. By the Late Middle
Ages and the Renaissance the technology of metallurgy advanced to the
point where smaller, lighter, yet stronger rings were used to provide
greater mobility and protection. Smaller steel rings also made it possible to produce smaller chain-mail items such as gloves, foot covers, and
helmets.

Full Body Armor

Although mail and plate armor were used during medieval times, it
was not until the 14th to 16th centuries that complete body plate armor
became practical. The chain-mail design provided greater flexibility but
less protection than did the improved armor and helmets made of
stronger steel plates. Even though chain-mail armor was effective
against slashing sword attacks, it provided inadequate protection from
thrusting spears, javelins, crossbow arrows, or small arms bullets. By the
15th century Italy and Germany were the major countries that produced full armor. One type referred to as *cap à pied* (head to foot),
"white armor" was highly sought throughout Europe. It was named
because of its highly polished finish, which not only made it attractive
but also provided some protection from rust. Full body armor included

articulated breastplates and plates to protect arms, legs, shins, and feet, and a full helmet with only a small hole for breathing and a slit for eyesight, or else a visor that could be lowered during battle. Knights and noblemen decorated their steel body armor with etched designs. (See Figure 8.4.) Their *surcoats* were also elaborately designed, as were their shields.

The reintroduction of gunpowder and the development of both handheld guns and larger bombardment weapons during the Renaissance greatly transformed warfare. The crews that fired bombardments into enemy-troop concentrations or fortifications were exposed and vulnerable as they reloaded their mortars. Since this process required some time to accomplish, large, heavy, hinged shields called *mantlets* were crafted as protection from the returning fire. By the late Renaissance the use of body armor became less important, due to the rapid improvement of gunpowder and handheld firearms. As a result it became necessary to increase the amount and thus the weight of body armor to protect soldiers from bullets and grenades. Consequently, this reduced the soldiers' mobility and hampered their ability to fight. Second, since bullets were more effective over greater distances than most other weapons during the Renaissance, it became imperative to increase mobility in order to survive and fight another day. Thus, armor became less cumbersome. Even so, some body plate armor was still in use after the Renaissance in the 17th and 18th centuries, including a type of ancient helmet still used during the First World War by some European countries. Today, modern soldiers wear several types of lightweight head and body armor that provide limited protection. Body armor in the form of small steel plates, fiberglass, boron and carbon fibers, or nylon fabrics sewn into garments that cover the vital chest, abdomen, and groin areas was first used in World War II. Since then, the personal bulletproof vests worn by both military and civilian law enforcement officers have been greatly improved with a new material called *Kevlar,* which is both lightweight and bulletproof. Kevlar-type material has also replaced steel in some current military-type helmets.

Gambeson

This was either a quilted or leather garment worn under the mail shirt designed to provide additional protection for the knights. A quilted gambeson was often the only protection provided to foot soldiers.

Hauberk (Saxon byrnie)

This was a rather long coat of mail worn by armed horsemen. Its lower back was split to allow for a more comfortable fit when riding horseback. This design has persisted to the present inasmuch as almost all modern men's suit coats retain the split at the bottom-rear.

Head Mail

A hood-type chain-mail covering, it resembled a ski mask with an opening for the eyes, while covering the entire head down to the throat and neck area.

Helmets (helm)

The chain-mail hood-like device of the 9th to 12th centuries later developed into a more open pot-like helmet to provide not only protection, but also a better view of the enemy.

The earliest type of helmet known appeared circa the 7th to the 11th centuries in Rome. It was called the *Spangen* helmet and was first invented by the predecessors of Rome. It was made of several small plates of iron bound together with rivets fastened onto a cap-like headpiece. Some museum examples include semiprecious stones and gold designs. Another early version in the 13th century resembled a bowl fastened to the top of the head. It was known as the *Chapel de Fer,* or "iron cap" helmet. By the end of the 13th and early 14th centuries the *Barrel* or *Great Helmet (Heaulm)* was invented, followed by the *Sugarloaf,* the typical medieval helmet. Many other types of helmets were designed to reduce the weight and improve protection. For example, the *Sallet,* a 15th-century helmet, had a flared tail and sides to protect the upper shoulders and neck of the warrior. Occasionally, chain mail was attached to the sides and back of helmets for neck protection. Later versions were constructed with a protruding nosepiece that evolved in design to cover the entire face except for an eye slit. This design often restricted vision, and thus a hinged visor that could be raised for a better view was developed soon after. By the 16th century the invention of a pivoted visor was used with the *Armet* helmet, which evolved into the 17th-century *Close Helmet* and became the first helmet made and used internationally. However, the helmet became so heavy that it required considerable strength to wear it. As a result, many warriors only donned their helmets during actual battles.

Figure 8.4 Armor of the Middle Ages and Renaissance
This photograph shows both a suit of full body armor and two types of shields
from the Late Middle Ages and Renaissance period. Most preserved suits of armor
indicated that warriors of this period were rather short men.
(Photograph by the author.)

As knights approached each other, they would raise their visors not
only to better see whom they were meeting but also to expose their face
as a salute to friend or enemy. This action of raising the visor may be the
origin of the modern military salute, where the right forearm is bent
45° with the hand brought sharply to just above the eye as a sign of
recognition and/or respect by a lower-grade soldier for an officer of a
higher rank. Ancient swordsmen and jousters also held up their swords
or lances in a salute to their opponent before doing battle or entering a
jousting contest.

Mail Gloves

Mail gloves were developed during the early Late Middle Ages after
craftsmen perfected techniques for producing fine chain mail that

could be formed into practical, flexible mail gloves to protect the hands of fighters. Some gloves had an open palm covered with leather to provide a more secure grip on weapons.

Shields

Most likely, the first shields were fashioned by ancient hunters as blinds to conceal themselves as they stalked their prey. Over the centuries this concealing shield was adapted for a new use as a defensive shield for protection from an attacking enemy. Originally, the 14th century soldiers' war shields were made of wood, covered in gesso (a mixture of plaster of paris and glue), that formed a relief design that could be painted. They were somewhat unwieldy but light, weighing only about 10 pounds or less. As body armor improved and provided better protection, shields became smaller and somewhat specialized. Foot swordsmen who marched into battle in close formation used a shield called a *hoplon*. It was a rather heavy, three-foot-in-diameter, circular, convex-shaped shield constructed of bronze and wood (and later, lightweight steel). The soldier supported his shield by placing his left forearm through a metal and rope attachment on the rear side of the shield. As spears were used more frequently in war, shields became more oval-shaped to afford the soldier greater room to throw the spear without removing his protection. When battles were fought with short swords and other handheld weapons, the shield was designed as a semi-cylinder or half-tube that protected the sides of the soldier as well as the front. Knights used small shields during tournaments but larger shields for protection from arrows during battle. From the time of the Crusades into the Renaissance, knights and noblemen decorated their shields with colorful representations of either the ladies for whom they fought or their family crests.

Spurs

The foot stirrup for horsemen was invented in China in the 2nd century B.C.E. No doubt, some form of sharp jabbing device to "spur on" the rider's horse was devised soon after. During the 14th century C.E. two types of spurs were in use. One used a spike attached to a *Y* that fit around the back of the heel with a strap. The other type consisted of a rowel (a sharp toothed wheel) attached to the *Y*. Spurs were effectively used by knights and other horsemen and also became a badge of honor. If a knight was dishonored, his spurs were broken and his shield was hung upside down.

Surcoat

A cloth garment that knights wore over the armor or mail. Often colorful, it was designed to identify the order of the knight.

Trousers

Mail pants were not worn; rather, cloth or leather material was utilized to pad the trousers of soldiers. Later, knights had full-frontal armor that covered their thighs, lower leg (shin) region, and feet. Leg armor was sometimes open in the back but jointed to permit motion.

Fortifications

There were (and still are) two main reasons to build fortifications: (1) to delay and possibly repel an intruder by placing obstacles in his path; and (2) to provide a shield that would afford the defender an advantage and allow him to utilize available weapons to reduce the advantage of the attackers. It is not known when humans constructed the first fortifications to protect their villages. Archaeologists have discovered postholes in the ground at the sites of prehistoric villages that they believe could have been used to erect timbers to support either walls for buildings or fences consisting of interwoven branches. These structures either kept their domesticated animals inside the compound or kept wild animals and/or their enemies out. The types and styles of fortifications have evolved over the centuries. During the Middle Ages and Renaissance they were designed to protect villages and even entire cities. Today man-made underground bunkers and command centers, as well as natural underground fortresses such as mines and caves, serve as fortifications for some countries.

In a military sense, fortifications are wood, stone, concrete, metal, or earthen structures that provide protection against attack. In essence, there are just two types of fortifications.

Field Fortifications

This somewhat temporary type of fortification has been employed since the warring days of the ancient Greeks and Romans. They are often hastily erected deterrents either in preparation for or during an impending battle and may or may not be effective obstacles in preventing an enemy from engaging in close-quarter combat. If they are used in conjunction with natural barriers, such as mountain passes, rivers, shorelines, or other obstacles, they can slow down an enemy, create a stalemate, or provide the fortifiers with a battlefield advantage. In the United States, westward-moving wagon trains would use the wagons and other objects as a temporary fort when attacked by Native Americans. Various types of temporary fortifications are still used in warfare, and thus the term "circle the wagons" during an emergency or argument is often applied.

Permanent Fortifications

These include a variety of well-known ancient and medieval walls, forts, and secure compounds.

The Romans built walls around their cities, including the cities they occupied in England. But these walls never compared with the Great Wall of China, which began as earth mounds and embankments constructed in northern China in the 4th century B.C.E. It soon became obvious that horses could jump over low earthen mounds or walls, thus the necessity of building fortifications higher. It was always much easier to fight men on foot than on horseback. Several centuries later this rudimentary wall of China was complemented with 500 miles of walls that joined other sections to form a barrier 1,300 miles long. It was designed to keep the northern nomads and Mongols from advancing south. In the early centuries of the Christian Era the Great Wall was strengthened and towers were added. During the 15th century C.E. its walls were reconstructed of stone, 40 feet high and 32 feet thick at the base, with towers placed two arrow shots apart. This formable barrier, as with most fortresses, did not protect the Chinese as the northern Manchurians gained control of China in 1644. (Sooner or later, no fortress is impregnable to a determined enemy.) The final wall is 2,150 miles long and much of it is in disrepair, while some sections have been rebuilt and open to tourists. It is one of the few pre-20th-century man-made objects visible from space.

The city forts of ancient Greece, Rome, and Egypt were known as *citadels*. They utilized many designs over the centuries, but all had some form of gates to control and limit access to the municipalities. (The term *citadel* derives from the Latin word *civitas*, meaning "city.") Throughout history, wood timbers were used to construct fortifications, mainly because they were inexpensive, but they were easily destroyed by fire. The type of permanent fortifications that were built during the Late Middle Ages and Renaissance depended on the availability of local construction material, such as stone, rocks, trees, and soil. Often two rows of wood boards or timbers were erected several feet apart. The space between them was then packed with earth, bricks, stones, rocks, or adobe, along with vertical pilings combined with horizontal planks. If available, rivers, seashores, cliffs, mountain gorges, and other natural topographical features were incorporated into the fortifications. However, with the development of improved gunpowder and improved French-type siege cannons, these walls provided inadequate protection. Later, the advantage tilted the other way when it was discovered that great earthworks could be constructed inexpensively outside the walls for increased protection. These earthen mounds or embankments often incorporated natural barriers or moats (from the Middle English word *mote,* or *motte,* meaning "mound") that returned the advantage to the fortifiers. The ditches resulting from excavation of dirt for the mounds outside the fortress walls were sometimes filled with water and were also referred to as "moats." By the mid-1400s the design of forts began to change. The high, vertical, flat walls of forts were vulnerable to bombardment because they could be breached by direct hits. This resulted in the construction of lower, broad-based, sloping walls that incorporated an earthen rampart at the base to provide better protection from projectiles. However, the sloping, low-walled rampart fortress could be scaled more easily by the enemy, which, in turn, required an increase in the defenders' firepower. The defending soldiers were stationed on parapets or battlements on top of the walls that connected towers. These gun platforms were known as *boulevards.* Rounded towers deflected cannon fire more easily than high, flat walls and also provided the defenders a safe place from which to fire their weapons at the attackers. Strategic openings along the parapet, called *crenellated* or firing positions, provided protection for defending archers, crossbowmen, and later those firing long guns. Platforms were erected atop the inside of the wall, or walkways were constructed on existing roofs of smaller buildings that were built against the inside of the fort. These offered an excellent view of the enemy and enabled defenders to dislodge ladders used by the attackers attempting to go over the top. In addition, a series of openings or "loopholes" were incorporated into the walls of the fort or along the fort's upper edges, where defenders shot crossbow arrows or guns at advancing enemies. (See Figure 8.5.)

When technical advances or improved techniques of war occurred on one side, corresponding changes were improvised on the other, thus restoring a balance or equilibrium that frequently shifted between the assailants and the besieged. As such, the advancement of war was, in some sense, the "mother of invention" for all kinds of technical devices during the Middle Ages and Renaissance. This improvement of war weapons also contributed to the science of the times, just as the science and technology of the times contributed to humans' abilities to conduct war. This concept of balance in offensive and defensive weapons and

Figure 8.5 "Loophole" for Firing Crossbows
A fortress "loophole" for crossbow use in the wall of Windsor Castle, Windsor, England.
(Photograph by the author.)

military techniques still exists today, as evidenced by offensive nuclear ballistic missiles and a proposed defensive missile-shield system to counteract this type of threat. The Maginot Line, a chain of defensive fortifications built on the northeastern border of pre–World War II France, was designed as a defense to stop and repel a possible German tank attack. It was never completed and proved completely ineffective, as dirt was just piled up over the obstacles providing a path for the tanks. In addition, the German tanks bypassed the Maginot Line to the north and rolled on to capture France without much of a fight by the French armed forces. This led to the expression "Maginot mentality," which refers to any military strategy that is exclusively defensive and therefore flawed, as well as any plan or action whose time has past.

Figure 8.6 Pentagon Shape of Fortification for Castle
An artist's depiction of Leonardo da Vinci's pentagonal-shaped castle fortification.

As a result of improvements in long-barreled cannons and other offensive weapons, the balance shifted in favor of the attackers until the late 1400s and early 1500s, when a new concept for fortifications was invented. It was called the *trace italienne,* better known as the *bastion* system. The bastion incorporated all the technical advantages of past fortifications, in addition to some important innovations. The bastion was a pentagonal-shaped stone fortification with sloping walls, often having semidetached extension forts with many angled sides that deflected artillery shells. Leonardo da Vinci (1452–1519) is credited with inventing the polygon-shaped (many-sided) fortress that incorporated massive outworks (mounds, etc.). (See Figure 8.6.)

The bastion-type fortification also included a series of ditches and artificial earthen slopes that reduced the effectiveness of artillery by scattering individual shots. Although the *trace italienne* evolved over 100 years, it is credited with holding off attackers for long periods, thus ending the era of offensive sieges. The military advantage then shifted to the fortifiers until the late 1400s and early 1500s when static or stationary defenders could no longer delay attackers. Improved grain-sized gunpowder and long-bore cast-iron cannons that were developed during the Renaissance again shifted the balance, as the attackers' guns held the advantage and sieges became longer. (During the Middle Ages and Renaissance sieges lasted about two months, on the average. In that time the defenders were either starved out, the walls were breached, or the attackers gave up. In some cases, the defeat of one side over the other was caused by factors other than war, such as the bubonic plague or other diseases.) Over the centuries the spread of the *trace italienne* over Europe tended to restore equilibrium, but it did not necessarily restore long-lasting peace. Wars, both land and naval, in one or several European nations were common during the Middle Ages and Renaissance periods. Several major wars were: the series of Crusades (1096 to 1291), the Hundred Years War (1337–1453), the War of the Roses (1455–1485), the Dutch Revolt (1554–1648), and later, the Thirty Years War (1614–1648).

Castles

Castles might be considered a special type of fortification, as they were more like fortified households of royalty and noblemen or protected churches. Most larger, permanent fortifications encompassed towns and cities. There are two general periods when castles were constructed, the pre-gunpowder era and the post-gunpowder years.

The earliest known castles were built in the first few centuries of the Christian Era. One of the best known was the castle at Axum, Ethiopia, which was constructed between the 1st and 5th centuries C.E. Four square stone towers were constructed around the ruling monarch's palace. These towers were about 35 to 40 feet tall and were built on a stepped foundation. They were connected by stone walls that provided protection for several hundred years. It seems that this design was imported to the Western world when the Muslims conquered north Africa and brought this castle-fortress concept to Europe.

Many early castles were fortified with walls of wooden pilings that were combined with earthworks. Castles, as with larger fortifications, constructed with wood were subject to rot and decay over time, or to fire during a siege. This design became obsolete as siege weapons improved. The typical Normandy castle encompassed a *motte* (moat) and *bailey* and was erected on a mound of earth often protected by water on one or more sides. Builders constructed the castle itself from the excavated material from the surrounding ditch. Protective dry ditches were at times lined with pikes made of sharpened tree trunks, with one end of the trunk buried while the sharp end pointed toward the enemy. If water was available, the ditch was filled with water to form a moat. (Note: During the 2003 war with Iraq, the Iraqi defenders dug moats around Baghdad and other cities and filled them with burning oil in an attempt to improve their defenses.) The *bailey* was the area directly outside the castle's walls and behind the ditch moat. This strip of land just outside the walls provided housing for the castle's household staff in times of peace. During wartime a defensive garrison was stationed on the bailey closest to the moat to enable the defenders to shoot arrows at anyone trying to attack the castle. Often an additional ditch and earthen rampart surrounded by a **palisade** was constructed. By the Late Middle Ages and early Renaissance (post-gunpowder), castles were built of stone and masonry works. The main turrets were high, rounded structures that deflected artillery shells and other missiles, while the moat and bailey prevented the enemy from approaching the castle. Added security was developed to prevent entry to the castle itself. Higher walls were built that provided crossbowmen the advantage of shooting downward from galleries or platforms toward the approaching enemy. If the enemy attempted to breach the wall, hot oil, tar, stones, and arrows rained down upon them. Entrance gates were narrow, with a drawbridge over the ditch located just outside the gate area. Heavy wooden or iron doors protected the entrance. In addition, a *portcullis,* a large, sliding, wooden, grille-like gate structure was suspended inside the wall just over the gate opening. The portcullis was positioned so it could be dropped rapidly if the moat and doors were breached. These defensive measures were effective to the extent that most sieges either failed or lasted until starvation or disease resulted in a truce. Most small villages had at least one castle belonging to royalty or to a local noble-man. During wartime this local castle was used as the village fortress. The construction of castles continued into the 17th century. Most of the well-known castles, such as Windsor Castle, were originally built as wooden stockades in the 9th or 10th centuries. Windsor Castle is located on 13 acres of high ground on the north bank of the Thames River several miles west of London, England. Currently it consists of sev-eral stone and masonry building complexes connected by the large stone Round Tower built on the highest point. Windsor has been the residence of the royal family since Saxon times of the 9th century, and it is currently the official home of the British royal family. Henry II rebuilt the original wooden castle using stone in the 11th century. He also added outer walls and stone towers. In the 13th century Henry III completed the walls and built the royal chapel. However, it was not until the Renaissance that upper stories of the fortress were converted into residential quarters for royalty. These quarters have been remodeled several times since then. Currently, this well-preserved castle is a national historical relic and a popular tourist site. (See Figures 8.7 and 8.8.)

The United Kingdom has preserved many of their castles and fortifi-cations from medieval times, the most famous being the Tower of Lon-

Figure 8.7 Windsor Castle—Main Tower
(Photograph by the author.)

don, which was long used as a prison, a garrison for troops, and a mint to produce coins.

Gunpowder

It is generally accepted that gunpowder was first invented in ~850 C.E. by the Chinese, who used it primarily for fireworks and ceremonies, not for defensive or offensive weaponry. Even so, the Chinese developed a rudimentary type of flamethrower referred to as a "fire lance" (Figure 8.9) and a type of grenade or percussion bomb (Figure 8.10) that later was adapted by Islamic forces to use against Christian Crusaders. The Chinese were also the first to invent a crude prototype cannon (Figure 8.11).

Figure 8.8 Windsor Castle—Round Towers
(Photograph by the author.)

As with several other early Chinese inventions, such as printing and the compass, gunpowder did not play an important historical role in the cultural evolution of the Chinese people. However, Europeans conceived an entirely different role for gunpowder than did the Chinese. Europeans placed an emphasis on its destructive use, whereas the Chinese saw its primary use for ceremonies and entertainment, with military applications being secondary, at least in the Early Middle Ages.

There is little dispute that the invention or discovery of gunpowder by Chinese alchemists and the importation of it to the West as a means of conducting war at a distance was one of the greatest influences on history from the time of the Renaissance up to the present. Modern manufactured gunpowder is specialized according to its intended use. Both the shot size, caliber of the weapon, and powder size (grain) are standardized. In addition, many other types of explosives, such as nitro compounds—for both firearms and noncombat explosive uses, such as blasting out tunnels, quarries, and mines—have been invented or developed and continue to be improved upon. (See Chapter 7 for more on gunpowder.)

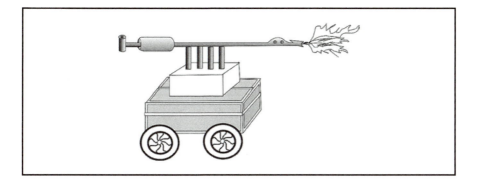

Figure 8.9 Chinese Fire Lance
Represents a design for an ancient "fire lance."

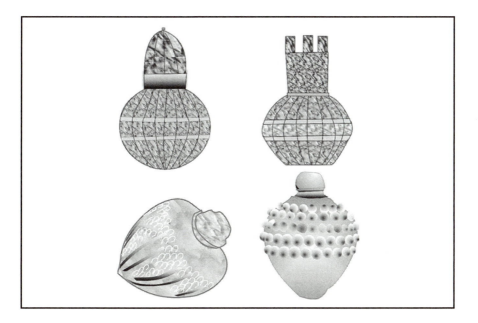

Figure 8.10 Early Types of Islamic Grenades
Several early types of Islamic hand grenades.

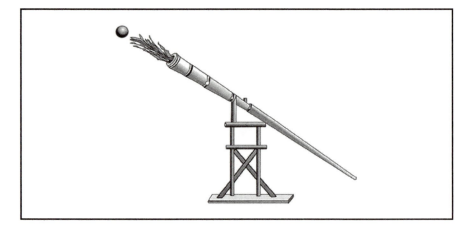

Figure 8.11 Early Chinese Cannon
An example of an early Chinese cannon that required a *Y* ground post to support it.

Weapons of War

From the early prehistoric periods when humans first lived in groups, there have been conflicts with other groups that could be considered a form of war. And when at war, both sides attempted to utilize any weapon they believed would provide an advantage over the other. This section presents the inventions, discoveries, and development of weapons of war from the Middle Ages through the Renaissance.

Siege Weapons

In the previous section on fortifications, weapons and techniques used to lay siege to these defensive structures were described. Following is an account of some of the offensive siege weapons:

(1) *Catapults*—The introduction of catapults, dating back to the 4th century B.C.E., became a major siege weapon in several countries. Dionysius of Syracuse designed a large arrow-firing bow-type catapult in ca. 399 B.C.E. It could hurl 13-foot arrows twice as far as a strong bowman. Other catapults were designed to hurl large rocks weighing several hundred pounds into fortresses. By the Late Middle Ages and early Renaissance the ancient catapult was transformed into an advanced counterpoise engine called the *trebuchet*.

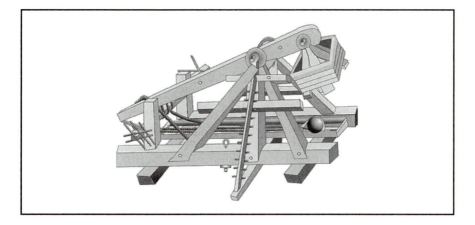

Figure 8.12 Sling-Type Trebuchet
A large sling-type trebuchet of the Middle Ages. Note the counterweight that provides the force to sling the ball which is placed under the weight. When released, the ball is propelled forward toward the enemy.

Using the principle of a giant lever it worked similar to a giant see-saw. A **pivot** in the arm of the trebuchet was located at about one-fourth of its length. This short end of the trebuchet supported a large stone weight called the counterpoise. A spoon or sling-like attachment was affixed to the other long end of the trebuchet's beam. (See Figure 8.12.)

A winch was used to lower the long end, thus raising the counterpoise weight on its shorter end. A missile, such as a stone or ball of flaming tar, was placed in the spoon or sling at the end of the long end, and when released, the trebuchet hurled the missile into the fortress. A trebuchet with a 50-foot-long firing arm could hurl a 300-pound stone a distance of almost 1,000 feet. This would require a counterpoise weight of several tons. Smaller trebuchets were designed to rapidly fire smaller stones that could harass the archers on the parapets of fortifications.

(2) *Greek Fire*—As the name implies, this siege weapon was developed by the ancient Greeks as both an offensive and defensive weapon. It was updated as a siege weapon during the Medieval Period and the Renaissance specifically to set fire to wooden fortresses by using trebuchets. While the original formula for Greek fire is unknown, it was most likely a mixture of several puri-

fied, light metals that, when immersed in water, produced hydrogen gas and burst into flame—namely, metallic sodium, potassium, or lithium. These metals, and possibly some other ingredients, were suspended in an oil-like base. It was most effective when used against approaching water vessels, since when wet, it would ignite by releasing hydrogen from the water and could only be extinguished by smothering it with sand to cut off the supply of oxygen in the air. The Crusaders in 1204 C.E. may have used Greek fire during the defeat of Constantinople; the Turks used it in 1453 when they recaptured Constantinople (present-day Istanbul). Versions of Greek fire have been used ever since in warfare, for instance, as flamethrowers of World War II, as napalm in the Vietnam War, and as a means to clear out caves in search of terrorists in the more recent conflict in Afghanistan.

(3) **Battering Ram**—The ancient Assyrians are credited with the invention and use of the battering ram. The Assyrian battering ram contained an iron head that could be quite effective in breaking down wooden fortresses. Later, the Greeks added a pendulum device onto a tall wood frame that could be repeatedly swung to destroy a fortress's wooden door. During the early Renaissance rams were designed not only with wheels to move them into position, but with a roof-like structure to protect the soldiers. A variation was the "siege engine," a tall structure that could be rolled on logs up to the fortress wall. It consisted of different levels (floors) connected by ladders. (See Figure 8.13.)

However, there were two reasons why the battering ram and siege engine never became major siege weapons. (See Figure 7.7 for a type of siege ladder.) First, defenders built ditches and moats that deterred the ram from getting close to the walls. These defenses exposed the attackers to archers as they filled in the ditch with dirt in order to approach the walls or door. The other reason was that defenders constructed thicker and higher ramparts and walls. Today, law enforcement SWAT teams sometimes use short but heavy handheld battering rams to burst open doors when making an arrest.

(4) **Tunneling**—Tunneling under the fortress wall was used in countries where the ground permitted this type of underground excavation. If done in secret, it was somewhat effective. One technique was to shore up the tunnel under the wall and, when completed, burn the wooden supports. This caused a section of the wall to collapse. Sometimes the tunnel provided an egress inside the fortress providing entrance for the attackers. Tunneling was used with varying degrees of success until the development and extensive use of black powder in the mid-1400s. When

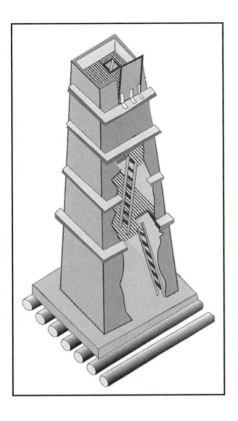

Figure 8.13 Siege Engine
The ancient siege engine provided a platform for attacking soldiers to gain entrance over the fortress walls. It was not very effective because it was very heavy and difficult to move into position.

a tunnel's construction was detected, defenders would explode a charge in it, driving back the attackers.

(5) *Cannon Fire*—By far, the most effective short-term siege weapons were cannon and artillery fire. These will be addressed later.

(6) *Time*—The most effective long-term siege weapon was time—just waiting and starving out the enemy. Other techniques, with varying degrees of effectiveness, were treachery, bribery, trickery, spies, and inducing diseases.

Large Guns

The Chinese invented gunpowder and used it for both pyrotechnics and, to a lesser extent, in small weapons to fire Roman candle-like "fire-lance" rockets. (See Figure 8.9.) The fire lance could be considered a gun but not in the same sense as a true gun, since its bore, charge, and projectiles were not standardized, nor was it a very effective weapon—essentially, it was just a Roman candle-type fireworks tied to a spear. It acted not only as a missile but also a flamethrower that continued to burn for about five minutes. Using large volleys of these proto-guns gave fortress defenders only a short-term advantage. Fire lances evolved over the centuries from bamboo tubes to metal barrels that, in addition to Roman-candle devices, could eject broken pottery, rocks, and scraps of metal. (Self-contained bullets had not yet been invented.) It took 450 years for the Chinese fire lance to reach Europe, where fortress defenders tied fire lances to poles that could be raised above the walls. In this way, they could then be fired at an enemy attempting to breach the parapets while not exposing the defender. This version of the metal fire lance, called a *mattock gun,* spewed out flames and several rock projectiles. By the late 1200s the Chinese also developed bombs that could be catapulted over several hundred feet. The powder in these bombs contained over 75 percent saltpeter in the mix, causing the bombs to explode upon impact with the ground. They also invented a long bamboo tube that might be considered a small cannon. The front end was elevated and braced by a *Y* support, while the back end was inserted in the ground. A charge of black powder was placed about halfway down the front opening, followed by stones or bits of metal. The powder was ignited through a small fire hole about halfway back from the front end. See Figure 8.11 for an artist's conception of this early cannon. It is assumed that the concept for this crude Chinese cannon filtered through Muslim countries to the west during the Middle Ages.

The first known description of a stationary gun in Europe was written in a 1327 manuscript that depicted a vase-shaped cannon resting on a platform, loaded with gunpowder and with a large spear-like arrow inserted in the open end. It was ignited through a *touch-hole* and could not have been very effective, since it would be rather bulky to move about and aim. (See Figure 8.14.)

Soon after, the larger bronze gunpowder-type cannon originated in Europe, not China, during the early 14th century. The evolution of these large guns spread throughout the known world. By the mid-1330s cannon technology traveled from Europe back to Asia, mainly to Arabia and China. (The first reference to bronze gunpowder-type cannons in China was during the late 1300s.) During the 15th century the Chinese developed cannons called "eruptors," made from cast iron, that fired hollow cast-iron shells filled with gunpowder consisting of a high concentration of saltpeter. When a series of these "bomb shells" landed inside a fortress, exploding with much fire and shrapnel, the advantage again returned to the attackers. A 7-foot 10-inch long, 5.5-inch caliber

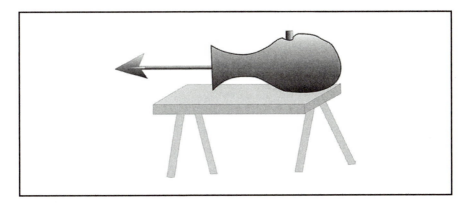

Figure 8.14 Spear-Shooting Cannon
An early type of cannon that shot arrows instead of cannonballs. Note an extra-large firing chamber was required to prevent it from exploding.

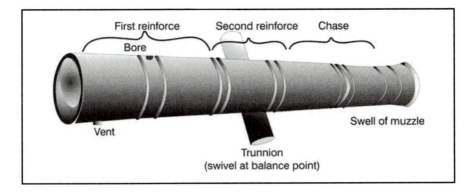

Figure 8.15 Chinese Bronze Cannon
The Chinese bronze cannons were the first true cannons designed to withstand the force of a large gunpowder charge that was capable of sending a cannonball over long distances. Note: the trunnion at its balancing point allowed its elevation to be adjusted.

bronze cannon was cast in China during the Ming Dynasty (c. 1644 C.E.). It is now a museum piece. (See Figure 8.15.)

And by the 1500s the manufacture of cannons spread to the Moghal Empire in India, the Ottoman Empire (present-day Turkey), and other regions of the Old World and Middle East. The cannons that the Turks used in the recapture of Constantinople in 1453 were huge. In fact, they were so big that they had to be cast on site since they could not be moved once formed. Foundries were built outside the fortifications to cast these bronze cannons. Some cannons invented during this period were over 10 feet long and weighed from 15,000 to 20,000 pounds. An enormous amount of gunpowder could be used to send 20- to 22-inch stone balls several hundred yards.

By the early 1400s European gun makers experimented with different methods of casting lighter, more mobile cannons. At first they adapted the "bell-casting" method to produce cannons of desirable shapes. Basically, this method used vitrified clay in a vertically hollowed-out mold constructed to the outside measurements of the cannon. Then a solid cylindrical post was suspended in the center of the upright mold to form the smooth bore for the powder and shot. Once constructed, a molten mixture of copper, tin, or zinc, and at times other metals, was poured into this vertical mold to form a bronze cannon. These *muzzle-loading* cannons required that the powder be inserted in the open end and then tamped down into the closed end of the barrel, until finally the ball was placed in front of this charge. A hot wire device inserted into a small hole where the powder was located ignited the charge. The technique was improved to include a stronger metal alloy for the breech area where the pressure from exploding gunpowder is greatest. Later, the *breech-loading* cannon was invented, in which the barrel and gunpowder sections were cast separately. This required a system of wedges to close the hinged breech before firing after the rear end of the cannon was loaded. At first, this did not work very well, as the superheated gas escaped around the juncture between the breech and the rear end of the barrel, thus reducing the pressure needed to project the cannonball. By the 1500s breech-loaded cannons were also constructed with wrought iron. Since this technique had the advantage of producing stronger, smaller, lighter, and more rapid-loading cannons, they could be used for naval warfare. Numerous methods of sealing the breech more securely were soon developed, further improving its firing power. From the rear-end breech, the cannonball was inserted first, followed by the gunpowder that was ignited through a torch hole. Breech-loading cannons also made possible the later invention of shells made from cylindrical brass cases that contained gunpowder, with the missiles located at the front end of the shells. This self-contained shell/missile greatly improved the firing rate of cannons. Or a large explosive cannonball was inserted in the breech followed by bags of gunpowder that could be ignited. Placing the powder in bags made the loading process easier and faster. As cannons became lighter and more mobile, they were cast with mounting pivot posts called *trunnions* as part of the barrels. This provided a means of attaching the cannon to a wagon or its own set of wheels. The trunnion pivots also provided a means of changing the barrel's elevation, thus increasing the cannon's efficiency.

By the 1400s improved methods of manufacturing cast-iron and wrought-iron cannons were devised. Of some note is the fact that cast iron was first produced in Europe in Scandinavia in the 8th century C.E. Even so, it was not generally produced in Europe in any quantity until the late 14th century. However, records indicate that the Chinese invented cast iron in the 4th century B.C.E., but they did not use iron for their cannons until about 1,000 years later.

Throughout England and Europe cannons were often referred to as *bombards,* or *lombards.* The long-barreled cannons were named *basilisk,* after a mythical fire-breathing dragon. The term *cannon* derived from a "cannonade," or battery of artillery. Iron was the preferred metal in the manufacture of cannons because of economics—that is, it soon became much less expensive to prepare than brass or bronze. However, its main problem, particularly at sea, was rusting of the internal barrel. (Rust is still a problem today for all guns made of steel. Cleaning after each firing, or using stainless steel to manufacture the weapon, can solve the rust problem.) Other methods soon replaced the use of cast-iron cannon barrels. One method cast the barrel in two pieces (lengthwise) and then welded the halves together, thus eliminating the need for the center part of the mold. Barrels were also constructed of a sheet of wrought iron that was beaten and rolled around a cannon barrel-shaped form. The seam was then welded. Another technique involved placing longitudinal strips of wrought iron on a wooden barrel frame and then welding the seams. This was followed by placing a series of closely spaced rings around the longitudinal iron staves of the barrel. This provided added strength. These methods were somewhat of a turning point in the use of cannons both on land and on sea. The lighter cannons could be moved rapidly by horse carts from place to place to form an artillery battery for attacking fortresses. By this time most warships had several cannons of different sizes on board. The next advance in cannon technology did not occur until the mid-1700s, when the barrels of cast-iron cannons were bored out after the cannon was cast as a solid in a mold. This greatly improved the accuracy of the diameter or caliber of the weapon.

As previously mentioned, this increased firepower and accuracy provided the attacker with an advantage, which was soon met by countermeasures such as ramparts, thicker, sloping walls, and rounded towers. The response to improved cannon fire also resulted in another innovation, the star-shaped stone/masonry structure called a bastion, or *trace italienne*, that provided an advantage in responding to the attackers with small-arms fire as well as small cannons to a position on the walls. Since new manufacturing methods produced cannons light enough to place on board ships, naval warfare was also transformed.

As the technology for large weapons developed, so did the need for special gunpowder and a means for improving the accuracy of weapons. It was soon discovered that a low percentage (about 30%) of saltpeter in the powder mix was preferred for charging cannons. Also, the use of large grain (corned) powder slowed the burning rate for cannon powder. If larger percentages of saltpeter and fine-grained powder were used for ancient cannons, they had a tendency to explode. The concept was that slower-burning powder to reduce the point of high-pressure gases reached a peak and thus slowly accelerated the more massive projectiles for cannons. (The heavier the projectile, the greater is its inertia at rest [Newton's first law of motion], which means that it resists movement to a greater degree than does a small revolver slug.) If the powder contained too high a proportion of saltpeter or the grain was too fine, the resulting accumulation of super-hot gas built up and peaked so fast that the inertia of the massive cannon shot resulted in the breech area of the gun exploding. This problem was partially solved by constructing cannons with a larger, heavier rear portion of the barrel (the breech) where the powder is exploded. By the time cannons were made of cast or wrought iron, their forward barrels were thinner and longer, which resulted in greater accuracy. Even so, smooth-bore cannons were notoriously inaccurate.

Ballistics

In the mid-1500s Niccolo Tartaglia (ca. 1501–1557), an Italian philosopher, attempted to solve the problem of cannon accuracy. He did this by experimenting with various methods for determining the flight paths of projectiles. He invented a quadrant that could measure small angles. In his 1546 book *Quesiti e Inventioni Diverse* (Various Queries and Inventions), Tartaglia described a system consisting of a two-arm quadrant. One arm was inserted into the mouth of the cannon, and the second arm was adjusted as perpendicular to the ground; thus, the angle of elevation could be determined by means of using geometry to calculate the trajectory of the cannonball. However, he was unable to gauge the velocity of the cannonball leaving the barrel. Therefore, his calculations were not accurate. This ballistic problem was not solved until the 1700s, when Benjamin Robins, who invented the ballistic pendulum, also invented a means of measuring muzzle velocity, which determined the effectiveness of different types of gunpowder and shot size. His book describing this method was the first scientific book published on ballistics. Another problem of cannon accuracy was not solved until Galileo's experiments with motion demonstrated that all projectiles followed a parabolic path when launched. When any object is projected—for example, a stone thrown, an arrow or bullet shot, or a cannonball fired—no matter its weight, how hard it is thrown or shot, or how great a distance it covers, it follows a parabolic path as it is affected by gravity. Aristotle's and other ancients' concepts of motion did not include this phenomenon mainly because they never bothered to observe the path of thrown missiles. Robins's ballistic pendulum was the first device that accurately measured the muzzle velocity of both large and small guns and thus the effective power of a given charge of gunpowder in relation to the size of the shot. Today, this information, along with meteorological data (e.g., wind and humidity) is required to determine the accurate trajectory of artillery fire.

The type of propellant and shells used for ancient cannons and handheld guns, as well as modern weapons, is important to achieve the intended purposes of the weapons. During the Late Middle Ages and early Renaissance stone-cut cannonballs were preferred to cast-iron projectiles. Even though they were larger, the stone balls were lighter and thus could be launched over greater distances than those of iron. Also, the larger stone projectiles could make a larger hole in fortress walls than a comparable iron one of the same weight but much smaller. The disadvantage was the cost. Forming perfect spherical stone balls, as compared to casting spherical iron ones, required considerable amounts of time for skilled stonemasons. Ancient iron cannons cost about one-third less than bronze cannons, but they were also heavier. Nonetheless, iron cannons could be reduced in size and weight if stone balls were used instead of cast-iron shot since less powder was required for stone shot. Over the years both accuracy of large and small weapons improved greatly. For example, the smooth bore of barrels was replaced with spiral grooves to create a spin on projectiles, which greatly improved accuracy. Today, the barrels of most types of firearms contain spiral grooves. (Shotguns are the exception.) Both the caliber and types of explosive propellants have been improved and standardized, as has the quality of metal used in the weapons. As cannons evolved, and knowledge of ballistics became more sophisticated, so did the types of cannonballs and shells. During the 1400s a cylindrical, hollow, cast-iron metal shell that contained gunpowder was fitted with an exposed fuse. It required lighting the fuse and getting out of the way before the cannon fired. If the shell became stuck in the barrel then the whole cannon would explode. Exploding shells were effective but dangerous to handle. These limitations were overcome somewhat by placing these types of exploding shells in short mortars, that were somewhat like stubby cannons. Another technique was to heat iron cannonballs red-hot and then fire them into a fortress, starting a fire. However, this too was dangerous, so a clay plug was inserted between the gunpowder and the hot ball to prevent a premature explosion. A thin-walled canister-type shell was sometimes loaded with grapeshot (small iron balls), or other shrapnel-type material. The shell casing was designed to separate, spreading multiple projectiles over a large area. This was very effective against approaching troops. Another ingenious device was the chain shot, which consisted of two cannonballs connected with a chain and which would slowly spin as it approached either foot soldiers or cavalry troops or ships. It was an ideal weapon to knock down attacking horsemen or the masts of enemy ships.

Handheld Weapons

There are two basic types of handheld weapons, those that are mainly dependent on human muscle power for their use and those dependent on gunpowder or other explosives as the major source of launching projectiles. Following is a short history of the invention and development of early handheld weapons, including the human, muscle-powered type and the gunpowder-powered type.

There are a number of muscle-powered weapons that were used by humans long before the period of the Middle Ages and Renaissance. (Records indicate that by 1500 B.C.E. the sharpened metal ax and sickle evolved into a type of sword. Later, straight as well as curved swords were cast in bronze with handle areas thickened for a firm grip.) Included among these ancient handheld weapons were the following:

(1) The handheld *ax*, one of the oldest weapons.

(2) The handheld *sling*, used since Biblical times.

(3) *Spears* and *javelins* were updated in the 15th century as the *Halberd* (or Halbard). This was a 5- or 6-foot hybrid pike-type combination ax and spear.

(4) *Swords* were invented before 500 C.E. However, the Romans devised a distinctive short bronze model only 20 to 24 inches long, known as *gladius* or "jabbing" weapons used in close combat. These were followed by swords with longer, double-edged slashing-type blades. As metallurgy techniques improved, so did the quality of steel used in these long swords. The Middle Ages and Renaissance was the age of knights and their exploits. The legend of Excalibur is the story of young King Arthur being the only person who could pull a special sword out of a stone to which it was fixed so tightly by magic that no one else could release it. He won the crown by this achievement. The Excalibur (which supposedly means "cut-steel") of the 15th-century legend was said to be not only a magical sword but also one made of superior steel.

(5) *Bow and Arrow*—The exact date when the bow was invented is not known, but one is pictured in a cave painting that dates to about 30,000 B.C.E. No doubt, the first bow was a springy branch of a tree that was bent and strung with a thong of animal hide. This design had limited tension and thus thrust. Even so, the *English longbow*, made of elm or yew wood in the 12th century C.E., which required some strength to use, became a formable offensive as well as defensive weapon of the Middle Ages. Today, similar longbows are fashioned from lemon wood or Osage. The *reflex bow's* ends curve outward (forward), which gave it extra power. This was followed by the *composite bow*, which was reinforced in early times by using thin strips of wood or leather. Today, special plastic compounds such as fiberglass or carbon fibers are laminated to the outer surface of the bow. Since the curve of composite bows is reversed, it is sometimes known as the *recurved bow*. This design provides the greatest mechanical advantage of all bow designs, except the crossbow.

The bow seems like a simple weapon, yet it involves complex physical mechanics. Potential energy is stored in the bow itself as the bowman draws back on the string, bending the bow. When the string is released, the tension in the bow converts this stored potential energy to kinetic energy in the string and arrow with enough force (mass of the arrow × the arrow's velocity) for the arrow to be a deadly weapon. The only limit to the effectiveness of the longbow is how much pull the archer can input to the string and bow, thus determining the trajectory, kinetic energy, and killing power of the arrow. This is referred to as the "weight" of the bow and is usually expressed in pounds of pull on the string for the length of draw, which is dependent on the length of the arrow. The longer the bow's arms and the thicker the bow's wood (as well as the type of wood) also determined the amount of potential energy (tension) stored in the bow. The longer the bow's arms, the slower it changed potential energy to kinetic energy. In other words, the arrow was projected with a lower velocity while at greater total energy than a short bow whose arrows achieve a higher velocity but with less total energy. The weight of different arrows as well as the length and stiffness of bows determined both the range and killing power of a longbow weapon. Archers of the Middle Ages and Renaissance developed enlarged right shoulder muscles and shortened bones in their left arms by pulling very strong bows over many years.

(6) *Crossbows*—The crossbow may have originally been designed as a trap to kill game. The bow was cocked so that when an animal knocked against a stick, it would fire an arrow and kill the animal. This design remained unchanged since then, except when used as a handheld weapon. Then it is held horizontally. The first record of a weapon-type crossbow was discovered in a 5th-century B.C.E. Chinese grave. Basically, it is a short horizontal bow mounted on a stock (similar to a gun stock) that contains a groove on its top surface in which to rest the arrow. Later developments provided a mechanical lever that gave a great mechanical advantage when cocked. When a trigger released the tension in the cocked bow, a large arrow could be shot over great distances. The Chinese perfected and used the crossbow as a weapon, and in the Late Middle Ages an aiming grid-sight was included. (See Figure 8.16.)

There is a 17th-century record of a Chinese artillery-type crossbow that could fire six bolts at a time with a range of several hundred yards. Another catapult-type bow, which required several men to operate, used multiple arrows and could kill several men at a time at a distance of almost one mile. In the 1600s the Chinese also developed a repeating weapon that might be considered as the first machine gun. The arrow-resting groove was designed to stack one arrow above the other in a **magazine.** As one arrow was fired, the next would drop into firing position. Tests made with this rapid-firing crossbow determined that 10 to 12 arrows could be fired within seconds and that 100 men could discharge over 2,000 arrows in about 15 seconds. As the technology pro-

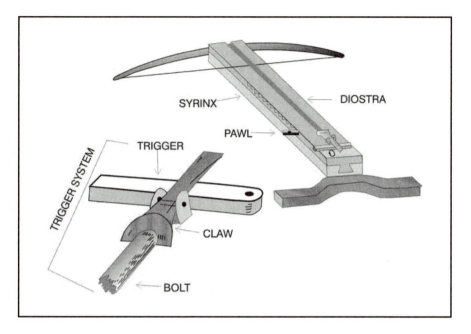

Figure 8.16 Ancient Crossbow
An example of a crossbow used during the Late Middle Ages and early Renaissance before small arms (rifles and pistols) were in general use.

gressed in Europe during the Middle Ages, crossbows were capable of piercing armor, particularly mail-type armor. Another advantage for the crossbowmen was that their cocked crossbows could be carried as they advanced. Regular longbowmen had to take time to nock (insert) their arrow into the string and draw the bow before firing. By the late 1200s in Europe, crossbows were manufactured of "crossbow steel," a highly refined spring-steel alloy of its day. By the 1300s large, portable crossbows were mounted on wheels similar to cannons. They were designed with a draw force of up to 1,000 pounds when using a rack-and-pinion type of windlass.

Even though a crossbow arrow could pierce body armor, the development of gunpowder and guns rendered crossbows less effective, since guns shot missiles over greater distances and reduced personal contact with the enemy to an even greater extent. (Ever since the use of explosives and other advanced weapons in combat, the conduct of war has become less and less of a personal, one-on-one engagement.) An interesting transition from the crossbow to the gun was a hybrid weapon developed by the Italians. They invented a crossbow with a bow-tube that held either arrows or small bullets. The ones that fired bullets were called *muschettae* ("gadflies") by the Italians, which is the origin of the English word *musket*.

(7) ***Handheld Guns*** are basically of two types: long barrels (muskets, rifles, shotguns) or short barrels (pistols, revolvers).

A Chinese stone carving dated to 1128 C.E. depicts a small demon holding a handheld cannon that appears to be about 2.5 or 3 feet long, with a bulbous closed end and a fluted open end. It shows a discharge of gunpowder shooting out fire and a ball projectile. The first true long-barreled, handheld gun was also invented in China sometime after the 13th century C.E. It is sometimes referred to as a small cast-iron cannon since it has a pivot mount. One well-preserved example is 39 inches long, with a 1-inch caliber bore. Both of these examples were more like light cannons than handheld firearms.

(a) The first true handgun was developed in China about 1250 or 1280 C.E. One such artifact was found in Manchuria. It was made of bronze, about a foot long, and weighed approximately eight pounds. It was typical of early guns in that the firing chamber was thicker to withstand the pressure of exploding gunpowder. It is assumed that this version of a handheld gun was brought to Europe by travelers and explorers even before Marco Polo made his journeys. The idea of light portable weapons came rather late because most armament makers believed that the larger the gun, the more damage it could do. Consequently, in the early 13th century massive cannons were manufactured. The need for more portable weapons was obvious, but this created a problem since at the time the standards of gun making required that the weapon's weight be related to the size of its bore in order to control kick-back recoil. (This formula also made guns heavy.)

(b) An early solution was to construct handguns by the stave method used for *bombards*. (See cannons.) In 1521 the Spanish introduced a new weapon that was somewhat smaller than the *arquebus*. (The arquebus had a 40-inch long muzzle with a small bore and a "stock" at the breech end.) It was too heavy to be considered a true handheld gun for foot soldiers to carry into battle. Even though this newer Spanish weapon, called a *musket,* was light enough to carry in battle, it required a Y-shaped ground support for its muzzle. By the early 17th century, small guns were cast in single pieces by similar methods used for cannons. They were heavy, smooth bored, and muzzle loading, and designed as portable weapons.

(c) During the 15th and early 16th centuries Leonardo da Vinci (1452–1510) invented or sketched several important types of military hardware. He is credited with designing the breech-loading cannon, rifle-barreled firearms, the wheel-lock pistol, and a cannon that used steam instead of gunpowder. By Leonardo's time in history, many improvements had been made to the loading and firing of the muzzle-loaders. First, they were made lighter so they could be raised and fired from the shoulder. Even so, the *Doppenlhacken* used for many years in fortress defense looked like a giant elephant gun. The matchlock design, premixed cartridge packets, better fuses (saturated in saltpeter), standard-size balls, reduced weight, and designs that allowed a single soldier to load and fire them were some of the improvements. In 1515 the first flintlock-firing, handheld, long-barreled gun was used in battle. It had a spring-loaded cogwheel that, when released by a trigger, would strike the flint igniting the gunpowder.

(d) By the late 16th century the term *musket* described a class of hand-held portable weapons used by the Spanish Army. It weighed twice as much as the arquebus but fired a larger shot over 400 yards. A single gunner could operate a musket because its barrel was supported on a forked *Y*. A typical musket had the following characteristics. Its bore was approximately .80 caliber. It was between 115 and 140 centimeters in length, weighed between 7 and 9 kilograms, with lead bullets weighing about 50–70 grams. This type of musket required large, well-built men, called *musketeers,* to carry them into battle. Besides its portability, the musket had the advantage over the arquebus in that it used heavier bullets, meaning that at close range a musket ball could pierce armor. It was not until the 17th and 18th centuries that smooth-bore, muzzle-loaded gun barrels were *rifled* (spiral-grooved) to impart a spin on the bullet—thus the term *rifle.* Rifling the barrel greatly improved the gun's accuracy. From the 18th century to modern times, breech-loading rifles have used brass cartridges that are inserted in the breech areas of the guns. The cartridge has the bullet at the front end, is filled with a high explosive powder and with a cap at the rear end that, when hit by the firing pin, explodes the powder. The first breech-loading rifles were known as "needle guns" due to the firing pin's needle-like shape.

(e) The *wheel-lock* mechanism to fire a gun was invented in the early 1500s. It used a small, spring-wound steel cogwheel that, when released, would strike a flint to ignite the powder. (This device was similar to modern cigarette lighters that feature a spring that turns the serrated wheel to spark the flint.) This invention led to an improvement known as the *flintlock* that, in turn, led to guns small enough to be considered true portable firearms, that is, pistols. The pistol could quickly be put into play. It also could be concealed which, even today, gives an advantage to the weaker combatant, or the citizen versus the criminal. One reason small arms and pistols were not manufactured as early as cannons was that serpentine powder, ideal for large-bore cannons with heavy balls and shells, was not adequate for smaller-bore guns. If this mixture of gunpowder was used in small arms, the bullets would only be propelled a few yards. It was not until the invention of corned or "crumb" powder with a higher percentage of saltpeter in the mix that small arms became possible. A charge that explodes faster and that propels the bullet with greater velocity made pistols that used smaller-caliber bullets possible. The reason smaller-caliber guns can use highly explosive powder is that they have a more effective ratio of barrel thickness to bore diameter than do cannons. In addition, improved steel and casting methods made the manufacture of pistols efficient.

A better understanding of the principles of chemistry and physics of weaponry has developed since the Renaissance. The next innovation for light, handheld guns was the development of shells in brass cases with the igniting charge and cap at one end and the slug (bullet) at the other. The famous 6-shooter was invented by Samuel Colt in 1835. It used a revolving chamber holding six shells that could rotate after each time the trigger released a hammer. The result was the firing pin hitting the cap that, in turn, exploded the shell's charge. The development of automatic handguns was not far behind the revolver. Other major inventions and discoveries for handheld weapons over the years included new and improved types of gunpowder (e.g., **cordite,** a smokeless gunpowder), specialized shells and bullets, rifling the barrels for greater accuracy, increased rapid firepower, and increased portability.

(8) *Other Types of Weapons*—Following the invention (or discovery) of gunpowder, its multiple uses for both pyrotechnics and weapons of warfare increased rapidly from the Middle Ages through the Renaissance. Several examples, not usually classified as guns, but which are still deadly, follow:

(a) *Rockets:* There is evidence that the Chinese used gunpowder to fire several types of rockets, and by the 1300s C.E. they were firing two-stage, solid-fueled rockets. Most of these were in the form of fireworks. During the 1200s the "earth rat" was a type of pyrotechnic designed to skid across the ground as the gas from the burning powder escaped from the rear end of the tube. This was later adapted as a "water rat," as one of these devices was attached to a small water ski-type board. It was used to scoot across water during celebrations. There is a common story that a salvo of these "earth rats" was fired along the ground toward advancing horse-mounted warriors and so frightened the horses that the tide of battle was turned. Obviously, someone had the idea that if you could mount one of the "rats" to a board, you could also mount one to an arrow. Soon after, the Chinese attached rockets to arrows with poison tips. Soldiers placed a group of arrows in an open basket-like launcher, called a "bee's nest," that held a salvo of about 5 or 6 arrows at a time. These arrows and launchers were light in weight, thus enabling each soldier to carry many into battle in wheelbarrows (which the Chinese also invented). During battle, multiple batteries of these launchers could fire as many as 32,000 rocket-arrows at a time, hurling as many as one million rocket-arrows during one battle. Over the next several hundred years the rocket was improved by adding stabilizing fins and a longer cylindrical body that held more gunpowder. By the 1200s small lead weights were placed behind the feathers on the arrow's shaft to provide a balance to the heavy rocket tube at the front of the device. This counterbalance weight kept the tail end down and maintained the rocket's flight path while increasing its distance. By the 1300s the open rear ends of rockets were constricted to increase the flow-velocity of the escaping gases and thus the internal forward force on the rocket. Later, in the late 1700s, this became known as the *Venturi* effect, which utilizes the *Bernoulli* principle and also creates the lift on an airplane wing. (Fluid pressure is lowered as the fluid's [air] speed increases.) Sir Isaac Newton's (1642–1727) third law of motion explains the phenomenon of why the greater the mass and velocity of gas escaping from the rear of the rocket increases the internal force propelling the rocket forward. ("When two bodies interact, the force exerted on body 1 by body 2 will be equal to [and opposite] the force exerted on body 2 by body 1.") In other words, the greater the force (velocity and mass) of the escaping gas out the back end of a rocket, the greater is the opposite force being exerted internally on the front end (nosecone) of the rocket. (Newton's third law of motion is the reason why rockets can fly in outer space, where there is no air. Many people in the past, and even today, mistakenly believed that the exhaust from a rocket engine needed something to push against to make it go forward.)

It is assumed that the story of Chinese rockets was imported to the West by travelers such as Marco Polo just two centuries after they were first invented in the East. Rockets used in European battles during the 14th and 15th centuries were tube-like structures that used black powder rather than the feathered sticks popularized by the Chinese. Sir William Congreve was among the first to experiment with the powder formula and construction of rockets. He used metal tubes that were strong enough to support a large charge as well as explosive or incendiary warheads. It is believed that "the rockets' red glare, the bombs bursting in air..." that Francis Scott Key saw from prison during the War of 1812, which inspired his "Star Spangled Banner," were Congreve rockets. During the 17th and 18th centuries there were some successes with using rockets in battles, but they soon took a backseat to more powerful weapons. With the development of cannons and lighter handheld guns, rockets were neglected as tools of war until the 20th century. The German V-1 and V-2 rocket bombs of World War II and the later ICBMs (Inter-Continental Ballistic Missiles), nuclear tipped and developed by several nations, are now the preferred types of war rockets. Of course, the so-called space age could not have been possible without the Western improvements in rocket technology, particularly liquid-fueled rockets and **telemetry.**

(b) *Bombs:* Soon after the Chinese discovered gunpowder they invented exploding "bombs"—that is, firecrackers. Although the Chinese knew about fireworks display-type gunpowder in the 9th century, the first record of gunpowder-type bombs was made in 1040 C.E. Tseng Kung-Liang published a paper describing three types of gunpowder bombs: (1) an explosive device to be hurled by a catapult; (2) a bomb constructed with hooks that continued to burn after it secured itself upon landing on wooden structures; and (3) a poison-smoke ball of chemicals mixed with gunpowder used in chemical warfare. By the 1100s the very explosive "thunderclap bomb," containing a high concentration of saltpeter, was developed. It was constructed by using an 18-inch to 24-inch length of dry bamboo about 1.5 inches in diameter that was filled with about 3 or 4 pounds of black powder. Shrapnel consisting of small pieces of porcelain was mixed with the powder as well as wrapped around the tube. A ball consisting of a gunpowder mixture was formed around the tube with a small portion of the tube exposed at each end. A red-hot iron was used to explode the bomb. By the early 1200s an improved version of a more modern type of bomb was invented. It was called the "thundercrash bomb" and required an even higher percentage of saltpeter because it was encased in metal, as are most modern bombs that require more powerful charges. Large bombs were often launched, after igniting the fuse, by placing them in the slings of trebuchets that could hurl them great distances. (See Figure 8.12.) These cast-iron bombs were exploded in air, which effectively spewed out deadly shrapnel and fire. The Chinese also developed many other types of special-purpose bombs, including those known as the *fire-oil* bomb, *magic-fire* bomb, *dropping from heaven* bomb, *flying magic* bomb, *flying sand* bomb, and the *wind-and-dust* bomb. Today, bombs are custom designed for specific purposes, including bunker busting and "smart bombs," which are somewhat like guided missiles.

(c) *Grenades:* Most of the ancient types of bombs were large and required a fuse to be lit before launching that, in turn, required good timing. Many of the designs for larger bombs, including the thunderclap type, were adapted to smaller hand-launched grenades. These also required a fuse to be lit before throwing at the enemy, making them dangerous to handle. Hand grenades were also used for hunting in ancient China. (See Figure 8.10.) Grenades have not changed significantly over the centuries. Many modern grenades have a timing mechanism that is activated by pulling a pin before being heaved. In addition, most grenades have the same pineapple-design outer container that, when exploded, spews out shrapnel.

(d) *Mines:* The Chinese also adapted the bamboo-type bomb into a deadly mine. Their invention consisted of a 9- or 10-foot hollowed-out length of bamboo filled with gunpowder and metal shrapnel. One end was sealed while the other contained a fuse. The bamboo tube was coated with wax and buried several feet (one meter) deep. When an enemy soldier tripped a trigger mechanism, the fuse ignited and exploded the mine. When a series of these mines were lined up, a chain reaction occurred with devastating effect. Sometime later, the Chinese also invented sea mines that were floated on a board that was kept just below the water's surface by weighted stones. An ox bladder encased the mine to keep it dry. A slow burning **joss-stick** fuse ignited the charge when it reached its target as the stream's current slowly sent them into the enemy's ships. It was not until the 16th century that land and sea mines were improved and utilized as weapons of war in the West. Today, there are many different types of mines, both land and sea, thousands of which are currently spread over many countries and are still in place, even after hostilities have ceased. An unintended consequence is that thousands of these land mines have not been removed and thus are a great danger to civilian populations living in these regions.

(e) *Flamethrowers:* The first record of a simple type of flamethrower using gunpowder dates back to 675 c.e. A more advanced model that used Greek fire was developed in 904 c.e. By the

10th century, attacking ships regularly used bundles of dried, oil-soaked reeds that, when the wind blew in the in the appropriate direction, were hurled into enemy warships. At times these dried, oil-soaked reed devices, as well as rudimentary rockets, were altered to become "fire arrows." However, these altered weapons were not true flamethrowers. In 1137 c.e. the Chinese invented what could be called a true flamethrower or "fire lance." (See Figure 8.9.) It was a rectangular brass tank supported on a table-like structure, and later on wheels. The brass box had several pipes leading up from the top to a horizontal tube with an igniter on one end and the flame nozzle on the other. The tubes were ingeniously connected inside the tank to provide a continuous flow of fluid once started. The fuel was either gasoline or kerosene stored in glass jugs carried into battle, so the brass tank could be reloaded from a side opening. Evidently, the Chinese knew how to fractionally distill crude oil to produce lighter, more volatile fractions of crude oil. Later, other countries attempted to develop flamethrowers, but their models had to be shut down for repumping to build up pressure between firings. The Chinese invented the double-acting piston bellows-type pump, which assured a continuous pressure to provide a steady flow of burning fluid. Modern flamethrowers have not been much improved since their initial invention, except they are now portable, hand-held, and can be operated by an individual soldier.

(f) *Poisons:* Since the beginning of human warfare a great variety of chemical weapons and biological weapons, in many forms of poisonous and toxic solids, liquids, gases, and diseases (bacteria and viruses), have been used in organized combat with varying degrees of effectiveness. As far back as 400 b.c.e. there are records of biological organisms being used in warfare. Examples include the following incidents: Scythian archers in Mesopotamia dipped their arrow tips in the blood and bile of decomposing bodies, thus creating poisonous weapons. In the 6th century b.c.e. the Assyrians poisoned enemy wells with rotting rye-grain ergot (a deadly fungus). In 184 b.c.e. Hannibal's army catapulted pots of poisonous snakes into the enemy's fortress. This act contributed to his victory. In the mid-1300s c.e. the Tartars hurled bodies of bubonic plague victims and several cartloads of excrement over the fortress walls of the Crimean city of Kaffa, resulting in surrender by the defenders. It is believed by some historians that this act spread the deadly bacteria and was the cause of the devastating black plague in Europe that wiped out about one-third of the population (approximately 25 million people). During the late Renaissance there are records of one group poisoning the wine of their enemies by spiking it with blood from leprosy victims, although it is doubtful this was effective. Even as late as the Civil War in the United States in the 1860s, blankets and clothing of Southern soldiers who died from smallpox were sold to Northern troops by Southerners as biological weapons against Northern soldiers. The Chinese also applied poisons to the tips of their fire arrows. The double-acting piston bellows that the Chinese developed for flamethrowers was also utilized to shoot poisonous gases toward the enemy, assuming the wind was blowing in the right direction. (Teargas was first used in China in about the 2nd century C.E. The Chinese also burned dried mustard to produce a type of poisonous mustard gas in the 4th century, which was not exactly the same as the mustard gas used in World War I by the Germans.) By the 6th century the Chinese drove chariots equipped with bellows that spewed quicklime into the air as they raced through enemy ranks, in order to irritate the enemy's eyes. These devices, along with gunpowder "smoke bombs" and horses with burning rags tied to their tails that became so frightened they charged the enemy, provided crossbowmen and foot soldiers an advantage in battle. By the 12th century C.E. the Chinese perfected teargas, and by the 14th and 15th centuries many varieties of gunpowder that produced great smoke clouds had been developed. One type that was fired from a proto-gun was the *poison-fog magic-smoke eruptor,* which produced poisonous smoke. Another mixture contained a substance that, as it burned, would stick to the clothing and skin of the enemy (similar in effects as napalm). In the 1500s the Chinese developed a poison teargas formula called "five-league fog" composed of 28 percent saltpeter, 28 percent sulfur, and 44 percent charcoal. It most likely also contained arsenic, sawdust, and human or animal excrement. This mixture burned slowly producing a cloud of debilitating smoke. Another, called the "soul-burning" fog, had a high percentage of saltpeter and less sulfur and charcoal, but contained arsenic sulfide and deadly animal poisons that were spread from the resulting high explosion.

By the 16th century the Chinese, as well as the Europeans, employed many kinds of minerals, plants, or animal poisons in their antipersonnel bombs. In the early 1500s Leonardo da Vinci designed plans for attacking the enemy with fumes from burnt feathers, sulfur, arsenic, and other elements, as well as toad and tarantula venoms mixed with the saliva of rabid animals. During World War I, chlorine, mustard, and phosgene inorganic chemical poison gases were dispersed. Fortunately, since then, poisonous gases have been outlawed internationally. However, they have been used in isolated instances. Today, organic nerve gases (sarin), demonic biological germs (smallpox, anthrax, botulism, cholera, Glanders [equine disease], plague, typhoid, and tularemia), and advanced poisonous chemical formulas are being developed and stockpiled as weapons in a number of nations, including most Western countries. There are several reasons why chemical and biological terrorist agents are seldom used: (1) The agent must be either highly toxic or infectious. (2) The virulent strain of the desired organism or infectious agent must be obtained. (3) It must be easily produced and cultured in large quantities. (4) It must be stable enough to be maintained and withstand transportation to the site. (5) It must also be stable enough to withstand heat and the blast of the bomb explosion or mechanical spraying. (6) It must be dispersed in the most effective particle size. (7) It must be spread in a large geographic area in order to cause mass infection. One reference lists 119 recorded historical events of the use of bioterrorism (not chemical gases, but biological agents) from the 6th century B.C.E. to November 2, 2001 C.E. Unfortunately, many chemical and biological poisonous agents are rather easy to concoct, and the threat of their use by terrorists or by armies in future warfare will, no doubt, remain as long as human beings continue to make war against each other.

Conclusion

Many of the inventions as well as innovative technologies for weapons of war were developed in the East (China). During the Middle Ages these Eastern tools of war passed through the Arabic regions as they were transported to Europe during the Islamic invasion of the West. The West then further developed and improved these weapons, and in time, some improved weapons were transferred back to the East. The Islamic Arab countries contributed substantially to some of the sciences, such as astronomy, alchemy, and medicine, but they made few original contributions to the weapons of war. The Muslims did little more than adapt the Eastern weapons of war and transfer them to the West. One of their major contributions was the Arabic translation of Greek knowledge that was brought westward with their invading armies. The retranslation of these Arabic books to Latin in Europe (and later into vernacular languages) hastened the beginnings of the European

Renaissance and the later Age of Enlightenment, but the Arabs did not add much to the art of weapons and war.

Humans are what humans do, and our history reflects that war and conflict are the norm—with some exceptions of mutual cooperation. It is said that peace is just an interlude between wars, and that war is just an extension of politics. Over the past several thousands of years humans have spent more time at war than we have spent living in peace. Even in modern times of peace there are always several minor wars being fought somewhere on Earth. There has never been a time when war has not been waged someplace on our planet. Many theories abound as to why this is so. Evolutionary geneticists explain aggression as an inherited trait for survival and along with some learned experiential factors as part of our "human nature," but exploration of this topic is beyond the scope of this book. What is known, however, is that scientific experimentations, discoveries, inventions, tools, and technical knowledge have enabled humans to control their environment and improve their lives in general. The very same science and technology that can improve humanity's condition also provides methods that lead to its destruction, if we choose to do so. Humans are responsible for wars—not science and technology.

Summary

"Invention Breeds Invention"

Ralph Waldo Emerson (1803–1882)

Just as "Invention Breeds Invention," so are "Discoveries Followed by More Discoveries," and "Experiments Lead to New Experiments." Emerson's insight is especially true for the Renaissance and the period of Enlightenment that followed. Just about every invention, discovery, or experiment from the fall of Rome (476 C.E.) through the Renaissance and beyond has some historical precedence. For instance, the ancient Greeks developed relatively high levels of technology as well as a lasting body of literature and political systems. Centuries later the Nestorian Christians (a cult of Homeric-age Christians who followed Nestor, the wise Greek) translated the Greek texts into Arabic during the 9th and 11th centuries C.E. The knowledge gained from these translations resulted in a new type of Muslim who studied these astronomy, alchemy, medicine, and mathematics texts. The Arabs' absorption of Classical Greek-Roman culture resulted in more scientific discoveries and innovations than in all previous Arab history. This scientific knowledge and a similar philosophy based on reason was not assimilated in Europe, in either thought or culture, until the Late Middle Ages and early Renaissance, when these texts were retranslated into Latin—the language of the educated few of Europe. The Medieval Period or Early Middle Ages, between the years 400 to about 1200 C.E., has been correctly referred to as the "Dark Ages," as far as science, technology, and discovery were concerned. Little if any in the way of significant scientific progress was accomplished during the Medieval Period in Europe.

However, between the years 800 to 1300 C.E. Islamic Arabs developed a widespread intellectual civilization. This new type of Arab intellectualism was referred to as *Falsafab,* which might loosely be translated as a "philosophy" that expressed a desire to live rationally according to one's view of reality within the cosmos. As Muslims explored the natural sciences, they also applied Greek metaphysics to their Islamic religion of that period. This revival of rational intellectualism arrived in the West shortly after the Muslim conquest of countries reaching from the western border of China, to northern Africa, and finally westward to southern Spain. The translations of Greek science, mathematics, medicine, and technologies into Latin during the 10th to the 13th centuries led the West out of the darkness, so to speak. Arabic intellectual rationalism influenced Western thought, bringing about Scholasticism (the European philosophy that unsuccessfully attempted to meld reason with religion). Up to this time there was no consistent religious belief system related to the nature of the existing world. (There still is little consistency among major religions as to their beliefs related to the origin and nature of the universe.) Before the Renaissance superstitions abounded. Mathematicians, physicians, and philosophers practiced the occult. Astrology and alchemy retarded rational inquiry into the nature of things, while the uneducated struggled to eke out an existence. This "darkness" began to fade in Europe as Arabic rational intellectualism was absorbed. Ironically, at the same time that western Europe emerged from the Dark Ages and into the Renaissance, the Spanish Muslims were driven from Europe. A decline in their science and technology, in specific, and their level of rational philosophy, in general, soon followed—a process that continued over the next centuries and into the present. Historians do not agree on the exact cause or causes for this decline in Islamic rationality, but they do agree that it began in 1492 when the Spanish monarchs Ferdinand and Isabella conquered Granada, the last Islamic stronghold in Spain. This date may be significant because it is the same year that Columbus discovered the New

World, which, along with the printing press, signified the height of the European Renaissance.

Just as the term "Dark Ages" is appropriate for the Middle Ages, the term "Renaissance," meaning "awakening," is apropos for the period following the Middle Ages from about the 14th to the 16th centuries. This period of renewed interest in classical knowledge led to significant changes, first in Italy and then slowly spreading to northern Europe. Both land and sea trade routes were established between merchants of the Far East and western Europe. More significantly, new concepts of science and technology spread westward from the East. Two of the most important technologies imported from the East were the printing press and manufacture of paper, both inventions of ancient China. The reinvention of this technology during the mid-1400s provided a source of books and reading materials that heretofore had been restricted to the clergy and the wealthy. This increase in knowledge spurred literacy. The advent of the Renaissance saw a weakening of the Roman Catholic Church's influence and hold on both the sources and philosophy of learning. Superstitions and Scholasticism slowly gave way to more rational approaches to understanding the nature of the world as new ways of thinking developed. The evolution of universities as influential centers of education and philosophies began in the 13th century. Notable professors, such as Robert Grosseteste, Albertus Magnus, Roger Bacon, Thomas Aquinas, and Alexander of Hales, among others, advanced the revival of learning. For example, Roger Bacon, although not an experimenter himself, was one of the first to recognize that experimentation and mathematics were essential to the understanding of nature. He was aware that a systematic approach to investigation (i.e., the scientific method) was required to gain practical knowledge so that man could understand and control nature.

Many discoveries, inventions, concepts, and ideas that are now considered products of Western civilization were, in fact, imported from the Far East and reinvented or significantly improved by the newly educated generations of European scholars and scientists. One major difference between the East and West from the Renaissance until today is the change in attitude of Western culture toward the use of science and technology, rather than religion, to explore the nature of the universe for the improvement of the human condition. Emerson's quote, "Invention Breeds Invention," also applies to the period of post-Renaissance Enlightenment when many older technologies and inventions were further improved and exploited. In addition to the printing press, moveable type, and paper, several other examples are the following:

1. The European importation from the East of the magnetic compass, as well as methods of constructing larger, more seaworthy ships, equipped with rudders, watertight bulkheads, and efficient sails. Improved sailing vessels and navigation led to exploration and discovery of new lands.

2. The formula for gunpowder and its use as an explosive, as well as blueprints to construct and improve offensive and defensive strategies.

3. The casting of brass cannons with molds to form the bore was replaced by boring the hole in a solid brass tube. The heat generated by this process led to the theory of latent heat that later was applied to the development of efficient steam engines, as well as the theories of thermodynamics and entropy.

4. Leonardo da Vinci's many drawings of mechanical devices that he made during the Renaissance have become the applied science and technology of today, for example, the airplane, parachute, and submarine.

5. Paracelsus, the Renaissance alchemist/physician, challenged the established tenets of the medical profession based on many of Claudius Galen's 1,500-year-old misconceptions. Paracelsus considered the human body from a chemical point of view and thus is credited with inventing iatrochemistry, that is, treating specific diseases with specific chemicals and drugs. He is also a well-known practitioner of homeopathic medicine.

6. The use of ancient magnifying crystals led to the development of ground lenses for spectacles, later resulting in the invention of the microscope and telescope. Over the years improvements in these instruments enabled the eyes of scientists to study all aspects of our world—from the minutest particles to the very distant edges of the universe.

Classical science in mechanical terms originated with the ancient Greeks and Romans, and later these terms continued to be used by Arabic and European scientists. During the Renaissance, natural philosophy became more abstract with the renewed interest in rational approaches to investigating science and mathematics. While this emphasis on rationality and the search for reality continues into the 21st century, it is becoming more evident that science may be reaching a stage that will require a closer association with a metaphysical type of philosophy to complete our understanding of the interconnectedness of the whole of nature.

GLOSSARY

ascension
In astronomy, it refers to the rising of a star above the horizon.

attenuate
In medical terminology, it refers to an attempt to lessen the virulence of a disease caused by a microorganism, e.g., syphilis, smallpox, often through the use of heat treatments, vaccination, etc.

autopoiesis
The self-organization and self-maintenance of an organism. Possible system for the origin of life.

bestiaries
A collection of Middle Age or medieval allegorical fables concerning actual or imaginary animals, i.e., morality tales.

cordite
A nitrogen-based explosive, smokeless gunpowder, resembling brown twine, used in medieval weaponry.

cupping
An ancient medical practice that involved either the application of heated cups to the patient's skin in an effort to draw blood to the surface, or making small incisions in the skin, whereupon a cup containing a small piece of lint was burned to produce heat to draw blood when applied to the surface of the skin.

dead reckoning
A navigational term referring to the use of instinct or supposition, often based on previous experience, to determine one's geographic position. Ancient sailors used this method when celestial observations— i.e., those using the moon, sun and stars as positional markers—were not possible.

declination
Also called magnetic variation. In astronomy, it is the horizontal angle in any given location on the celestial sphere between true north and magnetic north.

dicots
A shortened version of the biological term *dicotyledons*, which refers to the majority of flowering plants (angiosperms) that have embryos with two seed leaves. Dicots have broad leaves with net-like venation. Examples are most forest trees, beans, potatoes, broccoli, cabbages, roses, geraniums, and hibiscus.

diurnal
Refers to an event that occurs on a daily basis, or one that occurs regularly during daylight hours. Opposite: nocturnal.

eclectic
From the Greek word *eklektikos,* meaning selective. It refers to an attitude or tendency to choose the best from a diverse set of systems, sources, or trends, e.g., a Renaissance person.

ecliptic
The great circle in which the plane of the Earth's orbit around the sun intersects the celestial sphere. In other words, it is the apparent annual path of the sun across the sky.

escapement
In timekeeping devices, it is the mechanism that provides both the energy impulses and rate of movement of the wheel that drives the clock.

fletched
An archery term meaning that feathers have been attached to an arrow.

geodesy
The geological science that determines the size and shape of the earth, as well as the precise location of points on its surface.

germ theory
The hypothesis that states that infectious diseases in both animals and plants are caused by simple organisms, i.e., microscopic bacteria.

gimbal rings
An ancient invention consisting of two rings mounted on axes at right angles to each other holding an object (e.g., candleholder, compass) that will remain suspended in a horizontal plane regardless of any intervening motion, such as that produced by an oceangoing ship.

gnomon
An ancient astronomer's timekeeping tool. It is a pole or stake, placed in the ground in an open, flat area, which casts a shadow from sunrise to sunset. Gnomons assisted astronomers in determining the time of day and the time of year, as well as the solstices (the longest and shortest days of the year).

hydraulic
Refers to something that is operated or affected by the action of a pressurized fluid of low viscosity, usually water.

hydrostatics
The study of liquids at rest (equilibrium) (e.g., liquids contained in dams, storage containers, and hydraulic machinery).

joss stick
A stick of incense burnt before a Chinese idol or image. However, in medieval Chinese weaponry, it characterized a type of solid fuse.

kinematics
The study of the motion of objects, excepting the effects of mass and force.

lateen
A nautical term referring to the use of a triangular sail on a long yard (pole) that is secured at an angle to the shorter mast (vertical pole).

macroevolution
An evolutionary term referring to the large-scale phylogenic changes of species, including extinction, that occur over eons of geological time, i.e., evolution of one species into another, such as dinosaurs to birds.

magazine
In military parlance, it is a place to store ammunition.

metaphysical
Refers to beliefs that are based on abstract or speculative interpretations rather than objective observations. A preponderance of the writings of the Middle Ages and Renaissance were filled with metaphysical reasoning, characterized by elaborate images.

metrological
Refers to the science (system) of measurement.

microevolution
An evolutionary term referring to changes that occur in the outward characteristics of species over generations, i.e., evolution within a species.

monocots
A shortened version of the biological term *monocotyledons,* which refers to flowering plants (angiosperms) that have embryos with a single seed leaf (cotyledon). Typically, monocots have narrow leaves, smooth edges and a parallel vein structure. Examples are grasses, cereal grains, bananas, palms, orchids, lilies and tulips.

palisade
Geological: A line of extended high rock cliffs, usually along a river, but occasionally rising above streams or lakes. Military: A defensive structure constructed of pales (pickets) that form a barrier.

parallax
The apparent change in direction and/or position of an object viewed through an optical instrument (e.g., telescope), which occurs by the shifting position of the observer's line of sight.

pivot
A short staff or rod on which another object rotates. Medieval weapons, such as the trebuchet, were affixed with pivots.

pole star
Also called Polaris, or the polar star. It is a guiding star around the northern celestial pole that never rises or sets. It was used by navigators of antiquity, the Middle Ages, and the Renaissance before the invention of the compass.

projection
In cartography, it refers to the system or grid of intersecting lines that illustrates the Earth, all or in part, as a plane, i.e., a globe projected onto a level or flat surface.

relief
The raised projection of figures or objects from a flat surface. Reliefs were part of ancient mapmaking.

schistomiasis
Also known as bilharzia. It is a serious disease caused by a blood fluke (schistosoma) that penetrates snails that inhabit the soil of moist or tropical regions. The parasitic larva emerges from the snail into the soil and enters the human body, typically through bare feet.

sines
Mathematical devices deriving from the Latin word *sinus,* meaning "curve." It is the function of the acute angle in a right triangle that is the ratio of the *opposite* side to the hypotenuse. Conversely, a cosine is the function of an acute angle in a right triangle that is the ratio of the *adjacent* side to the hypotenuse.

slag
The residual material, mostly silicon, from the smelting of iron ore in a blast furnace.

sluices
Natural or manmade waterways (channels, canals, or waterwheel mills) whose flow is regulated by a system of gates or valves.

spontaneous generation
An ancient and persistent theory of life that espoused that a "life force" was present in all inorganic matter, including air, which could cause life to occur spontaneously. It was not refuted definitively until the 18th century.

telemetry
The measurement and transmission of data via wire or radio signals from distant sources, e.g., space vehicles, to a recording and/or display station on Earth.

theodolite
A precision surveying tool equipped with a small telescope that is used for measuring horizontal and vertical angles.

type metal
An alloy, usually comprised of tin, lead, and antimony, that is used for manufacturing metal printing type.

SELECTED BIBLIOGRAPHY

Print Media

Ackerknecht, Erwin H. *A Short History of Medicine*. Baltimore, Md.: Johns Hopkins University Press, 1982.

Allstetter, William, ed. *Science and Technology Almanac*. Westport, Conn.: Oryx Press, 2001.

Armstrong, Karen. *A History of God: The 4,000-Year Quest of Judaism, Christianity and Islam*. New York: Ballantine, 1993.

Asimov, Isaac. *Asimov's Chronology of Science and Discoveries*. New York: Harper & Row, 1989.

———. *A Short History of Biology*. Westport, Conn.: Greenwood Press, 1964.

Barnes, Jonathan, ed. *Complete Works of Aristotle*. 2 vols. Princeton, N.J.: Princeton University Press, 1984.

Barnes-Svarney, Patricia, ed. *The New York Public Library Science Desk Reference*. New York: Macmillan, 1995.

Beckmann, Petr. *A History of Pi*. New York: Golem Press, 1971.

Britannica Atlas. Chicago: Encyclopaedia Britannica, 1996.

Bunch, Bryan. *Handbook of Current Science and Technology*. New York: Gale, 1996.

Cantor, Norman F. *The Civilization of the Middle Ages*. New York: HarperCollins, 1993.

———. *The Medieval World: 300–1300*. New York: Macmillan, 1963.

———, ed. *The Encyclopedia of the Middle Ages*. New York: Viking, 1999.

Carey, John, ed. *Eyewitness to Science*. Cambridge, Mass.: Harvard University Press, 1995.

Cobb, Cathy, and Harold Goldwhite. *Creations of Fire: Chemistry's Lively History from Alchemy to the Atomic Age*. New York: Plenum Press, 1995.

Cohen, Elizabeth S., and Thomas V. Cohen. *Daily Life in Renaissance Italy*. Westport, Conn.: Greenwood Press, 1946.

Crombie, A.C. *The History of Science: From Augustine to Galileo*. New York: Dover Publications, 1995.

Daintith, John, ed. *Biographical Encyclopedia of Scientists*. 2 vols. London: Institute of Physical Publishing, 1994.

Derry, T. K., and Trevor I. Williams. *A Short History of Technology*. New York: Dover Publications, 1960.

Dyson, James. *A History of Great Inventions*. New York: Carroll & Graf Publishers, 2001.

Ford, Brian J. *Images of Science: A History of Scientific Illustrations*. New York: Oxford University Press, 1993.

Garin, Eugenio. *Science and Civic Life in the Italian Renaissance*. New York: Anchor Books, 1969.

Gatti, Hilary. *Giordano Bruno and Renaissance*. Ithaca, N.Y.: Cornell University Press, 1999.

Giblin, James Cross. *From Hand to Mouth, On. How We Invented Knives, Forks, Spoons, and Chopsticks, and the Manners to go With Them*. New York: HarperCollins, 1987.

Giles, Frances, Joseph Giles. *Cathedral, Forge, and Waterwheels: Technology and Invention in the Middle Ages*. New York: HarperCollins, 1994.

Gimpel, Jean. *The Medieval Machine: The Industrial Revolution of the Middle Ages*. New York: Penguin Group, 1976.

Griffith, Samuel B. *Sun Tsu: The Art of War*. London: Oxford University Press, 1963.

Hall, Bert S. *Weapons and Warfare in Renaissance Europe: Gunpowder, Technology and Tactics*. Baltimore, Md.: Johns Hopkins University Press, 1997.

Hall, Marie Boas. *The Scientific Renaissance: 1450–1630*. New York: Dover Publications, 1964.

Hawking, Stephen W. *A Brief History of Time*. New York: Bantam Books, 1988.

Hogben, Lancelot. *Mathematics for the Millions*. New York: W. W. Norton, 1971.

Holmyard, E. J. *Alchemy*. New York: Dover Publications, 1990.

Jacob, James R. *The Scientific Revolution: Aspirations and Achievements, 1500–1700*. Amherst, N.Y.: Humanity Books, 1999.

Jaffe, Bernard. *Crucibles: The Story of Chemistry: From Ancient Alchemy to Nuclear Fission*. New York: Dover Publications, 1976.

James, Peter, and Nick Thorpe. *Ancient Inventions*. New York: Ballantine Books, 1994.

Konstam, Angus. *Historical Atlas of Exploration: 1492–1600*. New York: Checkmark Books, 2000.

Koyre, Alexandre. *From the Closed World to the Infinite Universe*. Baltimore: Johns Hopkins University Press, 1957.

Krebs, Robert E. *Basics of Earth Science*. Westport, Conn.: Greenwood, 2003.

———. *The History and Use of Our Earth's Chemical Elements*. Westport, Conn.: Greenwood, 1998.

———. *Scientific Developments and Misconceptions through the Ages*. Westport, Conn.: Greenwood, 1999.

———. *Scientific Laws, Principles, and Theories*. Westport, Conn.: Greenwood, 2001.

Lewis, Bernard. *What Went Wrong?: Western Impact and Middle Eastern Response*. New York: Oxford University Press, 2002.

Lindberg, David C. *The Beginnings of Western Science: 600 B.C. to A.D.* Chicago: University of Chicago Press, 1992.

Margotta, Roberto. *The History of Medicine*. New York: Smithmark Publishers, 1996.

Mason, Stephen F. *A History of Science*. New York: Simon & Schuster, Macmillian, 1962.

McClellan, James E., III, and Harold Dorn. *Science and Technology in World History: An Introduction.* Baltimore: Johns Hopkins University Press, 1999.

Menocal, Maria Rosa. *The Ornament of the World: How Muslims, Jews, and Christians Created a Culture of Tolerance in Medieval Spain.* New York: Little Brown, 2002.

Miller, David, et. al. *The Cambridge Dictionary of Scientists.* New York: University of Cambridge Press, 1996.

Motz, Lloyd, and J. H. Weaver. *Conquering Mathematics: From Arithmetic to Calculus.* New York: Plenum Press, 1991.

Nordenskiold, Erik. *The History of Biology: A Survey.* New York: Tudor Publishing, 1928.

North, John. *A Norton History of Astronomy and Cosmology.* New York: W. W. Norton, 1995.

Pannekoek, A. *A History of Astronomy.* New York: Dover Press, 1961.

Porter, Roy, ed. *The Cambridge Illustrated History of Medicine.* Cambridge: Cambridge University Press, 1996.

Read, John. *From Alchemy to Chemistry.* New York: Dover Publications, 1995.

Roberts, J. M. *History of the World.* New York: Oxford University Press, 1993.

Sagan, Carl, and Ann Druyan. *Shadows of Forgotten Ancestors: A Search for Who We Are.* New York: Ballantine Books, 1992.

Sarton, George. *Six Wings: Men of Science in the Renaissance.* New York: World Publishing Company, 1957.

Science and Technology Department of the Carnegie Library of Pittsburgh, comp. *Science and Technology Desk Reference.* Detroit, Mich.: Gale, 1996.

Scientific American Science Desk Reference. New York: John Wiley & Sons, 1999.

Silver, Brian L. *The Ascent of Science.* New York: Oxford University Press, 1998.

Simonis, Doris, ed. *Lives and Legacies: An Encyclopedia of People Who Changed the World: Scientists, Mathematicians, and Inventors.* Phoenix, Ariz.: Oryx Press, 1999.

Singer, Charles. *A History of Scientific Ideas: From the Dawn of Man to the Twentieth Century.* New York: Barnes and Noble, 1996.

Struik, Dirk J. *A Concise History of Mathematics.* New York: Dover Publications, 1987.

Temple, Robert. *The Genius of China: 3,000 Years of Science, Discovery, and Invention.* London: Prion Books, 1999.

Teresi, Dick. *Lost Discoveries: The Ancient Roots of Modern Science—From the Babylonians to the Maya.* New York: Simon & Schuster, 2002.

Tiner, John Hudson. *Exploring the History of Medicine: From the Ancient Physicians of Pharaoh to Genetic Engineering.* Green Forest, Ark.: Master Books, 1999.

Usher, Abbott Payson. *A History of Mechanical Inventions.* New York: Dover Publications, 1982.

Vauchez, Andre, ed. *Encyclopedia of the Middle Ages.* 2 vols. Chicago: Fitzroy Dearborn, 2000.

Webster, Charles. *From Paracelsus to Newton: Magic and the Making of Modern Science.* New York: Barnes & Noble, 1982.

White, Michael. *Leonardo: The First Scientist.* New York: St. Martin's Press, 2000.

Wightman, W. P. D. *Science and the Renaissance: An Introduction to the Study of the Emergence of the Sciences in the Sixteenth Century.* New York: Hafner Publishing, 1962.

Wilber, C. Keith. *Revolutionary Medicine.* Chester, Conn.: Globe Pequot Press, 1980.

Williams, Trevor I., et al. *A History of Inventions: From Stone Axes to Silicon Chips*. New York: Checkmark Books, 2000.

Woolley, Benjamin. *The Queen's Conjurer: The Science and Magic of Dr. John Dee, Adviser to Queen Elizabeth I*. New York: Henry Holt, 2001.

Digital Media

CD-ROM Disks

Academic Press Dictionary of Science & Technology. San Diego: Academic Press, 1996.

Britannica CD. Chicago: Britannica, 2003.

Comptom's Interactive Encyclopedia. SoftKey Multimedia, Inc., 1996.

Grolier Multimedia Encyclopedia. Danbury, Conn.: Grolier Interactive, 1998.

Microsoft Encarta. New York: Funk & Wagnalls, 1994.

Oxford English Dictionary. London: Oxford University Press, 2000.

Internet Web Sites

American Trade Association. "The Spices of Antiquity; Spices of the Middle Ages; and An Arab Monopoly and The Age of Discovery." http://www.astaspice.org/history/history.htm.

Ancient and Modern Science. "Biology: Part III." http://www.wisdomworld.org/setting/biology.html.

Annenberg/CBP. "Renaissance: Out of the Middle Ages." http://www.learner.org/exhibits/renaissance/middleages.html.

Bartleby.com. *Columbia Encyclopedia*. "Renaissance." http://www.aol.bartleby.com/65/re/Renaisnc.html.

Botany. "Traditions and Innovations." http://www.ibiblio.org/expo/vatican.exhibit/exhibit/g-nature/Botany.html.

Botany Online. "Renaissance and History—First Scientific Descriptions." http://www.biologie.uni-hamburg.de/b-online.htm.

Christopher Columbus. "Early Career and the First Voyage." http://www.carmensandiego.com/products/time/columbusc10/plans.html.

Claudius Ptolemy. http://www-groups.dcs.st-and.ac.uk/~history/Mathematicians/Ptolemy.html.

The Concepts of Evolution. "Aristotle's *Scala Naturae*." http://www.sp.uconn.edu/-bi102vc/102su01/evolutionintro.htm.

Crystalinks. "Science in Ancient Greece." http://www.crystalinks.com/greekscience.html.

Dana, Peter H. "Map Projection Overview." http://www.colorado.edu/geography/gcraft/notes/mapproj/mapproj.html.

Discoverers Web. "Exploration in the Medieval Period." http://win.tue.nl/~engels/discovery/medieval.html.

Epact. "Scientific Instruments of Medieval and Renaissance Europe." http://www.mhs.ox.ac.uk/epact/indix.html.

Era of Elegance, Inc. "A Brief History of Science and Technology." http://www.erasofelegance.com/sciencehistory.html.

Filippi, Zivan. "Marko Polo and Korcula: Epilogue." http://www.korcula.net/mpolo/mpolo10.htm.

Fort Raleigh National Historic Site. "Heritage Education Program." http://www.nps.gov/flra.htm.

Garo, Laurie A. B. "Introduction to Map Projections." http://www.uncc.edu/~lagaro/cwc/mapproj/intro_mp.html.

Ghazanfar, S. M. "Islamic World and the Western Renaissance." http://www.cyberistan.org/islamic/ghaxil.html.

Harley, J. Brian, and David Woodward. "Cartography in Prehistoric, Ancient, and Medieval Europe and the Mediterranean." http://www.feature.geography.wisc.educ/histcart/series.html.

History of Science. "History of Biology." http://biology.clc.uc.edu/courses/bio104/hist_sci.htm.

Hobbs, Christopher. "Botanical Taxonomy—A Historical Summary." http://www.healthy.net/asp/templates/article.asp/PageType=Article&ID=863.

Hooker, Richard. "Reformation: The Northern Renaissance." http://www.usu.edu:8080/~dee/REFORM/NORTHER.HTM.

"Humanism." http://www.wsu.edu:8080/~dee/REN/HUMANISM.HTM.

"Johann Gutenberg, c. 1400–1468." http://www.home.vicnet.net.au/~einstein/gutenberg.htm.

Kessler Corp. *Inventors Museum.* "Inventing Printing." http://www.inventorsmuseum.com/Printing.htm.

Lanius, Cynthia. "Mathematics and History of Cartography: What are Maps?" http://math.rice.edu/~lanius/pres/map:html.

Lawton, Chris. *History of Astronomy.* http://www.homepages.tcp.couk/~carling/astrhis.html.

Maptown. "Reference: Types of Maps." http://www.maptown.com/reference maptypes.html.

Marx, Irma. "Travelers of the Silk Route." http://www.silk-road.com/artl.html.

"Mathematicians through 1500." Clark University. http://aleph0.clarku.edu/~djoyce/mathhist/Europe.html.

Medicine Transformed. "Renaissance." http://www.ibiblio.org/expo/vatican.exhibit/exhibit/f-medicinebio/Medicine-2.html.

"Medieval Europe—Mathematics and the Liberal Arts." http://www.math.truman.edu/~thammond/history/MedievalEurope.html.

Memorial University of Newfoundland. "Before Cabot—the Saga of Erick the Red." http://www.heritage.nf.ca/exploration/precabot.html.

Millersville University. "Columbus, Explorations, and Spice Trade." http://www.Marauder.millersv.edu/~columbus/data/art.

Mulcahy, Karen. "Map Projections." http://www.everest.hunter.sunny.edu/mp.html.

National Aeronautics and Space Administration (NASA). "Landsat 7 Gateway." http://www.landsat.gsfc.nasa.gov/.

Pickering, Keith A. "Columbus, Navigation, Maps." http://www.1.minn.net/~keithp.htm.

Plague and Public Health in Renaissance Europe. http://www.Jefferson,village,virginia.edu/osheim/plaguein.html.

The Road to Modern Evolutionary Biology. "Influence of the Greeks." http://www2.evansville.edu/evolutionweb/history.html.

Rosenberg, Matt. "Geography and Exploration." About.com. http://www.geography. minigeo.com/library/weekly.htm.

Saunders and Cooke. "History of the Astrolabe (Quadrant)." http://www.saunders andcooke.com.html.

Smith, Gene. "A Brief History of Astronomy." http://casswww.ucsd.edu/public/ tutorial.History.html.

Sorkin, Adam. "Inventions and Technology (Renaissance)." http://www.library. thinkquest.org/3588/Renaissance/University/Inventions.html.

St. Michael's Library. "Medical Chronological Table: Surgery Through the Ages." http://www.st-mike.org/library/smlibmdc.html.

Taxonomy. "The Real Poop." http://www.cod.edu/people/faculty/fancher/ Taxon.htm.

Tuck, Jim. "Jeronimo De Aguilar: The Marooned Priest Who Speeded the Conquest." http://www.mexconnect.com/mex_history/jtuck/jaquilar.html.

"Tycho Brahe and Johannes Kepler." http://www.csep10.phys.utk.edu/astr161/ lect/history.html.

UNESCO. "Silk Road." http://www.unesco.kz/natcom/turkestan/e05silkroad. htm.

University of St. Andrews, Scotland. http://www-gap.dcs.st-and.ac.uk/~history. html.

World Book Online. "Fractions, Zero, Decimal System, π, Number Systems." http:// www.aolsvc.worldbook.aol.com/wbol/wb/Auth/jsp/wbArticle.jusp.

Zahoor, A. "Muslem Scientists and Explorers." http://www.users.erols.com/ zenithco.html.

NAME INDEX

Abu-Bakr Mohammed ibn-Zakaria. *See* al-Razi

Abu Kamil Shuja, 129

Abul Hasan Ali al-Masu'di, 41

Abu'l-Wafa, 135

Adelard of Bath, 9, 22, 184, 143

Aesculapius, 90

Aguilar, Jeronimo de, 52

Ahmes the Moonborn, 128

Albategnius. *See* Al-Battānī, Abu Abdullah

Albert the Great. *See* Albertus Magnus

Albertus Magnus (Saint Albert the Great), 72–74, 78, 157, 174, 185, 289

Alcmaeon of Croton, 110

Alcuin of York, 142–43

Aldrovandi, Ulisse, 79–80

Alexander of Hales, 289

Alexander the Great, 13, 35

Alhazen, 20–21, 136, 171, 172, 174

Alpini, Prospero, 75

Alvarado, Pedro de, 52

Anaxagoras of Clazomenae, 156

Anaximander, 54, 68, 81–82

Anaximenes of Miletus, 156

Apianus, Peter, 59

Apollonius of Perga, 125, 128, 196

Aquinas, Thomas, 78, 157, 289

Archimedes of Syracuse, 125, 126, 145, 146, 153, 156, 159, 162, 176

Archytas of Tarentum, 175

Aristarchus of Samos, 7–8

Aristotle, 6–7, 22, 35, 36, 46, 68–70, 77, 78, 81, 116, 145–46, 156, 158–59, 162, 163, 171, 175, 176, 185, 192

Arnald of Villanova, 96, 185–86

Arthur of Camelot, King, 275

Aryabhata I, 16–17, 130–31

Aryabhata II, 132

al-Asmai, 78

Aspdin, Joseph, 203

Atahuallpa, 18

Augustine, Saint, 82, 176

Avenzoar. *See* Ibn Zuhr

Averroës (Ibn-Rushd), 72

Avicenna (Ibn Sina), 72, 95, 107, 115, 135, 171, 184, 185

Bacon, Roger, 79, 136, 144, 157, 167, 172, 173–74, 207, 214, 232, 289

Baillou, Guillaume de, 97

Barentsz (Barents), William, 50

Bartholomew the Englishman, 185

al-Battānī, Abu Abdullah (Albategnius), 20, 30, 134

Bauhin, Caspar, 75

Beg, Ulugh, 21, 137

Bellovacensis, Vincentius, 78

Belon, Pierre, 80

Berengario of Carpi, 111

Bhaskara (Bhadkarcharya), 132–33
Bingen, Hildegard von, 72
al-Biruni, Abu Rayhan, 96, 135–36, 172
Bock, Hieronymus, 74–75
Boethius, Anicus Manlius Severinus, 176
Bohr, Niels, 192
Boyle, Robert, 180
Bradwardine, Thomas, 145–46
Brahe, Tycho, 23, 24, 25–26, 27, 225–26
Brahmagupta, 17, 131
Brahmah, Joseph, 210
Brancas, The, 113
Brunel, Olivier, 50
Brunfels, Otto, 71, 74
Bruno, Giordano, 23, 24–25, 70, 151
Buridan, Jean, 160, 162

Cabot, John, 48
Cabot, Sebastian, 48
Caesalpinus. See Cesalpino, Andrea
Caesar, Julius, 30
Calvin, John, 117
Calvin, Melvin, 76
Canano, Giambattista, 117
Cantimpratensis, 78
Cardano, Gerolamo, 148–49
Cartier, Jacques, 49
Cayley, George, 211
Celsus, Aulus Cornelius, 112, 113, 114
Cesalpino, Andrea (Caesalpinus), 75, 118
Champlain, Samuel de, 49
Chancellor, Richard, 49–50
Charles IX of France, King, 152
Chauliac, Guy de, 109
Chekov, Anton, 107
Chiao Wei-Yo, 198
Chopin, Frederic, 107
Chou Kung, 139
Christian IV of Denmark-Norway, 50
Chu Shih-Chieh, 139–40
Clavius, Christoph, 30
Cleopatra, 30
Colombo, Matteo Realdo, 118
Colt, Samuel, 280

Columbus, Christopher, 17, 23, 44–46, 58, 105, 167
Congreve, William, 280–81
Constantine, Emperor, 82
Constantinus Africanus, 96
Contarini, Giovanni Matteo, 60
Copernicus, Nicolaus, 7, 22, 23–24, 233
Cordoba, Francisco Fernandez de, 51
Coronado, Francisco Vásquez de, 52
Cortés, Hernando, 51–52
Crapper, Thomas, 210
Crateus, 70
Cueva, Doña Beatriz de la, 52
Cumming, Alexander, 210

Daedalus, 210
da Gama, Vasco, 46
Dalton, John, 180, 192
D'Armate, Salvino, 207
Darwin, Charles, 82
da Vigo, Giovanni, 112
da Vinci, Leonardo, 21, 109–11, 147, 160–61, 172, 198–99, 210–11, 232, 258, 278–79, 285, 290
Dee, John, 24, 59, 150–51
de Elcano, Juan Sebastian, 47
de las Casas, Bartolome, 52
del Ferro, Scipione, 147, 148, 149
della Porta, Giovanni Battista, 172
Democritus of Abdera, 70, 192
Descartes, René, 136, 152
de Soto, Hernando, 49
Diaz, Bartolomeo, 44
Digges, Thomas, 24, 151
Diogenes of Apollonia, 82
Dionysius of Syracuse, 264
Diophantus, 128
Dioscorides, Pedanus, 70, 74
Drebbel, Cornelius, 234
Dumas, Alexandre, 107

Edward III of England, King, 186
Ehrlich, Paul, 106
Einstein, Albert, 169
Elizabeth I of England, Queen, 151, 190, 209

Empedocles of Acragas, 68, 90, 156
Epicurus, 70
Erasistratus, 110
Eratosthenes, 14, 35, 45, 47
Eric the Red, 38
Eriksson, Leif, 38
Euclid of Alexandria, 110, 125, 128,
    133, 143, 145, 147, 150, 156, 170,
    176
Eudoxus of Cnidus, 6
Eustachio, Bartolomeo, 119

Fabricius, David, 26
Fabrizio, Girolamo, 82
Fa-Hsien, 37
Fallopio (Fallopius), Gabriele, 82, 118,
    120
Fermat, Pierre de, 161
Fernel, Jean Francois, 96–97, 104, 107
Ferrari, Ludovico, 148, 149, 150
Fibonacci, Leonardo Pisano, 143–44
Fior. See Tartaglia, Niccolo Fontana
Fludd, Robert, 234
Fracastoro, Girolamo, 7, 23, 104
Franklin, Benjamin, 31
Frederick II of Hohenstaufen, 78
Frederick IV of Palatine, 153
Frisius, Regnier Gemma, 60, 149
Fuchs, Leonhard, 75
Fust, Johann, 222

Galen, Claudius, 71, 77, 90–91, 93–94,
    98, 106, 107, 108–9, 110–11, 115,
    116–18, 290
Galilei, Galileo, 12, 24, 26–27, 160,
    162–63, 164, 176, 233–34, 273
Gaspar à Myrica, 60
Geber-Jabir ibn Hayyan, 183
Gensfelder, Reinhardus, 59
Gerald of Cremona, 184
Gerbert of Aurillac, 143
Gerhard of Cremona, 143
Gerson, Levi ben (Gersonides), 145
Gesner, Conrad, 79
Ghini, Luca, 74
Gilbert, William, 167–68

Goodyear, Charles, 120
Gregory V, Pope, 143
Gregory XIII, Pope, 30
Grosseteste, Robert, 173, 289
Guericke, Otto von, 177
Guerrero, Gonzalo, 52
Gutenberg, Johannes, 218, 221–23

al-Hakim, 136
Hansen, Armauer Gerhard Henrik, 102
Harington, John, 209
Hariot, Thomas, 59
Harrison, John, 28, 65, 149
Harvey, William, 118–19
Hatten, Ulrich von, 106
Hecataeus of Miletus, 54
Heinlein, Peter, 202
Helmont, Jan Baptista van, 76
Henry II of England, King, 260
Henry VII of England, King, 48
Henry II of France, King, 104
Henry IV of France, King, 152
Henry the Navigator, Prince, 43–44, 46
Heracleides of Pontus, 6
Heraclitus of Ephesus, 156
Herjulfsson, Bjami, 38
Hermes Trismegistus, 23
Hero (Heron) of Alexandria, 128, 170,
    239
Herophilus, 110
Hicates, 7
Hipparchus of Nicaea, 125
Hipparchus of Rhodes, 9, 21, 196
Hippocrates of Chios, 125
Hippocrates of Kos, 90, 112
Homer, 90
Hooke, Robert, 217
Hopkins, Samuel, 191
Hsuan-Tsang, 37
Hudson, Henry, 49
Huygens, Christian, 136, 163
Hypatia of Alexandria, 128, 196

Ibn-Batuta, 42
Ibn al-Haitham, Abu Ali Hassan. See
    Alhazen

Ibn-Masawayh, Jurjis ibn Bakhtishu
   Jibril Yuhanna, 94
Ibn-Maimon (Maimonides), 96
Ibn an-Nafis, 96, 117
Ibn-Rushd. *See* Averroës
Ibn Sina. *See* Avicenna
Ibn 'Umar, Abd al-Rahman, 20
Ibn Yunus, 129
Ibn al-Zarqala, 21
Ibn Zuhr (Avenzoar), 96
al-Idrisi, Abdullah Mohammad ibn al-
   Sharif, 41, 57
Ingrassia, Giovanni Filippo, 97
al-Istakhri, Ishaq Ibrahim ibn Muham-
   mad al-Farisi, 57
Ivan the Terrible, 50

al-Jahiz, 78
Janssen, Zacharias, 26, 217, 233
Jefferson, Thomas, 191
John of Ulynam, 190

Kan Te, 12
Kasparov, Gary, 200
Keats, John, 107
Kepler, Johannes, 21, 26, 27–28, 82,
   136, 163–64, 174
Key, Francis Scott, 281
al-Khalili, 136–37
Khan, Genghis, 21, 42
Khan, Kublai, 42
Khan, Nulagu il, 21
al-Khwarizmi, Abu Ja'far Muhammad
   ibn Musa, 133–34
al-Kindi, 134
Knox, Henry, 191
Koch, Robert, 107–8
Ko-Hung, 179
Krebs, Nicholas, 207

Lamarck, Chevalier de (Jean Baptiste
   Antoine de Monet), 68
Langenstein, Henry, 22
Lavoisier, Antoine-Laurent, 180
Lawrence, D. H., 107
Leeuwenhoek, Antoni van, 217–18

Leibniz, Gottfried, 153
Leucippus of Miletus, 192
Liang Ling-Tsan, 201
Li Chih, 139
Lightfoot, John, 83
Linnaeus, Carolus, 74
Lippershey, Hans, 26, 233
Liu Hsiang, 179–80
Lu Hui, 221
Lully, Raymond, 186
Luzzi, Mondino de, 109

Magellan, Ferdinand, 46–47
Magni, Einhard vita Karoli, 12
Mahavira (Mahaviracharya), 131–32
Maimonides. *See* Ibn-Maimon
Ma-Ling, 204
Marinus of Tyre, 54–55, 60
Mary of England, Queen, 150–51
Massa, Niccolo, 109
Matthias Corvinus of Hungary, King,
   146
Mendeleyev, Dimitry, 180
Mercator, Gerardus, 58, 60
Modena, Tommaso de, 207
Monet, Jean Baptiste Antoine de. *See*
   Lamarck, Chevalier de
Montezuma II, 527
Mo Ti, 171–72
Muhammad, 18, 39, 92
Müller, Johann. *See* Regiomontanus
Munk, Jens, 50–51

Nemorarius, Jordannus, 144–45
Newton, Isaac, 21, 81, 83, 151, 153,
   162, 164, 180, 192, 212, 281
Nicholas of Cusa, 22–23, 24, 146

O'Neill, Eugene, 107
Ortelius, Abraham, 58

Pacioli, Luca, 147
Paracelsus, 104, 114–16, 186–87, 290
Paré, Ambroise, 112, 224
Parme, Philibert de, 203
Parmenides of Elea, 5

Pascal, Blaise, 147, 161–62
Pauli, Wolfgang, 180
Petrarch of Anjou, 58
Petrus Peregrinus (Picard), 166–67
Philolaus, 7
Piri ibn Haji Memmed (Piri Re'is), 58
Pi Sheng, 221
Pitiscus, Bartholomeo, 153
Plater, Felix, 97
Platinus, 176
Plato, 82, 169
Pliny the Elder, 35, 46, 70–71, 74, 77, 212, 219
Poitiers, Diane de, 104
Pollio, Marcus Vitruvius. *See* Marcus Vitruvius
Polo, Marco, 11, 42, 46, 51, 221, 278, 281
Polo, Niccolo, 42
Ptolemy of Alexandria (Claudius Ptolemaeus): astronomy, 7, 8–9, 13, 20, 21, 22, 28–29, 128, 136, 137, 143; cartography, 54–55, 57, 58, 59–60; geography, 35–36, 46, 45, 47; optics, 170–71
Ptolemy I of Egypt, 35
Ptolemy II of Egypt, 35
Purbach, Georg von, 23
Pythagoras of Samos, 5–6, 125, 175, 176

Qin Jiushao, 139

Randolph, Edmund, 191
al-Razi (Abu-Bakr Mohammed ibn-Zakaria), 94–95, 183–84
Recorde, Robert, 149–50
Redi, Francesco, 84
Regiomontanus, 23, 146–47
Ripley, George, 186
Robert of Anjou, 58
Robert of Chester, 184
Robins, Benjamin, 273
Rondelet, Guillaume, 80
Roomen, Adriaan van (Adriannus Romanus), 153–54

Rutherford, Ernest, 180
Ruysch, Johannes, 60

Sacrobosco, Johannes de, 144
Santorio, Santorio, 234–35
Saul, King, 82
Scarlett, Edward, 208
Seneca, 170
Servetus, Michael (Miguel Serveto), 117
Shen Kua, 166, 172–73
Shou-ching Kuo, 12
Sixtus IV, Pope, 146
Spina, Alessandro, 207
Sridhara, 132
Sripati, 132
Stevin, Simon, 152–53, 161–62
Strabo, 46
Su Sung, 10, 201
Svarsson, Gardar, 38
Sylvester II, Pope, 143
Sylvius, Jakobus, 110, 119

Tagliacozzi, Gasparo, 113, 225
Tartaglia, Niccolo Fontana (Fior), 147–49, 161, 272–73
Thales of Miletus, 5, 54, 81, 125, 156
Theophilus, 212
Theophrastus, 70
Thoth, 23
Timur the Lame, 137
Torricelli, Evangelista, 234
Tseng Kung-Liang, 282
al-Tusi, Nasir al-Din, 21
Tutankhamen of Egypt, 205

al-Umawi, 137–38
al-Uqlidisi, 134

Varahamihira, 16, 131
Verrazano, Giovanni da, 48–49
Vesalius, Andreas, 110–11, 118
Vespucci, Amerigo, 51
Viète, François, 152, 153–54
Vitellius. *See* Witelo of Silesia
Vitruvius, Marcus, 175

Wallace, Alfred Russel, 82
Wang Ch'i, 168
Wang Chieh, 221
Wang Xiaotong, 139
Wan-Hoo, 211
Widman, Johannes, 147
William of Ockham, 145, 159–60
Willoughby, Hugh, 49–50
Witelo of Silesia (Vitellius), 174
Wright, Edward, 59

Xenophanes, 68

Yersin, Alexandre, 99
Yu Chao Lung, 171

al-Zahrawi, Abu-al Quasim Khalaf
    ibn'Abbas, 95
Zhang-Hen, 195
Zheng He (Cheng Ho), 42–43

Subject Index

Abacus, 193–94
Acupuncture, 89
Aether, 7
Alchemical processes, 180–82
Alchemy, 114, 177–87
Algebra. *See* Mathematics
Algorithmi de numbero Indorum (al-
    Khwarizmi), 134
Almagest (Ptolemy), 19, 22, 23, 36, 143,
    195
Amber effect, 167–68
Anatomy. *See* Medicine
Ancient Mathematics (Wang Xiaotong),
    139
Andalusia (Al-Andalus), 9, 19
Anothomia (de Luzzi), 109
Apollonian Problem, 154
Appendice Algebraique (Stevin), 153
Application of Statics (Stevin), 161
Arches, 194–95
Aristotle's Ladder of Life, 69, 77
Arithmetic. *See* Mathematics
Armillary sphere, 195–96
Armor, 246–53
Ars Magna (Cardano), 148
Arybhatiya (Aryabhata I), 130
Astrolabe, 28, 196–97
Astrology, 3, 12, 17
Astronomy: background and history,
    1–9; calendars, 29–31; Chinese, 3–4,

10–13; Earth's ecliptic, 20; Egyptian,
    4, 12; European, 4–5, 21–29, 146;
    Greek, 5–9; Indian, 4, 13–17,
    131–33, 135; instruments, 12, 14, 16,
    21, 25, 27–29, 225–27; Islamic,
    18–21; Kepler's three laws of plane-
    tary motion, 27, 163–64; length of
    year, 20, 30, 137; Mesoamerican,
    17–18; Mesopotamian, 3; Middle
    Ages and Renaissance, 9–29; optics,
    20; perigee of earth and sun, 20;
    planetary motion, 27; theories of
    universe, 6–9, 12, 22–26
Atomic theory, 192
Autopoiesis, 84
Ax, 275
Azimuthal map projections, 63–64
Aztec civilization, 17, 18, 52, 141–42,
    246

Ballistics, 272–74
Barents Sea, 50
Battering Ram, 266
Bernoulli principle, 281
Biological sciences: botany, 71–76;
    definitions, 68; evolution, 81–84;
    history and background, 67–71;
    zoology, 76–80
Black Death. *See* Plagues
Bombs, 282–83

Book of Animals (al-Jahiz), 78
Book of Healing (Avicenna), 72, 135
Book of Numbers (Gerson), 145
Book of Stars (Fracastoro), 23
Book of the Composition of Alchemy (trans. Robert of Chester), 184
Book of the Secrets of Secrets (al-Razi), 183
Book On Algebra (Shuja), 129
Book on Those Geometric Constructions That Are Necessary for a Craftsman, A (Abu'l-Wafa), 135
Botany, 71–76
Bow and arrow, 275–76
Byrne, 248

Calculus. *See* Mathematics
Calendars, 29–31, 140–42
Camera obscura, 20–21, 171–72
Canal lock, 198–99
Cannons, 199–200, 267–72
Canon (Avicenna), 72, 184
Canon Mathematicus (Viète), 152
Caravelae, 44
Carbon-14. *See* Dating
Cardinals' purple, 206
Carolingian Codice (Golden Gospels) (Alcuin of York), 142
Cartography: history, 53–55; introduction, 53; mapmakers, 54–55, 57, 58–59, 60; map projections, 59–65; maps, 53–54; Middle Ages and Renaissance, 55–59
Castle of Knowledge (Recorde), 150
Castles, 259–60
Catalogue of Stars (Beg), 137
Catapults, 264–65
Catoptrica (Hero), 170
The Causes of Plants (Theophrastus), 70
Celsius scale, 235
Cement, 203
Chain mail, 248–49
Chemistry: background and history, 155–57; practice of, 177–78
Chemotherapy, 116–17
Chess, 200
Christianismi restitutio, 117
Chronometers, 28, 65–66

Citadels, 255
Classification of animals and plants. *See* Taxonomy
Clocks, 11, 16, 200–202
Cochineal insect, 206
Comets, 12
Communia mathematica (Bacon), 144
Company of Merchant Adventurers, 50
Compass, 153, 165, 166
Concrete, 202–3
Conic map projections, 62–63
Contraception, 119–21
Corpus Hippocraticum (Hippocrates), 90
Crab Nebula, 12
Crossbow, 203–4, 276–78
Crusades, 259
Cuirasses, 246
Cupellation, 182–83
Cuttlefish, 206
Cylindrical map projections, 61–62

Dark Ages, 19, 36, 55, 288, 289
Dating, methods of, 83
De Algorismo (de Sacrobosco), 144
De Fabrica (Vesalius), 111, 119
De harmonicis numeris (Gerson), 145
De historia stirpium commentarii (Fuchs), 75
De immense (Bruno), 25
De inventione veritatis (Geber), 183
De investigatione perfectionis (Geber), 183
De Mineralibus (Avicenna), 184
De Mirabilibus Mundi (Magnus), 185
De Motu Cordis et Sanguinis in Animalibus (Harvey), 118
De Multiplications Specierum (Bacon), 173
De naturis rerum (Cantimpratensis), 78
De omni rerum fossilium genere (Gesner), 79
De piscebus marines (Rondelet), 80
De proportionibus velocitatum in motibus (Bradwardine), 145
De Re Anatomica (Colombo), 118
De Revolutionibus Orbium Coelstrum (Copernicus), 23–24

De Thiende (Stevin), 152
De Triangulis Omnimodis (Müller), 146
De vegetabilibus (Magnus), 72
Diamond Sutra (Wang Chieh), 221
Discourses on Two New Sciences (Galileo), 163
Disease: history of, 85–90; specific diseases, 98–108
Dissection. *See* Medicine
Divina Proportione (Pacioli), 147
Dream Pool Essays (Shen Kua), 166, 172
Dyes, 205–6

Earth: age, 83; circumference, 35, 36, 45, 47; declination, 167–68; ecliptic, 20; radius, 136, 172
Electricity, 164–68
Electromagnetic Spectrum, 169
Electromagnetism, 165–68
Electroweak Interactions, 168
Elements (Euclid), 125, 133, 143, 150
Elements of the Art of Weighing (Stevin), 152
Elixir vitae (elixir of life), 114, 177
English Muscovy Company, 50
Enquiry into Plants (Theophrastus), 70
Ephemerides (Regiomontanus), 146
Equatorial torquetum, 12
Equilibrium, 152–53
Ergotism. *See* St. Anthony's Fire
Evolution, 81–84
Excalibur, 275
Exploration, 36–53; explorers of the Middle Ages, 37–43; explorers of the Renaissance, 43–53
Eyeglasses, 206–8

Fahrenheit scale, 235
Fall of Rome, 9
Fibonnaci sequence, 143
Fire, 177
Five Treatises (Five Astronomical Canons) (Varahamihira), 131
Flamethrowers, 283–84
Flying buttresses. *See* Arches
Flying machines, 210–11
Flywheel, 212

Fork, table, 231–32
Fortifications, 254–60
Freemasons, 178
French Disease. *See* Syphilis
Furnace, blast, 197–98

Galls, 206
Gambeson, 250
Gb (Gilbert), 168
General History of Plants (Gerard), 75
Geography (Ptolemy), 36
Geography: background and history, 33–36; cartography, 53–66; exploration, 36–53; Silk Road, 38–39; spice trade, 39–41
Geometry. *See* Mathematics
Glass, 212–14
Gloves, mail, 251
Gnomon, 10, 14, 16, 200
GPS (Global Positioning System), 66, 227
Grand Unification Theory (GUT), 168
Great Pyramid, 13
Greek Fire, 265–66
Greenwich, England, 47
Gregorian calendar, 30–31, 151
Grenades, 283
Grounde of Arts (Recorde), 149
Guide through the Forest of Affairs (Shen Kua), 166
Gunpowder, 214–17, 261–62, 279–80
Guns, 267–72, 278–80

Halley's Comet, 23
Hansen's Disease. *See* Leprosy
Harmonics. *See* Sound
Hauberk, 249, 251
Havenvinding (Stevin), 153
Head mail, 251
Helmets, 251–52
Hematite, 206
Herbarum vivae eicones (Brunfels), 74
Hippocratic oath, 90
Historia animalium (Gesner), 79
Historia Naturalis (Pliny the Elder), 35, 70–71, 77
Homeopathy, 115, 187

Homocentrica sive de stellis liber (Fracastoro), 23
Hookes Law, 175
Hubble telescope, 233
Hudson Bay, 49
Human migration, 33–34
Hydrostatics, 152, 161–62

Iatrochemistry, 116–17, 186
Ilkhanic Tables (Nasir al-Din), 21
Illiad (Homer), 90
In Artem Analyticam Isagoge (Viète), 152
Indigo, 205
Ink, 222–23
Innovations, 192–93
Introduction to Mathematical Studies (Chu Shih-Chieh), 139–40
Inventions, 189–93
Islam, 18, 91
Islamic Atlas, 57

Javelin, 275
Julian calendar, 30
Jupiter (planet), 26, 233

Kelvin scale, 235
Kepler's Law of Areas, 27
Kepler's Laws of Planetary Motion, 27, 163–64
Kevlar vest, 250
Knights Templar, 178

Lady of the Camillias (Dumas), 107
Latitude, 28, 45, 47, 59, 60
Laudanum, 115
Leprosy, 101–2
Letters of patents, 48
Liber abaci (Fibonacci), 143
Liber de proprietatibus rerum (Bartholomew the Englishman), 185
Liber fornacum (Geber), 183
Liber quadratorum (Fibonacci), 144
Light. *See* Optics
Limonite, 206
Little Book of Alchemy (Magnus), 184
Lodestone, 165, 166
Longitude, 28, 45, 47, 59, 60

Madder (*Rubia tinctoria*), 205
Magnetism, 164–68
Mail gloves, 252–53
Maps. *See* Cartography
Mars (planet), 27
Mathematical Principles of Natural Philosophy (Newton), 212
Mathematical Treatise in Nine Sections (Qin Jiushao), 139
Mathematics: abacus, 143; algebra, 124, 129, 133–34, 150, 152–53; algorithms, 134; Arab, 133–38; arithmetic, 123–24, 137; background and history, 123–28; calculus, 153; Chinese, 127, 138–40; conics, 126; cubic equations, 147, 148–49, 150; decimals, 152–53; Egyptian, 127, 128–29; equations, 143–44; equilibrium, 152–53; European, 142–54; geometry, 125, 145, 146; hydrostatics, 152; Indian, 128, 129–33; Mesoamerican, 140–42; negative numbers, 139; numerals, 138; pi, 125–27, 153–54; quadratic equations, 131, 148–49; sexagesimal, 124; square root, 153; trigonometry, 129, 131, 134, 136–37, 145, 151, 153, 154; vigesimal, 140; zero, 127–28, 129–30, 140
Mayan civilization, 17, 18, 52, 140–42
Mechanics: definition, 158–59, 161–62; five simple machines, 163; formula for mechanical advantage, 144–45; work, 163. *See also* Motion
Medicine: anatomists, 108–11, 119; Chinese, 89; circulation of blood, 116–19; contraception, 119–21; dissection, 108–11, 119; Egyptian, 88; Eustachian tubes, 119; Fallopian tubes, 119; Greek, 89–91; Hebrew, 88–89; history of disease, 85–90; Indian, 89; Medieval and Renaissance, 91–93; Mesopotamian, 87–88; pharmaceuticals, 114–16; physicians, 91–97; primitive, 86–87; reconstructive surgery, 112–13, 224–25; specific diseases, 98–109; surgeons, 111–12; surgery, 111–12

Medulla Alchimiae (Ripley), 186
Mercator map projections, 60–63
Meteorologica (Aristotle), 35
Microscope, 217–18
Milky Way, 27, 173, 233
Mines, 283
Minoans of Crete, 34
Mirrors, 213–14
Mississippi River, 49
Motion: concepts and theories,
    158–64; forced, 7; laws, 192, 212,
    281; natural, 7; speed, 159; velocity,
    159
Muhammadanism, 18–19
Murex, 205
Musket, 278–79
Mycenaeans of Greece, 34
Mysterium cosmographicum (Kepler), 27

Natural History (Aldrovandi), 80
Natural History of Strange Fish (Belon),
    80
Nature and Differences of Fish (Belon), 80
Nebulas, 12
Newfoundland, 38
Nine Chapters of Mathematical Art (Lu
    Hui), 220
Northwest Passage, 48, 49, 51
Nürnberg Eggs, 202

Ockham's Razor, 145, 160
Odyssey (Homer), 90
Oeuvres (Paré), 224
Olmec civilization, 17, 165
Omicron Ceti, 26
On Arithmetical Rules and Procedures (al-
    Umawi), 137
On Conic Sections (Apollonius of Perga),
    128
On Learned Ignorance (Nicholas of
    Cusa), 146
On Sines, Chords, and Arcs (Gerson),
    145
On the Continuum (Bradwardine), 146
On the Nature of Things (Carus), 70
Opening of the Universe (Brahmagupta),
    131

Opticae thesaurus (Alhazen), 171
Optics, 20, 28–29, 168–74, 206–8
Optics (Euclid), 170
Optics (Ptolemy), 28, 36, 170–71
Opus majus (Bacon), 144, 173
Opus minus (Opus secundus) (Bacon),
    144, 173
Opus tertius (Bacon), 144, 174

Pantometria (Digges), 24
Paper, 218–19
Parsimony, law of. *See* Ockham's
    Razor
Pascal's hydrostatic paradox, 161–62
Patents, 190–91
The Pathway to Knowledge (Recorde),
    149–50
Perfect exhaustion theory, 126
Perspectiva (Bacon), 173
Perspectiva (Vitellius), 174
Phoenicians, 34
Pharmaceuticals, 114–16
Philosophers' Stone, 180, 183, 187. *See
    also* Alchemy
*Philosophiae Naturalis Principia Mathe-
    matica* (Newton), 164
Physics: alchemical processes, 180–81;
    alchemy, 177–87; background and
    history, 155–57; chemistry, 177–78;
    electromagnetism, 165–68; fire, 177;
    law of falling bodies, 162–63; mag-
    netism, 164–68; mechanics, 158–59;
    motion, 158–64; Newton's first law
    of motion, 162; optics, 168–74;
    sound, 174–77
Physiologus, 78
Pi, 125–27. *See also* Mathematics
Pillars of Hercules, 35
Plagues, 98–102
Plastic surgery. *See* Medicine, recon-
    structive surgery
Pleiades, 17
Plow, 219–20
Pneuma, theory of, 94
Poisons, 284–86
Pozzolana, 202
Practica geometriae (Fibonacci), 144

Precious Mirror of the Four Elements (Chu Shih-Chieh), 140
Pre-Columbian, 17
Prime mover, 7
Principles of Hydrostatics (Stevin), 161
Principles of Statics (Stevin), 161
Printing, 220–23
Propaedeumata Aphoristica (Dee), 151
Prosthetics, 223–25

Quadrant, 225–27
Quebec, 49
Quesiti e Inventioni Diverse (Tartaglia), 272–73

Rationalism, 125
Renaissance, 21, 289–90
Rhind papyrus, 125–26, 128
Rhumb line, 59
Rockets, 280–82
Rosicrucians, 178

Safflower, 206
Sage's Steps, 182
Salvarsan, 106, 186
Samarkand, 40, 137
Saturn (planet), 26
Scholasticism, 157–58
Science of Equations (al-Khwarizmi), 133
Sea Mirror of Circle Measurements (Li Chih), 139
Serendib, 41
Sextant, 28, 226–27
Shields, 253
Ships. *See* Exploration
Siege ladder, 227–28, 266
Siege weapons, 264–67
Sighting tube, 11
Silk Road, 38–39
Sinan, 165
Sling, 275
Smallpox, 101, 102–3, 285
Soap, 229
Sound, 174–77
Spear, 275
Spectacles. *See* Eyeglasses
Speculative Geometry (Bradwardine), 146

Speculum Naturae (Bellovacensis), 78
Speed. *See* Motion
Spice Islands, 47
Spice trade, 39–41
Spinning wheel, 229–31
Spontaneous generation, 83–84
Spurs, 253–54
Standard Model, 168
St. Anthony's Fire, 103–4
Stars: constellations, 18; maps, 21; novas, 12, 25; supernovas, 12
St. Lawrence River, 49
Stonehenge, 1–2, 5
Strait of Belle Isle, 49
Strait of Magellan, 47
Summa de Arithmetic, Geometria, Proportioni et Proportionalita (Pacioli), 147
Summi perfectionis magisterii (Geber), 183
Sundials, 200–201
Sunspots, 12, 26
Surcoat, 250, 254
Surgery. *See* Medicine
Swords, 275
Syphilis, 23, 101, 104–7, 186

Tabulae directionum (Regiomontanus), 23
Tannin, 206
Taxonomy, 68–70, 76–77
Telescope, 232–33
Ten Books of Surgery (Paré), 112
Tetrabiblios (Ptolemy), 36
Theatrum Orbis Terrarum (Ortelius), 58
Theories of the Moon (Abu'l-Wafa), 135
Theory of Everything (TOE), 168
Thermometer, 233–35
Thermoscope, 234
Tin islands, 34
Toilet, flush, 208–210
Toledo Tables (al-Zarqala), 21
Transmutation. *See* Alchemy
Trebuchet, 228, 264–65
Trephining (trepanning), 85–86
Trigonometry. *See* Mathematics
Triquetrum, 225
Trousers, 254

Tuberculosis, 107–8
Tunneling, 266
Tycho Star, 25
Tyre, 54, 205
Tyrian purple, 205

Universal Encyclopedia (Wang Ch'I), 168
Universal Medicine (Fernel), 96
Universe, theories of, 6–9, 12, 22–26

Velocity. *See* Motion
Venturi effect, 281
Verrazano Narrows Bridge, 49
Vitalism, 83

Waterwheel, 235–38
Weapons: ballistics, 272–74; body
    armor, 246–53; bombs, 282;
flamethrowers, 283–84; fortifications, 254–60; gases, 285–86;
grenades, 283; gunpowder, 214–17,
260–62, 279–80; guns, 267–72,
277–80; handheld, 274–80; mines,
283; origins, 245–46; poisons,
284–86; rockets, 280–82; siege,
264–67; types, 243–44, 262–64
Wheelbarrow, 238
Whetstone of Witte (Recorde), 150
Windmill, 239–41
Windrose, 166–67
Windsor Castle, 260
The Wounded Man (Paracelsus), 116

Zatopec civilization, 17
Zero, 127–28. *See also* Mathematics
Zoology, 76–80

About the Author

ROBERT E. KREBS is retired Associate Dean for Research at the University of Illinois Health Sciences. He is also a former science teacher, science specialist for the U.S. government, and university research administrator.